Additive Manufacturing of Metals

Additive Manufacturing of Metals

Editor

Gregory John Gibbons

MDPI • Basel • Beijing • Wuhan • Barcelona • Belgrade • Manchester • Tokyo • Cluj • Tianjin

Editor
Gregory John Gibbons
University of Warwick
UK

Editorial Office
MDPI
St. Alban-Anlage 66
4052 Basel, Switzerland

This is a reprint of articles from the Special Issue published online in the open access journal *Metals* (ISSN 2075-4701) (available at: https://www.mdpi.com/journal/metals/special_issues/additive_manufacturing_metals).

For citation purposes, cite each article independently as indicated on the article page online and as indicated below:

LastName, A.A.; LastName, B.B.; LastName, C.C. Article Title. *Journal Name* **Year**, *Volume Number*, Page Range.

ISBN 978-3-0365-0698-2 (Hbk)
ISBN 978-3-0365-0699-9 (PDF)

Contents

About the Editor

Gregory John Gibbons (Dr., CEng, FIMMM, MIMechE), the head of additive manufacturing (AM) at WMG, University of Warwick, has over 20 years of experience in AM research, with more than 60 peer-reviewed publications. He is responsible for the directorship of the AM group within WMG, establishing new research contracts and ensuring the delivery of existing projects. He has a wide teaching portfolio, including lecturing on undergraduate and post-graduate modules, supervising UG, M.Sc., Ph.D. and EngD students, and providing tutorship. His research interests include developing new applications and novel technologies for AM. Dr Gibbons is the co-chair of the High Value Manufacturing Catapult AM Forum, promoting AM from across the HVM centers to the industry.

Preface to "Additive Manufacturing of Metals"

Additive manufacturing (AM), through which products are digitally manufactured from their 3D design data in a layer-by-layer manner, is an exciting and rapidly growing family of manufacturing technologies, offering a unique capability to realize 'impossible' structures, and providing the flexibility to make low-volume production cost effective. Originally developed in the mid-1980s, for many it is seen as a polymer-based technology; however, with the developments in CAD/CAM and laser technology, AM has moved into the metal arena, and new metal AM technologies, materials and applications are emerging at an ever-increasing rate, and are among the most dynamic and exciting research frontiers in AM. The additive nature of AM also enables unprecedented process control over the microstructure and performance of metals, and ever-increasing demands from industry, for AM processed materials offering excellent mechanical properties, is driving research to elucidate the often complex relationships that exist between process parameters and material structure and properties. This Special Issue on the additive manufacturing of metals offers an insight into the cutting-edge research that is currently being carried out in metal AM, bringing together an interesting set of contributions that make it possible to understand some of the phenomena that are directly related to the AM processing conditions and the resulting microstructural and mechanical properties, as well as the impact that these have had on the industrial application of materials. Over two-thirds of this book follow the laser powder bed fusion (PBF) of a range of alloys, from steels through to titanium alloys, with PBF being the most widely adopted metal AM technology both in academia and in industry, having been the founding technology for metal AM in the mid-1990s. More recently, other methods have emerged that offer the potential for the larger-scale higher-rate AM of components. Principally directed energy deposition (DED), with its unrivaled range of applications, is leading as one of the most promising advanced AM technologies for the future, and the remainder of the book is devoted to exciting new developments in this technology. This book has strength in being a collection of articles from different authors, with the knowledge being brought by true specialists in each area and rigorously reviewed by their expert peers in their field. This book will be of interest and bring added value in terms of knowledge to the entire scientific community, both academic and industrial, providing insight into these exciting developments. The authors hope that their contributions will be of great benefit and an inspiration to those who study and research in this field, as well as to those who wish to exploit the benefits of metal AM in their research and for commercial enterprise.

Gregory John Gibbons
Editor

Article

Microstructure and Mechanical Properties of Low-Carbon High-Strength Steel Fabricated by Wire and Arc Additive Manufacturing

Laibo Sun [1], Fengchun Jiang [1], Ruisheng Huang [2], Ding Yuan [1], Chunhuan Guo [1,*] and Jiandong Wang [1]

[1] Key Laboratory of Superlight Materials & Surface Technology, Ministry of Education, College of Materials Science and Chemical Engineering, Harbin Engineering University, Harbin 150001, China; slb1984@126.com (L.S.); fengchunjiang@hrbeu.edu.cn (F.J.); yuanding@hrbeu.edu.cn (D.Y.); wangjiandong@hrbeu.edu.cn (J.W.)

[2] Harbin Welding Institute Limited Company, Harbin 150028, China; huangrs8@163.com

* Correspondence: guochunhuan@hrbeu.edu.cn

Received: 15 January 2020; Accepted: 29 January 2020; Published: 3 February 2020

Abstract: Wire and arc additive manufacturing (WAAM) is a novel technique for fabricating large and complex components applied in the manufacturing industry. In this study, a low-carbon high-strength steel component deposited by WAAM for use in ship building was obtained. Its microstructure and mechanical properties as well as fracture mechanisms were investigated. The results showed that the microstructure consisted of an equiaxed zone, columnar zone, and inter-layer zone, while the phases formed in different parts of the deposited component were different due to various thermal cycles and cooling rates. The microhardness of the bottom and top varied from 290 HV to 260 HV, caused by temperature gradients and an inhomogeneous microstructure. Additionally, the tensile properties in transversal and longitudinal orientations show anisotropy characteristics, which was further investigated using a digital image correlation (DIC) method. This experimental fact indicated that the longitudinal tensile property has an inferior performance and tends to cause stress concentrations in the inter-layer areas due to the inclusion of more inter-layer zones. Furthermore, electron backscattered diffraction (EBSD) was applied to analyze the difference in Taylor factor between the inter-layer area and deposited area. The standard deviation of the Taylor factor in the inter-layer area was determined to be 0.907, which was larger than that in the deposited area (0.865), indicating nonuniform deformation and local stress concentration occurred in inter-layer area. Finally, as observed from the fracture morphology on the fractured surface of the sample, anisotropy was also approved by the comparison of the transversal and longitudinal tensile specimens.

Keywords: wire and arc additive manufacturing; low-carbon high-strength steel; microstructure; mechanical properties; anisotropy

---

1. Introduction

Low-carbon high-strength steel has been applied in many fields such as shipbuilding, automobiles, mining instruments, and railways due to its unique characteristics of mechanical properties and weldability [1]. However, it is difficult to manufacture large scale fabrications with a complex structure by means of conventional methods, and it is also time consuming with high costs. When facing these problems, additive manufacturing (AM) may be a better choice as a promising technology [2].

AM refers to a technology that usually joins materials together by layers, which has been developed rapidly due to its high material utilization and high geometric freedom [3]. Compared with conventional technologies, AM is good at manufacturing complex components, especially those that can be performance-customized within a certain time [4,5]. When considering AM technique used for metallic alloys, wire and arc additive manufacturing (WAAM) offers better competitive advantages over other techniques such as a higher deposition rate, lower cost, and high buy-to-fly ratio of components that can be realized within a shorter delivery time. Meanwhile, WAAM can also overcome many difficulties associated with manufacturing special alloys [6–8], such that it has increasingly shown the great potential in the manufacture of large metal parts by an arc-based process [9]. The typical WAAM can be divided into gas tungsten arc welding (GTAW), plasma arc welding (PAW), and gas metal arc welding (GMAW) as heat sources. As a modified GMAW process, cold metal transfer (CMT) has some advantages, such as low energy input, high deposition rate, no spattering, and extremely stable arc, therefore this AM technique has become a popular as well as widely used technique. The excellent characteristics of CMT make it to be an ideal process for fabricating a large-scale part, which can overcome common troubles encountered during conventional welding process [10,11]. In this study, it was adopted as the heat source during the deposition process.

Although WAAM is widely employed in AM due to its advantages, several challenges remain to be addressed such as poor surface quality, inhomogeneous microstructure, and the anisotropy of mechanical properties caused by different thermal history [12–14]. As a result, researchers have paid even more attention to the analysis of microstructure evolution, mechanical properties, and fracture behavior in the process of WAAM. This analysis is vital for its application in the ship building industry. Tiago A. Rodrigues et al. [15] studied the microstructure and mechanical properties of a high-strength low-alloy (HSLA) steel fabricated by WAAM. The same microstructural constituents of ferrite, bainite, martensite, and retained austenite were obtained for all heat inputs. Average values for the ultimate tensile strength ranged between 700 MPa and 795 MPa. Dai Yili et al. [16] investigated the microstructure and mechanical properties of multi-directional pipe joints using WAAM and pointed out that the microstructure consisted of 71.8% ferrite and 28.2% pearlite, while the average grain size did not exceed 15 μm. The tensile strength of the forming part reached 562 MPa. Youheng Fu et al. [17] explored the microstructure and mechanical properties of the bainitic steel WAAM part post-treated by rolling, and illustrated that hybrid deposition and micro-rolling treatment provided a novel way for the full transformation of columnar dendrites to equiaxed grains in the production of multi-pass multi-layer specimens. The maximum tensile strength reached 1309 MPa after optimizing.

As aforementioned, numerous studies have been conducted on the microstructure and mechanical properties of high-strength steel. However, few researchers have addressed the fundamental aspects of the solidification behavior and microstructure evolution as well as local strain concentration near inter-layer zones using DIC technology. In this study, a low-carbon high-strength steel developed for ship building was deposited as a thin wall component. It also analyzed the surface quality, microstructure evolution, microhardness, and transversal and longitudinal tensile properties. Finally, the relationship between fractography and the anisotropy of tensile properties was revealed.

2. Materials and Methods

The experiments were conducted on a fixed substrate plate of 907 shipbuilding steel with dimensions of 150 mm × 300 mm × 10 mm. The alloy wire, called A-Fe-W-86, was developed for specific projects and used as a welding material with a 1.2 mm diameter. The chemical compositions of the tested materials are listed in Table 1.

Table 1. Chemical compositions of substrate and wire (wt.%).

Alloy	C	Mn	Si	Cr	Ni	Mo	Cu	V	Fe
907	0.12	1.00	0.80	0.64	0.67	–	0.42	–	Balance
A-Fe-W-86	0.05	1.60	0.38	0.58	2.55	0.58	≤0.10	≤0.02	Balance

During the deposition process, the CMT RCU 5000i (Fronius, Vienna, Austria) was used as a welding power supply and the welding wire was fed to the welding torch, which was kept stationary for each layer. The process parameters are listed in Table 2.

Table 2. Process parameters for deposition.

Process Parameters	Details
Wire feed speed	4.5 m/min
Travel speed	0.25 m/min
Voltage	152 V
Current	14.7 A
Shielding Gas	Ar (90%) + CO_2 (10%)
Flow of gas	15 L/min

The deposition started from the end point of the previous layer for each subsequent layer. The path strategy was chosen in order to ensure the thickness and width of the start and end portions similar to that of the central portion, thus avoiding significant deviation from the originally expected shape [18]. The schematic diagram is shown in Figure 1.

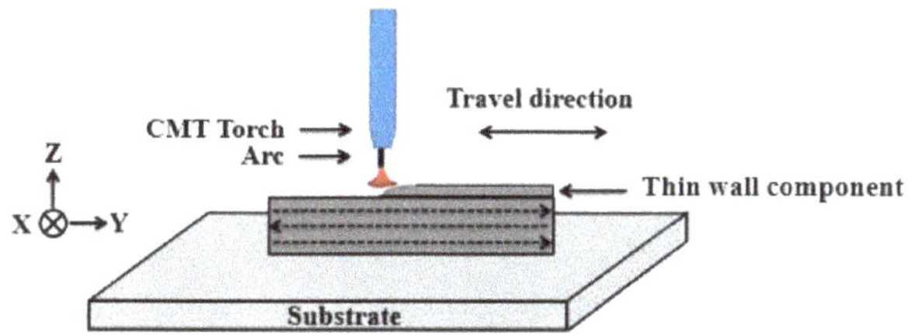

Figure 1. A schematic diagram for the experimental procedure.

The deposited component was prepared for the analysis of the microstructure and mechanical properties. For consistency, all specimens were adopted from the homogeneous and stable parts of the component. The specimens, which were used to analyze the surface quality, microstructure evolution, and Vickers hardness (HV 0.2) of the top and the bottom, were drawn from the cross section of the component by a ZwickRoell Indentec (ZwickRoell, Ulm, Germany) testing machine. The tensile experiments were carried out using a Shimadzu AG-X plus (Shimadzu Scientific Instruments, Shanghai, China) as the tester with a displacement rate of 0.02 mm/s. All tests were performed at room temperature. The specific sampling location is shown in Figure 2. Both transversal and longitudinal tensions were performed to obtain the data of strain evolution during the testing to reveal the anisotropy behavior. Before the tests, specimens were sprayed with a randomized speckle pattern that consisted of black micron size speckles on a white background to achieve high contrast. The size and surface treatment condition of the specimen is shown in Figure 3.

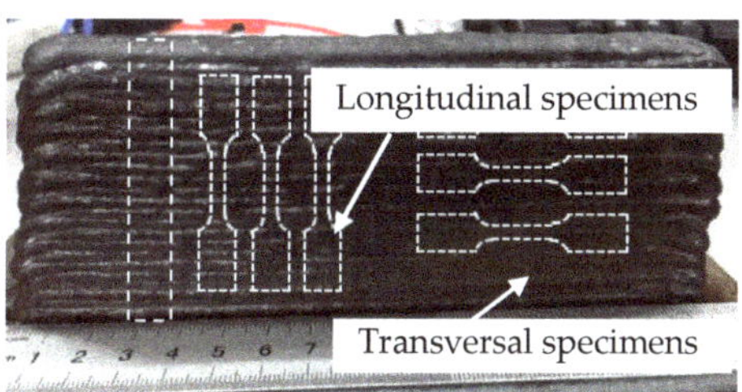

Figure 2. Schematic diagram for sampling location.

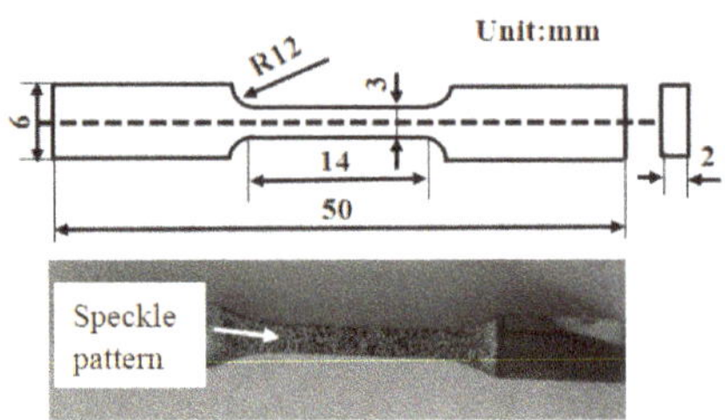

Figure 3. The size and surface treatment condition of the specimen.

During the testing, the strain and surface displacements were calculated by tracking the speckle pattern on the specimen surface. Images were captured using an iX i-SPEED 7 CCD camera (iX Cameras, Shanghai, China) at the rate of 10 fps (frames per second), with a pixel array of 350 × 750 pixels. Then, the data were analyzed by DIC software to obtain the strain field distribution [19]. The experimental setup is shown in Figure 4.

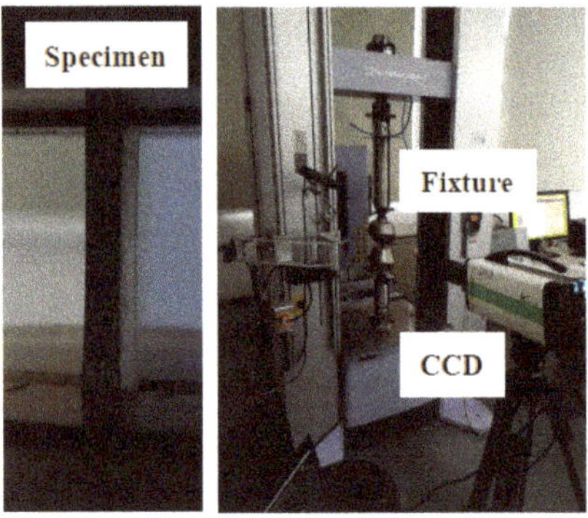

Figure 4. The tensile process and devices.

3. Results and Discussion

3.1. Surface Quality Assessment

It is well known that the components manufactured by WAAM have poor surface roughness and dimension accuracy. Additionally, cracks between the inter-layer adjacent region, lack of fusion and inclusions are usually visible. To meet industrial requirements, surface quality is a critical consideration in the manufacturing process [20].

Figure 5 shows the typical thin-wall cross section of the as-deposited component. It can be noted that the surface of the thin wall is uniform and well formed with good quality. The deposition strategy can improve the deposition quality by filling the arc craters that are frequently caused at the arc start and end points. The so-called step effect can be shown at two sides of the thin-wall, which results from the layer-by-layer deposition model. The cross section of the polished state is shown in Figure 5a, and it can be observed that the condition of the as-built deposition is steady, indicating that each part of the component shows the same size. Figure 5b,c demonstrate the surface conditions after the scanning electron microscope (SEM) analysis, which has a smooth surface and full fusion, although there are some oxides or impurities brought from the prepared specimen. At the same time, the juncture between layers is reported in Figure 5d,e. Optical microscopy (OM) analysis shows that cracks, inclusions, and pores are not found in the bonding areas. Generally, the deposited part is well formed without unfused effects, cracks, inclusions, and pores observed under the condition in the present experimental method.

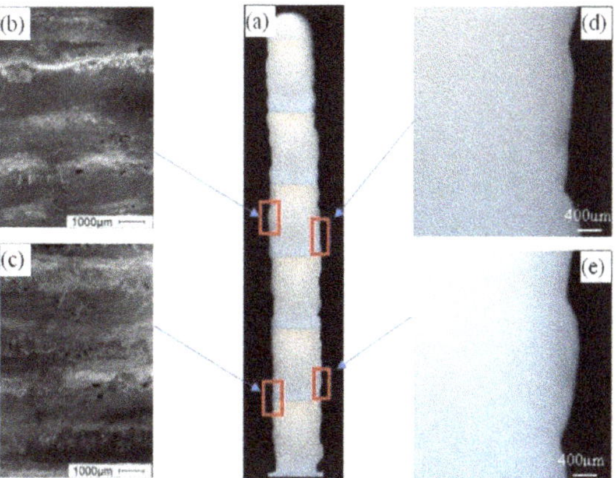

Figure 5. Assessment of the surface quality of the component: (**a**) Overall condition on cross section; (**b,c**) surface condition analyzed by SEM; (**d,e**) inter-layer adjacent region condition analyzed by OM.

3.2. Microstructure

3.2.1. Microstructure Evolution

In the process of WAAM, the heat cycle repeats over and over again. Some typical phenomena are caused by complex thermal cycles such as rapid heating, fast solidification in the fusion zone, various temperature gradients, a large amount of heat accumulated, reheated, and then recooled. The non-equilibrium thermodynamics governs the physical characteristics of microstructure formation in the WAAM process. Thus, analyzing microstructural characteristics and evolution is helpful for describing the manufacturing process [21]. Based on the thermal history illustrated above, the microstructure can be divided into three main morphologies: (1) columnar, (2) mixture of columnar and equiaxed, and (3) equiaxed. In other words, during the WAAM process, the microstructure can be divided into three different regions: (1) columnar grain zone, (2) boundary zone, and (3) equiaxed grain zone, as shown in Figure 6. The evolution of the microstructure is influenced by two crucial control parameters including the temperature gradient (G) and solidification velocity (SV) [22]. Based on solidification theory, the ratio of the G to SV determines the forms of the microstructure. When the G/SV is extremely large, the solidification microstructure contains planar grains, as shown in Figure 6. In this condition, the melt metal acts upon the former solidified layer and solidifies rapidly, G tends to be infinite, and VS tends to be zero. When the G/SV is a relatively high value, the major solidification microstructure is columnar grain, as shown in Figure 6; at that moment, the heat dissipation that comes from the substrate results in a rapid cooling speed and high temperature gradient. The melt metal solidifies quickly and then the columnar grain is obtained. In a general way, the core of the molten pool is finally solidified. The temperature gradient and solidification velocity tend to be low, resulting in smaller G/SV values, giving rise to the equiaxed grains as shown in Figure 6. In a word, heat is mostly transmitted through the substrate or formerly deposited layers, and different regions of the molten pool appear to have various temperature gradients and solidification velocities, which eventually influences the evolution of the microstructure.

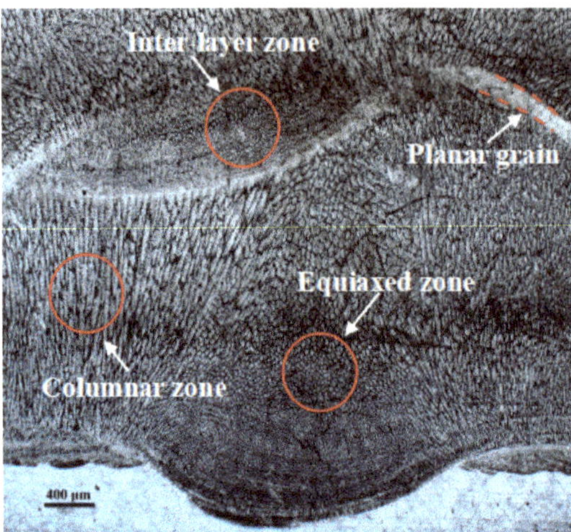

Figure 6. Three typical regions of microstructure in the process of WAAM.

During the deposition process, the temperature gradient dominates the directional growth of the microstructure and affects its morphology. The growth direction is usually perpendicular to the boundary of the molten pool [23]. The shape and growth of dendrites in different deposited layers vary as a function of heat dissipation, as shown in Figure 7. In this work, the component consists of deposited layers and its macrostructure at cross section is shown in Figure 7a. Region b is close to the substrate, representing a rapid cooling rate and large heat dissipation, which indicates a high temperature gradient. In this region, the heat dissipation of metal liquid is perpendicular to the substrate during the cooling and solidification processes. The direction of heat flow is therefore perpendicular to the interface, leading to the fact that solidification is directional. Thus, the structure of columnar grain formed is vertical to the fused interface. The typical and directional columnar microstructure in region b is shown in Figure 7b. Along with the deposition progress, the effect of heat dissipation from the substrate decreases by increasing the distance to the substrate, resulting in the decrease in the temperature gradient. The columnar microstructure formed appeares less typical than that in region b, and the details of the microstructure in region c are shown in Figure 7c. As deposition continued, region d is hardly influenced by the substrate, but is only affected by the thermal cycles between layers. At that time, the temperature gradient is relatively low, resulting in the inapparent directional growth of the microstructure. Simultaneously, region d undergoes a long period of heating, leading to a coarse microstructure. The conditions of region d are shown in Figure 7d. At the end of the deposition, the microstructure of regions e and f on the top of component are presented in Figure 7e,f, separately. In both regions, the columnar grain is rarely observed, while a uniform and fine microstructure appeare. To some extent, the characteristics of the microstructure are disclosed due to rapid cooling resulting from contact with the environment and no subsequent heating.

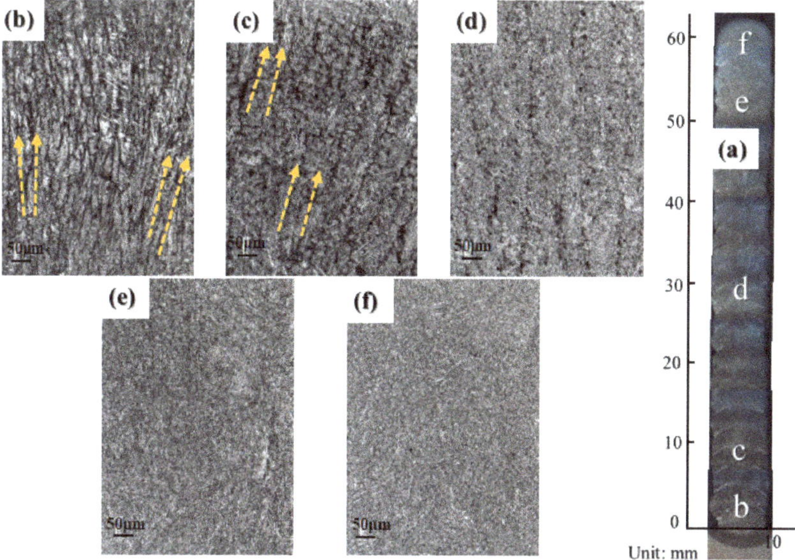

Figure 7. Microstructure evolution for different parts of the as-deposited component: (**a**) Macrostructure of the cross section; (**b**) Region b; (**c**) Region c; (**d**) Region d; (**e**) Region e; (**f**) Region f.

3.2.2. Phase Transformation

It is well know the phase transformation and microstructure development of the metal component fabricated by WAAM are completely different compared to that of conventional manufacturing. The WAAM fabricated components undergo a slower cooling rate and gradual/equilibrium thermodynamics processes, thereby obtaining more homogeneous properties. In this study, the phase transformation at different positions of (from bottom to top) the component is investigated and reported in Figure 8, in which Figure 8a shows the details of the typical phases of the bottom part. Here, the phase transformation analysis is based on phase morphology and the continuous cooling transformation diagram of a high strength steel with a similar chemical composition [24]. At the beginning of deposition, since the bottom layer is affected by the substrate and is then cooled rapidly, thus martensite transformation occures. It plays a tempering role under the influence of a long-term thermal cycle in the subsequent deposition layer, with the occurrence of tempered sorbite transformation. Along with the tempered sorbite transformation, the major phase transformation is bainite due to the intermediate cooling rate. Under the action of tempering, tempered bainite is achieved. Thus, the phase composition at the bottom is tempered bainite and tempered sorbite. As deposition continues, the part close to the bottom moves away from the substrate gradually, lowering the cooling rate and heat dissipation, thereby reducing the content of tempered sorbite. At this time, the phase composition at the bottom is tempered bainite and less tempered sorbite, as shown in Figure 8b. As deposition height increases heat accumulation approaches saturation state, while the heat input and output are in balance, which presents a lower temperature gradient and cooling rate, resulting in the fact that it could not achieve the transformation requirement of tempered sorbite. As a consequence, the phase composition at the middle is almost made up of tempered bainite, as shown in Figure 8c. When deposition arrives at the end, the melt pool emerges at a rapid cooling rate due to mass heat dissipation with the environment, which allows the tempered sorbite transformation to occur again [25]. At the same time, some ferrite transformation takes place due to the lack of subsequent heating. Under this circumstance, the phases on the top are composed of tempered bainite, tempered sorbite, and ferrite, as reported in Figure 8d.

In general, the complex phase transformation is caused by distinct thermal cycles and various cooling rates. This phenomenon eventually results in different phase transformation in diverse regions.

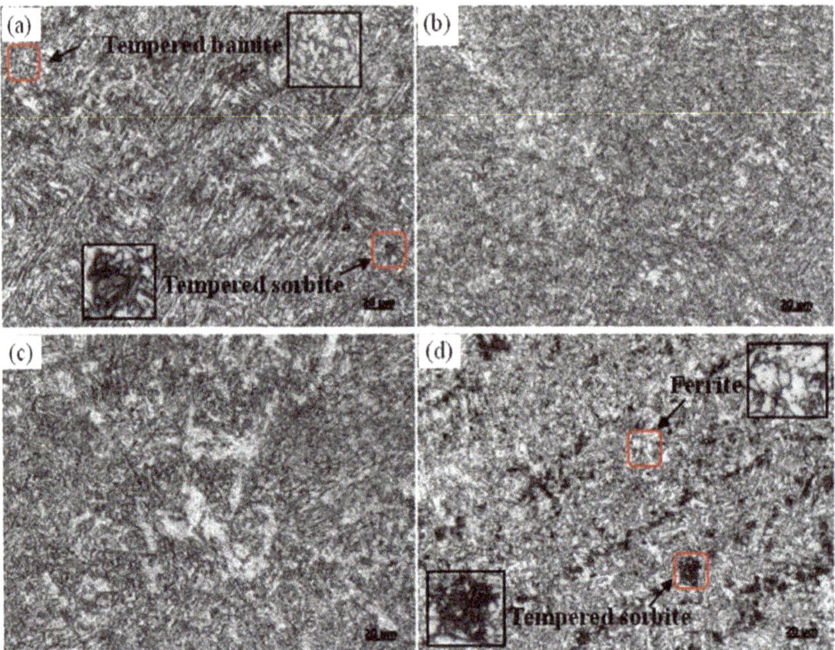

Figure 8. Phase transformation of different parts of the component: (**a**) bottom part; (**b**) part close to the bottom part; (**c**) middle part; (**d**) top part.

3.3. Microhardness

In this study, the average microhardness is measured from the bottom to the top, the changes in the microhardness of the vertical cross section of the as-deposited component are measured, the results are reported in Figure 9. It can be seen that the maximum value of microhardness is found to be around 290 HV near the bottom part. As the height of deposition increases, the microhardness decreases slightly and fluctuats between 280 HV and 270 HV. Then, the microhardness varies between 270 HV and 260 HV when measured near the top part. Generally, the deposited material goes through complex thermal cycles, which is expected to affect the microhardness [26,27]. It has been confirmed that the fluctuation of the microhardness comes from various thermal cycles and cooling rates in diverse parts of the component, and the microhardness values of multilayers are influenced by the heating at previous layers [28]. Furthermore, the compositions of the microstructure at the bottom, in the middle, and on the top also play a significant role on microhardness. As depicted from the previous analysis on phase transformation, the phase at the bottom is made up of tempered bainite and tempered sorbite, the phase in the middle is composed of tempered bainite, and the phase on the top is comprised of tempered bainite, tempered sorbite, and ferrite. In terms of the microhardness, tempered bainite + tempered sorbite > tempered bainite > tempered bainite + tempered sorbite + ferrite, hence, the microhardness of the components shows such a characteristic, which is in line with the microstructure results.

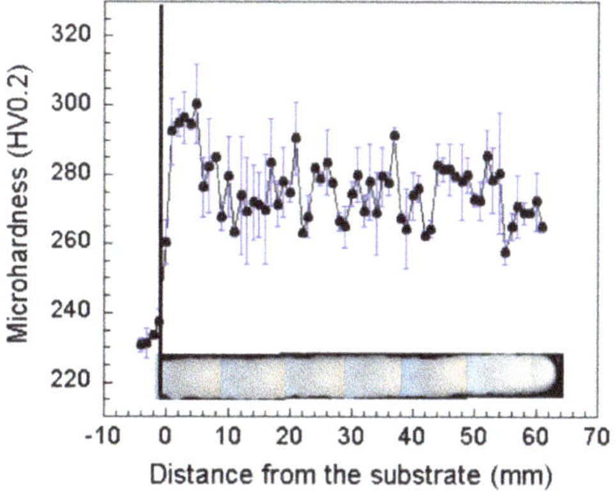

Figure 9. Microhardness along the longitudinal direction for the cross section.

3.4. Tensile Test and DIC Analysis

The results of transversal and longitudinal tensile tests are listed in Table 3. As summarized from the data, the transversal tensile property obtained is better than that of the longitudinally stretched, indicating an anisotropy behavior, which usually occurs in the process of WAAM [29].

Table 3. Transversal and longitudinal tensile properties.

Specimen Type	Ultimate Tensile Strength (MPa)	Yield Strength (Rp 0.2, MPa)	Elongation (%)
Transversal specimens	1007.6	818.0	12.6
	1020.3	825.5	12.4
	1025.5	831.6	12.6
Longitudinal specimens	968.2	743.2	10.2
	978.6	749.3	10.1
	982.8	755.1	10.5

To reveal the anisotropy mechanisms, a DIC analysis is performed on a typical group of specimens and eight typical points are selected to represent the strain evolution during the transversal and longitudinal tensile tests separately. The selected experimental points can be seen from the stress–strain curves, as shown in Figure 10.

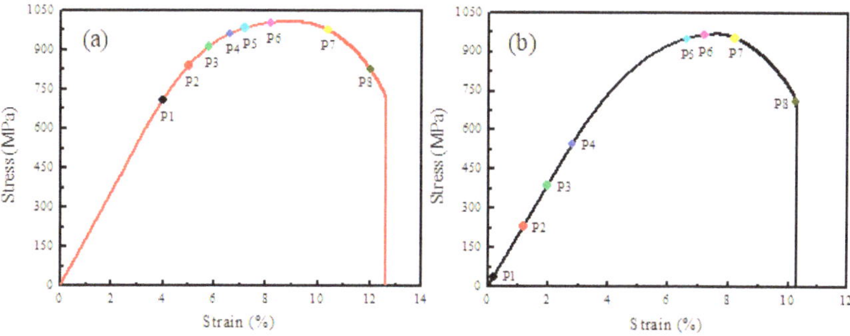

Figure 10. Transversal and longitudinal stress–strain curves and eight typical points selected. (**a**) Transversal curve; (**b**) longitudinal curve.

Figure 11 shows the strain evolution of the transversal and longitudinal specimens during the tensile testing. Each typical point selected in Figure 10 corresponds to the point in Figure 11. For the tensile test of the transversal specimen, strain evolution can be seen from P1 to P8, as shown in Figure 11a. Before the global strain reaches a certain extent, uniform strain distribution can be observed in the deformation area such as P1 and P2. This means that every area of the gauge section is involved in uniform deformation during the tensile process. As the tensile test proceeds, the applied load begins to increase. Local stress concentrations can be obtained from P3 to P6. Then, when necking takes place in the area of deformation, the local stress concentration intensifies and the macro stress reduces, as shown as P7 and P8. The strain evolution of the transversal tensile process is the same as that of the classic homogeneous material. In contrast, the strain evolution of the longitudinal specimen can be summarized by P1′ to P8′, as shown in Figure 11b. It can be seen that a high local strain occurred around several areas of the gauge section, shown as P1′ to P4′. This means that nonuniform strain distribution happens due to inhomogeneous stress. As the experiment continued, local stress concentration can be seen, but is markedly different from that of the transversal specimen. During this stage of deformation, the maximum strain existes while combined with local high strain near other areas, which can be observed as P5′ and P6′. At the end of the experiment, necking emerged, leading to stress concentration, which reduces the effect of the inhomogeneous strain distribution of the specimen, seen as P7′ and P8′. From the comparison of the strain evolution between the transversal and longitudinal specimens, the latter one shows inferior properties and a nonuniform strain distribution.

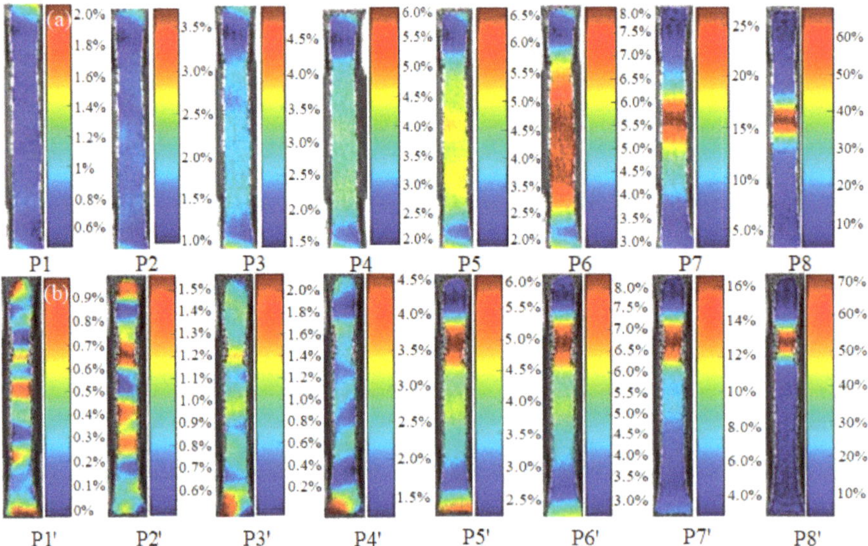

Figure 11. Strain evolution of transversal and longitudinal specimens during the tensile test. (**a**) Transversal specimen; (**b**) longitudinal specimen.

It is well known that the anisotropy of AM leads to different transversal and longitudinal behaviors and properties, mainly influenced by temperature gradients that are varied in successive layers and the penetration of the molten pool in the previous deposited layer [30–32]. As heat dissipation is higher in the forming direction than that in other directions, its solidified direction results in the anisotropy of grain morphology and mechanical properties. Additionally, longitudinal specimens are parallel with the building direction, implying that more inter-layer areas will be included within the test region. As a weakness, phase mutation and residual stress can usually be found near the inter-layer area, which results from the element interaction and large temperature gradient [33,34]. Researchers have also pointed out that the inter-layer area is considered as the weak link and that the strain tends to be focused in this region [35]. The cracks and material failure usually take place around these areas in the non-uniform specimen [36].

To give a further illustration of anisotropy, the strain cloud images of the transversal and longitudinal tensile specimens during the first 70 s of the experiment are shown in Figure 12. It can be seen from the differences in Figure 12b that they exhibit a nonuniform and intermittent strain distribution of the longitudinal tensile specimen because several interlayers are included, and that of the transversal specimen is relatively uniform, as shown in Figure 12a.

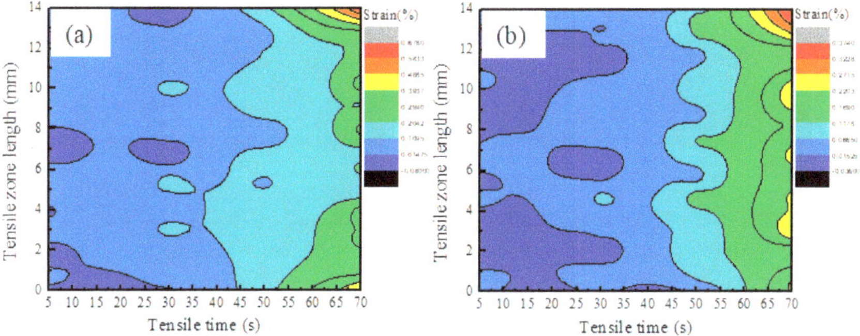

Figure 12. Cloud images of strain of transversal and longitudinal specimens. (**a**) Transversal specimen; (**b**) longitudinal specimen.

Additionally, after investigation on the initial stage of deformation, a comparison of the local strain distributed in the gauge section of the transversal and longitudinal tensile specimens is also made. Referred to as P5, P6, P7 and P5′, P6′, P7′, the results are shown in Figure 13. A high local strain region of the transversal specimen lay in the middle part and the local strain concentration reaches the maximum at P7, as shown in Figure 13a, while more than one region with a high local strain can be observed from the longitudinal specimens. With the experiment proceeding, the local strain difference increases and the local strain concentration is aggravated and reaches a peak at P7′, as shown in Figure 13b. Therefore, nonuniform local strain distribution of the longitudinal specimen is obvious, regardless of whether it is at the beginning or end of the deformation.

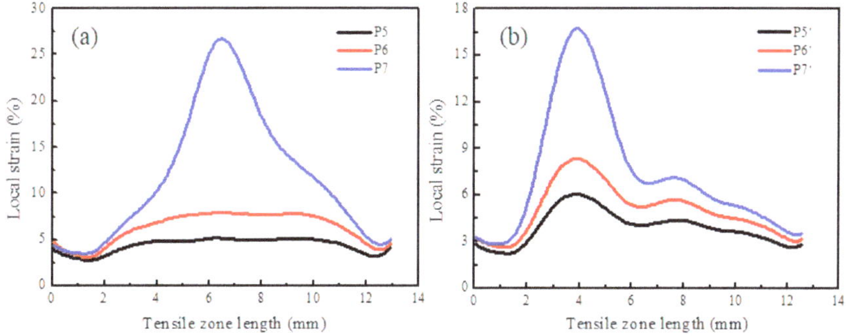

Figure 13. Comparison of the local strain distribution in the gauge section: (**a**) Transversal specimen; (**b**) longitudinal specimen.

The high strain zone is used to refer to all zones of strain concentration and treated as a significant consideration during the tensile test [37]. In this study, the strain distribution condition in the high strain zones before fracture on the transversal and longitudinal tensile specimens is shown in Figure 14.

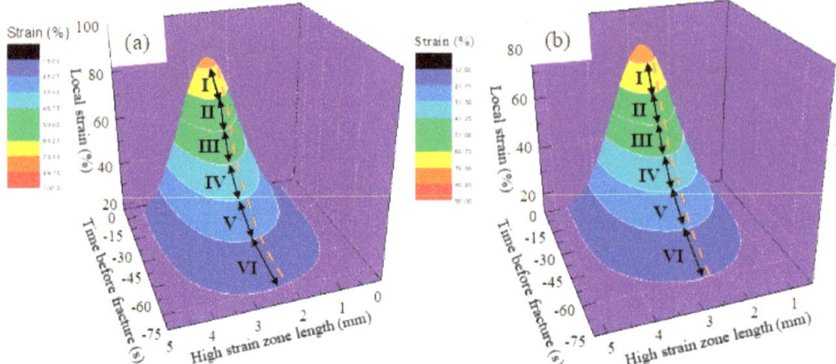

Figure 14. Strain distribution condition in high strain zones before fracture. (**a**) Transversal specimen; (**b**) longitudinal specimen.

Based on the strain distribution condition, high strain zones can be divided into six high strain gradients and a comparison of the duration of high strain gradients is made to reveal the differences between the transversal and longitudinal tensile specimens. The results are shown in Figure 15. It can be seen from Figure 15, that the duration in the high strain zone (from stage I to VI) of the transversal specimen is longer than that of the longitudinal specimen. Furthermore, as high strain varied from stage VI to I (close to fracture), the duration gap between the transversal and longitudinal specimens gradually expands, suggesting that in high strain zones, the transversal specimen performed better, and the higher strain stage the tensile test experiences, the more remarkable the duration superiority of the transversal specimen.

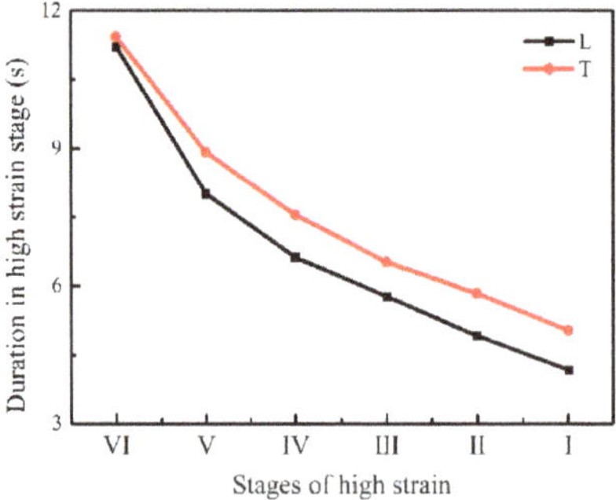

Figure 15. Comparison of the duration in high strain stages between the transversal and longitudinal specimen. T—Transversal specimen; L—longitudinal specimen.

3.5. Fractography Observation

Fracture property is one of the critical considerations in material design and plays a significant role in the engineering field. The longitudinal and transversal fracture morphologies are analyzed via SEM. The macroscopic fracture morphology of the tensile specimens is shown in Figure 16. It can be seen that several inter-layer lines are observed from the longitudinal specimen after fracture, as shown in Figure 16a. This phenomenon results from the inferior mechanical properties of inter-layer areas, which are treated as the weak link. It has been proven that strain tends to be concentrated in the weaker areas in the non-uniform regions and cause fracture in these regions [38]. This is consistent with the strain evolution and local strain analysis that has been previously mentioned, interpreting the anisotropic mechanical properties. The surface of the macroscopic fracture of the longitudinal specimen is shown in Figure 16b where a bumpy and undulating surface morphology can be observed. The surface of the macroscopic fracture of the transversal specimen is illustrated in Figure 16c where the surface is smooth and located at a certain angle with the axial tensile stress.

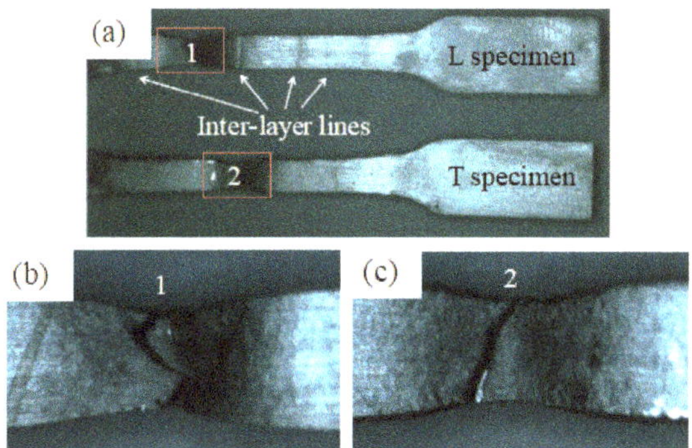

Figure 16. Macroscopic fracture appearance of tensile specimens. (**a**) Longitudinal and transversal specimens after fracture; (**b**) macroscopic fracture of longitudinal specimen; (**c**) macroscopic fracture of transversal specimen.

In order to give a further illustration of this phenomenon, a comparison between the inter-layer area and deposited area on the Taylor factor is conducted by electron backscattered diffraction (EBSD). It is generally accepted that the Taylor factor is an important parameter, which determines the stress required to activate a slip system. This means that the Taylor factor plays an important role in tensile behavior, followed by the relationship:

$$\sigma = M\tau_c \tag{1}$$

where σ is the applied stress; M is Taylor factor; and τ_c is the critical resolved shear stress on each of the activated slip systems [39].

The comparison results of the Taylor factor distribution are shown in Figure 17. Compared with the distribution of the Taylor factor in the deposited area shown in Figure 17b, a less uniform distribution of the Taylor factor in the inter-layer area can be observed in Figure 17a. This confirms that nonuniform strength distribution takes place in the inter-layer area. The statistics for the Taylor factor distribution are depicted in Figure 18. The standard deviation of the Taylor factor in the inter-layer area is 0.907, larger than that of the deposited area (0.865).

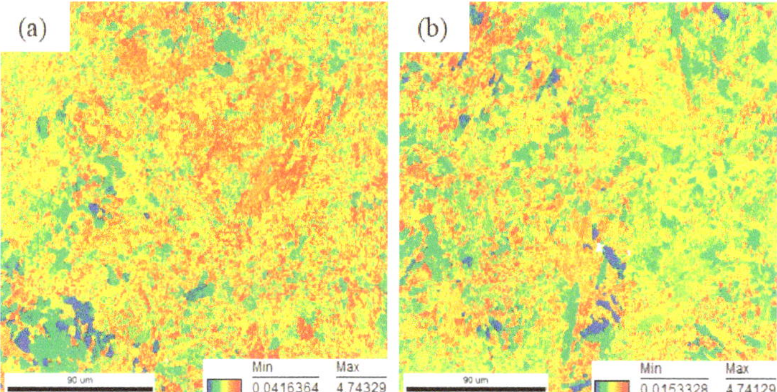

Figure 17. Comparison between the inter-layer area and deposited area on the Taylor factor: (**a**) Inter-layer area; (**b**) deposited area.

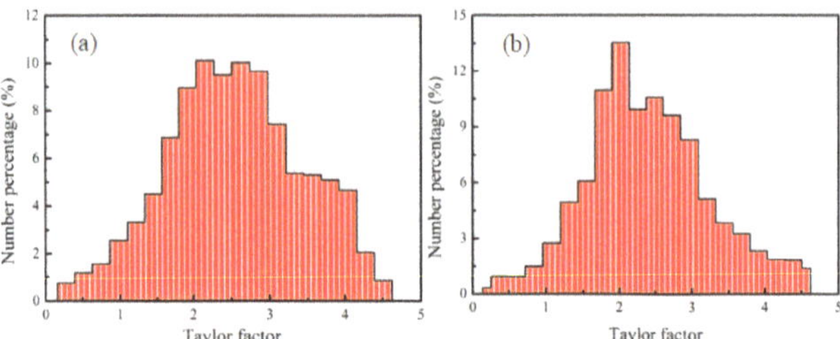

Figure 18. Statistics between the inter-layer area and deposited area with the Taylor factor: (**a**) Inter-layer area; (**b**) deposited area.

A higher standard deviation of the Taylor factor means less uniform strength distribution in the inter-layer area. During the tensile test, to maintain the consistence of deformation, the grains located on both sides of the boundary are forced to deform consistently. Thus, nonuniform deformation and local stress concentration take place around the boundary, resulting in zigzag surface features near the fracture of the longitudinal specimen, as shown in Figure 16b, where it is constant with the strain evolution and local strain analysis that has been previously mentioned, interpreting the anisotropic mechanical properties. The result mainly corresponds to the directional growth of the microstructure, which has been confirmed by previous investigations [40].

The fracture morphologies of the transversal and longitudinal specimens are shown in Figure 19. It can be seen from Figure 19a,b that the transversal and the longitudinal fracture surfaces both present as grey and fibrous, revealing ductile fracture. Additionally, a large number of equiaxed dimples with a relative uniform distribution are apparent on the fracture surface of the transversal specimen, shown in Figure 19c. In contrast, the mixed-rupture characteristics of quasi-cleavage and small dimples emerge on the fracture surface of the longitudinal specimen, as depicted in Figure 19d. Through the comparison of the appearance of dimple dimensions and depths of the transversal and longitudinal fracture, one can find the transversal are greater, which indicates that the transversal specimen has better ductility. Although the feature of quasi-cleavage is observed from the local part of the fracture on the longitudinal specimen, it can still show high ductility. Therefore, it is concluded from the result of the fracture analysis that the longitudinal mechanical property is inferior to that of the transversal one, which is in accordance with the above analysis.

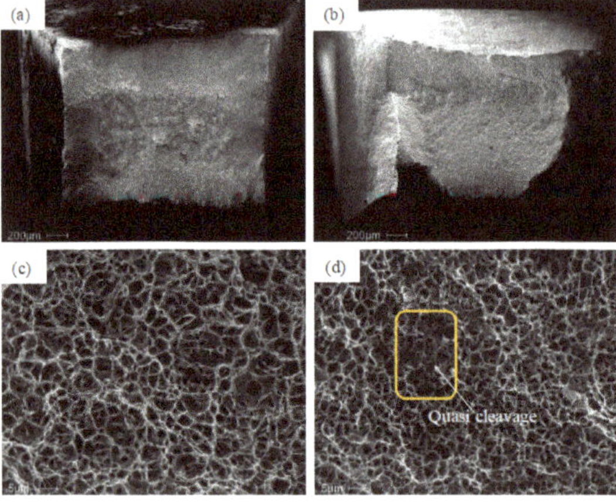

Figure 19. Fracture morphology of the tensile specimens: (**a,b**) Macroscopic fracture morphology of the transversal and longitudinal specimens; (**c,d**) typical fracture features of the transversal and longitudinal specimens.

During the process of fracture analysis, a large number of uniformly distributed particles are found in the dimples, as shown in Figure 20a. The particles may be instrumental in the formation of dimples [41]. In this study, most particles are smaller than 1 μm, implying that during the process of fracture, larger stress is required for the formation of microvoids, and that the growth rate of microvoids is lower [42]. Energy dispersive X-ray spectroscopy (EDX) is applied to analyze the composition of the particles, and the results are shown in Figure 20b. It can be seen that Fe, Ti, Al, O, Mn, and Si are the main elements of the particles, which can be identified as non-metallic inclusions such as Al_2O_3 or Ti_2O_3. The existence of the particles can be considered as nucleation points for dimples.

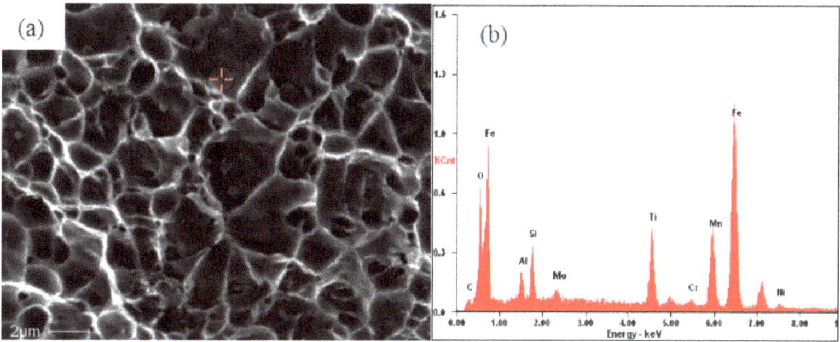

Figure 20. The distribution of particles and their EDX analysis: (**a**) distribution of second phase particles; (**b**) the EDX analysis.

4. Conclusions

In this investigation, a low carbon component with high strength was fabricated by WAAM without any visible detects and an analysis on its microstructure and mechanical properties was undertaken, which showed potential for industrial application. The following conclusions can be drawn:

1. The microstructure of the bottom part showed a directional growth of columnar grains, and as the deposition continued, the features of the directional growth of microstructure were weakened due to the temperature gradient variation, nearly disappearing in the top part. The formation of the phase at the bottom was tempered bainite + tempered sorbite, that at the middle part was tempered bainite, and the one at the top part was composed of tempered bainite + tempered sorbite + sorbite ferrite. The transformation of the microstructure resulted from the different thermal cycles and cooling rate.

2. The microhardness varied from the bottom to the top part due to the formation of different microstructures. The bottom part showed the highest microhardness, which was around 290 HV and dropped down from 280 HV to 270 HV in the middle part. Then, it emerged as a lower value at the top part, which fluctuated from 270 HV to 260 HV.

3. The tensile strength of the longitudinal specimen had an inferior performance to that of the transversal specimen, showing anisotropy behavior. Several areas with local strain concentration of the longitudinal specimen could be found at the beginning and end stage of the tensile test when investigated using a DIC method. Additionally, the duration in the high strain stages of the longitudinal specimen was shorter and deteriorated with increasing strain.

4. A ductile fracture was revealed in both the transversal and longitudinal fracture surfaces. Several inter-layer areas as weak regions could be observed from the surface of the longitudinal specimen after fracture, resulting in local strain concentration. It was confirmed that the longitudinal mechanical property was inferior to that of the transversal specimen, showing anisotropy characteristics. Some particles existing as inclusions such as Al_2O_3 or Ti_2O_3 were found on the fracture surface.

Author Contributions: Methodology, C.G. and D.Y.; writing-original draft preparation, L.S.; project administration and funding acquisition, F.J. and R.H.; data curation, J.W.; writing—review and editing, providing ideas, C.G and F.J. All authors have read and agreed to the published version of the manuscript.

Funding: This work was supported by the National Key R&D Program of China (No. 2017YFB1103701); the National Natural Science Foundation of China (No. 51671065); the Equipment Development Department for Commission of Science and Technology (No. 41423030504); the Applied Technology Research and Development Plan of Heilongjiang (No. GA18A403); and the Natural Science Fund of Heilongjiang Province (No. ZD2019E006).

Acknowledgments: Authors appreciate the help of Research & Development Centre and Quality Inspection Office, Harbin Welding Institute Limited Company.

Conflicts of Interest: The authors declare no conflict of interest.

References

1. Xue, Q.; Benson, D.; Meyers, M.; Nesterenko, V.; Olevsky, E. Constitutive response of welded HSLA 100 steel. *Mater. Sci. Eng. A* **2003**, *354*, 166–179. [CrossRef]

2. Guo, N.; Leu, M.C. Additive manufacturing: Technology, applications and research needs. *Front. Mech. Eng.* **2013**, *8*, 215–243. [CrossRef]

3. Frazier, W.E. Metal Additive Manufacturing: A Review. *J. Mater. Eng. Perform.* **2014**, *23*, 1917–1928. [CrossRef]

4. Dilberoglu, U.M.; Gharehpapagh, B.; Yaman, U.; Dolen, M. The role of additive manufacturing in the era of industry 4.0. *Procedia Manuf.* **2017**, *11*, 545–554. [CrossRef]

5. Cunningham, C.R.; Flynn, J.M.; Shokrani, A.; Dhokia, V.; Newman, S. Invited review article: Strategies and processes for high quality wire arc additive manufacturing. *Addit. Manuf.* **2018**, *22*, 672–686. [CrossRef]

6. DebRoy, T.; Wei, H.L.; Zuback, J.S.; Mukherjee, T.; Elmer, J.W.; Milewski, J.O.; Beese, A.M.; Wilson-Heid, A.; De, A.; Zhang, W. Additive manufacturing of metallic components–process. structure and properties. *Prog. Mater. Sci.* **2018**, *92*, 112–122. [CrossRef]

7. Ding, J.; Colegrove, P.; Mehnen, J.; Ganguly, S.; Sequeira Almeida, P.M.; Wang, F.; Williams, S. Thermo-mechanical analysis of wire and arc additive layer manufacturing process on large multi-layer parts. *Comput. Mater. Sci.* **2011**, *50*, 3315–3322. [CrossRef]

8. Ding, D.; Pan, Z.; Cuiuri, D.; Li, H. Wire-feed additive manufacturing of metal components: Technologies, developments and future interests. *Int. J. Adv. Manuf. Technol.* **2015**, *81*, 465–481. [CrossRef]

9. Williams, S.W.; Martina, F.; Addison, A.C.; Ding, J.; Pardal, G.; Colegrove, P. Wire + arc additive manufacturing. *J. Mater. Sci. Technol.* **2016**, *32*, 641–647. [CrossRef]

10. Gomez Ortega, A.; Corona Galvan, L.; Deschaux-Beaume, F.; Mezrag, B.; Rouquette, S. Effect of Process Parameters on the Quality of Aluminium Alloy Al5Si Deposits in Wire and Arc Additive Manufacturing Using a Cold Metal Transfer Process. *Sci. Technol. Weld. Join.* **2018**, *23*, 316–332. [CrossRef]

11. Kazanas, P.; Deherkar, P.; Almeida, P.; Lockett, H.; Williams, S. Fabrication of Geometrical Features using Wire and Arc Additive Manufacture. *J. Eng. Manuf.* **2012**, *226*, 1042–1051. [CrossRef]

12. Xu, X.; Mi, G.; Luo, Y.; Jiang, P.; Shao, X.; Wang, C. Morphologies, microstructures, and mechanical properties of samples produced using laser metal deposition with 316L stainless steel wire. *Opt. Lasers Eng.* **2017**, *94*, 1–11. [CrossRef]

13. Yang, D.; Wang, G.; Zhang, G. Thermal analysis for single-pass multi-layer GMAW based additive manufacturing using infrared thermography. *J. Mater. Process. Technol.* **2017**, *244*, 215–224. [CrossRef]

14. Hermans, M.J.M.; Ouden, G.D. Process Behaviour and Stability in Short Circuit Gas Metal Arc Welding. *Weld. J.* **1999**, *78*, 137–141.

15. Rodrigues, T.A.; Duarte, V.; Avila, J.A.; Santos, T.G.; Miranda, R.M.; Oliveira, J.P. Wire and arc additive manufacturing of HSLA steel: Effect of thermal cycles on microstructure and mechanical properties. *Addit. Manuf.* **2019**, *27*, 440–450. [CrossRef]

16. Dai, Y.L.; Yu, S.F.; Shi, Y.S.; He, T.Y.; Zhang, L.C. Wire and arc additive manufacture of high-building multi-directional pipe joint. *Int. J. Adv. Manuf. Technol.* **2018**, *96*, 2389–2396. [CrossRef]

17. Fu, Y.; Zhang, H.; Wang, G.; Wang, H. Investigation of mechanical properties for hybrid deposition and microrolling of bainite steel. *J. Mater. Process. Technol.* **2017**, *250*, 220–227. [CrossRef]

18. Zhang, Y.M.; Chen, Y.; Li, P.; Male, A.T. Weld deposition-based rapid prototyping: A preliminary study. *J. Mater. Process. Technol.* **2003**, *135*, 347–357. [CrossRef]

19. Krishnan, S.A.; Baranwal, A.; Moitra, A.; Sasikala, G.; Albert, S.K.; Bhaduri, A.K.; Harmain, G.A.; Jayakumar, T.; Rajendra Kumar, E. Assessment of deformation field during high strain rate tensile tests of RAFM steel using DIC technique. *Procedia Eng.* **2014**, *86*, 131–138. [CrossRef]

20. Aris, N.F.M.; Cheng, K. Characterization of the surface functionality on precision machined engineering surfaces. *J. Mater. Process. Technol.* **2008**, *38*, 402–409. [CrossRef]

21. Fayazfar, H.; Salarian, M.; Rogalsky, A.; Sarker, D.; Russo, P.; Paserin, V.; Toyserkani, E. A critical review of powder-based additive manufacturing of ferrous alloys: Process parameters, microstructure and mechanical properties. *Mater. Des.* **2018**, *144*, 98–128. [CrossRef]

22. Hunt, J.D. Steady state columnar and equiaxed growth of dendrites and eutectic. *Mater. Sci. Eng.* **1984**, *65*, 75–83. [CrossRef]

23. Zhong, Y.; Rannar, L.E.; Liu, L.; Koptiug, A.; Wikman, S.; Olsen, J.; Cui, D.; Shen, Z. Additive manufacturing of 316L stainless steel by electron beam melting for nuclear fusion applications. *J. Nucl. Mater.* **2017**, *486*, 234–245. [CrossRef]

24. Thompson, S.W.; Vin Col, D.J.; Krauss, G. Continuous cooling transformations and microstructures in a low-carbon. high-strength low-alloy plate steel. *Metall. Mater. Trans. A.* **1990**, *21*, 1493–1507. [CrossRef]

25. Ali, Y.; Henckell, P.; Hildebrand, J.; Reimann, J.; Bergmann, J.P.; Barnikol-Oettler, S. Wire arc additive manufacturing of hot work tool steel with CMT process. *J. Mater. Process. Technol.* **2019**, *269*, 109–116. [CrossRef]

26. Casati, R.; Lemke, J.; Vedani, M. Microstructure and Fracture Behavior of 316L Austenitic Stainless Steel Produced by Selective Laser Melting. *J. Mater. Sci. Technol.* **2016**, *32*, 738–744. [CrossRef]

27. Pradhan, P.; Robi, P.; Roy, S.K. Micro void coalescence of ductile fracture in mild steel during tensile straining. *Fatigue Fract. Eng. Mater.* **2012**, *19*, 51–60. [CrossRef]

28. Mazumder, J.; Schifferer, A.; Choi, J. Direct materials deposition: Designed macro and microstructure. *Mater. Res. Innov.* **1999**, *3*, 118–131. [CrossRef]

29. Carroll, B.E.; Palmer, T.A.; Beese, A.M. Anisotropic tensile behavior of Ti–6Al–4V components fabricated with directed energy deposition additive manufacturing. *Acta Mater.* **2015**, *87*, 309–320. [CrossRef]

30. Baufeld, B. Effect of deposition parameters on mechanical properties of shaped metal deposition parts. *Proc. Inst. Mech. Eng. Part B* **2012**, *226*, 126–136. [CrossRef]

31. Baufeld, B. Mechanical properties of Inconel 718 parts manufactured by shaped metal deposition (SMD). *J. Mater. Eng. Perform.* **2012**, *21*, 1416–1421. [CrossRef]

32. Wang, F.; Williams, S.; Colegrove, P.; Antonysamy, A.A. Microstructure and mechanical properties of wire and arc additive manufactured Ti-6Al-4V. *Metall. Mater. Trans. A* **2013**, *44*, 968–977. [CrossRef]

33. Ho, A.; Zhao, H.; Fellowes, J.W.; Martina, F.; Davis, A.E.; Prangnell, P.B. On the origin of microstructural banding in Ti-6Al4V wire-arc based high deposition rate additive manufacturing. *Acta Mater.* **2019**, *166*, 306–323. [CrossRef]

34. Zhang, J.; Wang, X.; Paddea, S.; Zhang, X. Fatigue crack propagation behaviour in wire+arc additive manufactured Ti-6Al-4V: Effects of microstructure and residual stress. *Mater. Des.* **2016**, *90*, 551–561. [CrossRef]

35. Liu, Y.; Dong, D.; Wang, L.; Chu, X.; Wang, P.; Jin, M. Strain rate dependent deformation and failure behavior of laser welded DP780 steel joint under dynamic tensile loading. *Mater. Sci. Eng. A* **2015**, *627*, 296–305. [CrossRef]

36. Zhang, J.; Zhang, X.; Wang, X.; Ding, J.; Traore, Y.; Paddea, S.; Williams, S. Crack path selection at the interface of wrought and wire + arc additive manufactured Ti-6Al-4V. *Mater. Des.* **2016**, *104*, 365–375. [CrossRef]

37. Jiang, D.; Williams, P.F. High-strain zones: A unified model. *J. Struct. Geol.* **1998**, *20*, 1105–1120. [CrossRef]

38. Liu, Y.; Huang, H.; Xie, J. Anisotropic deformation behavior of continuous columnar-grained CuNi10Fe1Mn alloy. *Acta Metall. Sin.* **2015**, *51*, 40–48. [CrossRef]

39. Bagherpour, E.; Qods, F.; Ebrahimi, R.; Miyamoto, H. Microstructure and Texture Inhomogeneity after Large Non-Monotonic Simple Shear Strains: Achievements of Tensile Properties. *Metals* **2018**, *8*, 583. [CrossRef]

40. Lin, Z.; Goulas, C.; Ya, W.; Hermans, M.J.M. Microstructure and Mechanical Properties of Medium Carbon Steel Deposits Obtained via Wire and Arc Additive Manufacturing Using Metal-Cored Wire. *Metals* **2019**, *9*, 673. [CrossRef]

41. Gurland, J.; Plateau, J. The mechanism of ductile rupture of metals containing inclusions. *Trans. ASM* **1963**, *56*, 442–454.

42. Argon, A.S.; Im, J.; Safoglu, R. Cavity formation from inclusions in ductile fracture. *Metal. Trans. A* **1975**, *6A*, 825–837. [CrossRef]

Article

Processing by Additive Manufacturing Based on Plasma Transferred Arc of Hastelloy in Air and Argon Atmosphere

Eva M. Perez-Soriano [1], Enrique Ariza [2], Cristina Arevalo [1], Isabel Montealegre-Melendez [1,*], Michael Kitzmantel [2] and Erich Neubauer [2]

[1] Escuela Politécnica Superior, Universidad de Sevilla, 41011 Sevilla, Spain; evamps@us.es (E.M.P.-S.); carevalo@us.es (C.A.)
[2] RHP Technology GmbH, 2444 Seibersdorf, Austria; e.ar@rhp.at (E.A.); m.ki@rhp.at (M.K.); e.ne@rhp.at (E.N.)
* Correspondence: imontealegre@us.es; Tel.: +34-95-448-2278

Received: 30 December 2019; Accepted: 28 January 2020; Published: 30 January 2020

Abstract: This research was carried out to determinate the effect of the atmosphere processing conditions (air and argon) and two specific thermal treatments, on the properties of specimens made from the nickel-based alloy Hastelloy C-22 by plasma transferred arc (PTA). Firstly, the additive manufacturing parameters were optimized. Following, two walls were manufactured in air and argon respectively. Afterwards, a determinate number of specimens were cut out and evaluated. Regarding the comparison performed with the extracted specimens from both walls, three specimens of each wall were studied as-built samples. Furthermore, a commonly used heat treatment in Hastelloy, with two different cooling methods, was selected to carry out additional comparisons. In this respect, six additional specimens of each wall were selected to be heat treated to a temperature of 1120 °C for 20 min. After the heat treatment, three of them were cooled down by rapid air cooling (RAC), while the other three were cooled down by water quenching (WQ). In order to study the influence degree of the processing conditions, and how the thermal treatments could modify the final properties of the produced specimens, a detailed characterization was performed. X-ray diffraction and microstructural analyses revealed the phases-presence and the apparition of precipitates, varying the thermal treatment. Moreover, the results obtained after measuring mechanical and tribological properties showed slight changes caused by the variation of the processing atmosphere. The yield strength of the extracted specimens from the two walls achieved values closer to the standards ones in air 332.32 MPa (±21.36 MPa) and in argon 338.14 MPa (±9 MPa), both without thermal treatment. However, the effect of the cooling rate resulted as less beneficial, as expected, reducing the deformation properties of the specimens below 11%, independently of the air or argon manufacturing atmosphere and the cooling rate procedure.

Keywords: additive manufacturing; plasma transferred arc; processing conditions; mechanical properties; microstructure; Hastelloy C-22

---

1. Introduction

In recent years, additive manufacturing technologies have become valued processes based on their competitive and comparative advantages in the production of specimens with complex geometries. Regarding the technique developments, one classification could then be done dividing into two big groups in light of the way the material is provided: Powder bed techniques or blown-powder/wire-feed techniques [1–7].

The technology of plasma transferred arc (PTA) is considered one of the most interesting additive techniques to process specimens through layer depositions [8–11] by blown powder. When the wire is employed, the technique could be considered as one of the wire arc additive manufacturing (WAAM) processes, in addition to tungsten inert gas/metal inert gas (TIG/MIG) [12]. The PTA is included in the recently named group 3D Plasma Metal Deposition (3DPMD) that belongs to the category of directed energy deposition processes [13–15]. This manufacturing process permits us to carry out the fabrication of specimens with larger dimensions than specimens produced by powder bed techniques. Furthermore, the production rate might be higher due to the employment of a higher feed rate in comparison to other additive techniques [16,17].

In order to research the possibilities of developing materials with interesting properties, this study was proposed. The flexibility of the employed technique allowed the realization of a detailed investigation of possible changes in the final specimen's properties as consequence of variation in the setup parameters. The selected material was Hastelloy C-22. This alloy presents excellent corrosion behavior as well as mechanical properties, being commonly employed in the industrial sectors [18–23]: Chemical, petrochemical, aerospace, etc.

The main goal of this research was the production of Hastelloy C-22 walls with good mechanical properties by PTA. In pursuing this goal, the aim of this study was threefold: (1) The manufacturing of walls made from Hastelloy with large dimensions by the optimization of processing parameters; (2) the determination of influence degree in the specimens of the manufacturing atmosphere, in air or argon conditions; and (3) the evaluation of thermal treatments on the final behavior of specimens. Therefore, a wall was fabricated in air conditions firstly, and a second wall was built in argon atmosphere subsequently. Thereafter, from the two walls, determined samples were extracted from marked positions to compare their final properties. Moreover, two thermal treatments were defined to evaluate if the specimens would suffer variation in their properties caused by their cooling rate.

2. Materials and Methods

The starting material employed was powder from Hastelloy C-22, supplied by Atomising Systems Limited (Sheffield, UK). This powder was produced by the conventional method of powder manufacturing known as plasma-atomization process. In Figure 1, the spherical morphology of the Hastelloy particles can be appreciated. Furthermore, the chemical composition of the manufactured powder was compared to the standard one; both are listed in Table 1. In the granulometry given by the manufacturer, d50 (average) was 82.74 μm, and d10 and d90 were 57.39 μm and 125.14 μm, respectively.

Table 1. Composition of Hastelloy C-22.

Element	Supplied Hastelloy C-22 [%wt.]	Standard Hastelloy C-22 [%wt.]
Ni	Bal.	Bal.
C	0.007	max. 0.015
Co	0.40	max. 2.50
Cr	22.11	20.00–22.50
Fe	5.00	2.00–6.00
Mn	0.65	max. 0.50
Mo	12.37	12.50–14.50
Si	0.91	max. 0.08
V	<0.02	max. 0.35
W	3.63	2.50–4.00

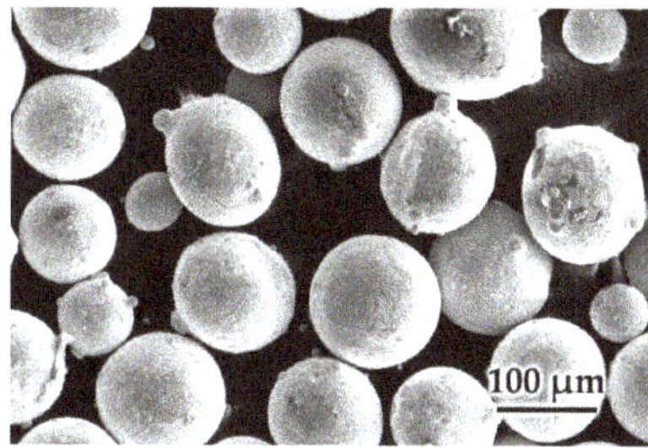

Figure 1. Circular backscatter detector (CBS)-SEM image of the starting powder of supplied Hastelloy C-22.

After the powder characterization, the production of the specimens was developed. In this research, the additive manufacturing equipment used for this research was based on PTA technology, self-made, and adapted (RHP-Technology GmbH, Seibersdorf, Austria). In the torch, the plasma was generated. Then, the feeding materials were melted with the plasma energy (Figure 2).

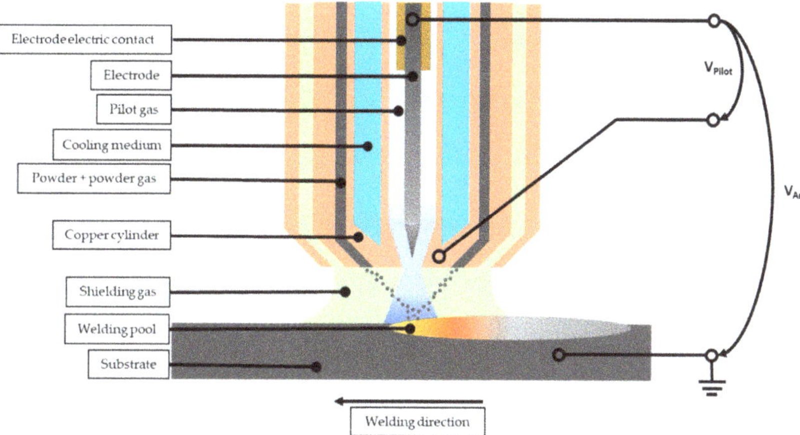

Figure 2. Scheme of the plasma transferred arc (PTA) torch.

As novelty in this research, an argon atmosphere was employed, in addition to an air conventional atmosphere. When using an argon atmosphere, the specimens were built inside a designed argon box. For ensuring the environmental conditions, an oxygen sensor (Oxy 3. Orbitec GmbH, Seligenstadt, Germany) was placed near the torch [24].

The torch had an internal cooling circuit where water ran, preventing its melting due to the plasma. The distance between the torch and the substrate or specimen during the fabrication was 10 mm. An argon plasma was induced by introducing the gas between the electrode and the copper cylinder (pilot gas, 1.5 L/min), applying a potential of 20 V. The pilot flame was employed to start the plasma arc between the electrode and the substrate connected to ground. Different electric currents could be applied to increase or decrease the intensity of the plasma arc energy.

At the same time, building in air atmosphere, there was another gas that acted as a shield (shielding gas, 15 L/min) preventing the seam from oxidizing during the process. It was applied by coupling an external copper cylinder to the torch. Several gases could be inserted as shielding gases, as argon pure or mixed with a 5 vol% CO_2 gas, depending on the material to be processed, and the properties to be reached. In this investigation, pure argon (99.99% purity, Air Liquide, Paris, France) was employed as shielding gas in the manufacturing of all the specimens.

The material was fed as powder to the plasma jet by aligned holes. For a better powder flowing through the ducts, it was injected with a pressurized gas (powder gas, 1.5 L/min), argon.

This additive manufacturing equipment could build large size components thanks to the torch fixed to an XYZ mechanical, which allowed the torch to move through the working table. This working table was made from aluminum profiles, cooled to ease the heat transfer through the system.

Firstly, a flow test was performed to check if the powder size and shape were optimal for the process. Values in g/min after using different engine units (U) of the rotating metering powder feeder were obtained during the flowability test (Figure 3). A clear linear trend was obtained as the value of motor unit's increase. This means that the powder had a good flow through the ducts of the system and there were no clogging problems.

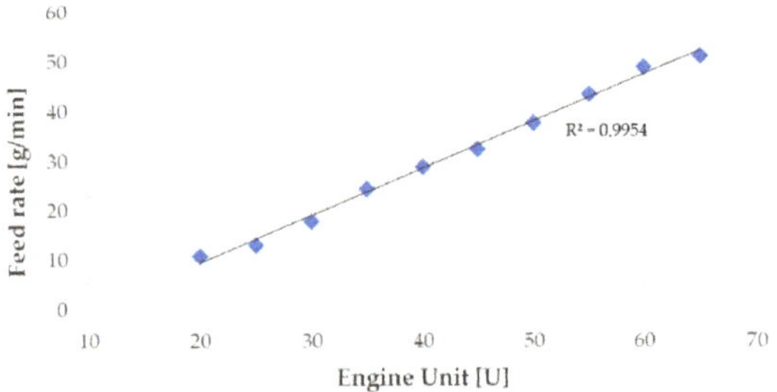

Figure 3. Flowability test of the Hastelloy C-22 powder on the PTA device.

After checking the flowability of the powders, a preliminary bead on a plate welding test was conducted with different parameters (see Table 2), to properly select the more efficient ones and to produce the best quality seam for a subsequent application to the final sample structure. Several parameters' combinations were tested in order to meet a range of processing conditions. By varying the most significant processing parameters (current 120–250 A, speed 100–400 mm/min, material feed rate 11–43.5 g/min), different single seams were welded on a steel AISI 1015 (Figure 4). The dimensions of this substrate were 200 mm × 300 mm × 10 mm. The surface was previously cleaned by brushing.

Table 2. Test processing parameters.

Layer	Current [A]	Travel Speed [mm/min]	Feed Rate [g/min]
1	120	200	13.5
2	140	200	13.5
3	180	200	13.5
4	220	200	13.5
5	120	200	11.0
6	120	100	11.0
7	140	100	11.0
8	160	100	11.0
9	140	300	29.0
10	180	300	29.0
11	220	300	29.0
12	250	400	43.5
13	140	300	11.0
14	180	400	29.0
15	180	400	24.5
16	140	300	13.5

Figure 4. Processing parameters test.

There were no initial geometrical specifications (seam width and height) predefined in this work. The final height resulted from the layer deposition, and the seam width was the obtained under the fabrication parameters.

By visual inspection, the parameters of the seams with better geometrical and surface quality were selected (6, 7, and 10 from Figure 4). Then, oscillation bead on plate welding tests were performed with these conditions (Figure 5). The oscillation movement helped to improve the quality of the different parts because the torch stayed longer on top of the hot spot after the melting pool. The oscillation parameters were: (1) The amplitude, 7.5 mm and (2) the overlapping, 2.5 mm. The oscillation amplitude resulted in 15 mm for the walls width. The shielding gas protected more during the very beginning of the cooling of the pool, reducing the oxidation due to high temperature. Considering the most promising parameters, seam number 14 from Figure 5, the constructions of both walls were conducted in air and argon. This sample was the candidate selected based on a visual checking of the first manufactured specimens, thus there were neither cracks nor pores.

Figure 5. Oscillation bead on plate welding test.

Obviously, the height of the resulting wall was set considering the height of each layer deposition. Setting the parameters listed in Table 3, the walls of dimensions 120 mm × 40 mm × 15 mm were manufactured. To reach this height of 40 mm, 16 layers were necessary, as can be appreciated in Table 3 and Figure 6. With the purpose of obtaining a proper comparison in the study of the effect of the manufacturing atmosphere on the material properties, the similar parameters, strategy, and the number of layers were used to build both walls, in air and argon atmospheres.

Table 3. Fabrication parameters for Hastelloy C-22 wall produced in air and in argon atmosphere.

Specimen	Layer Number	Current [A]	Travel Speed [mm/min]	Feed Rate [g/min]
Two Hastelloy walls produced in air and in argon	1	180	400	29.0
	2	160	500	24.5
	3–9	150	600	18.0
	10–16	150	700	13.5

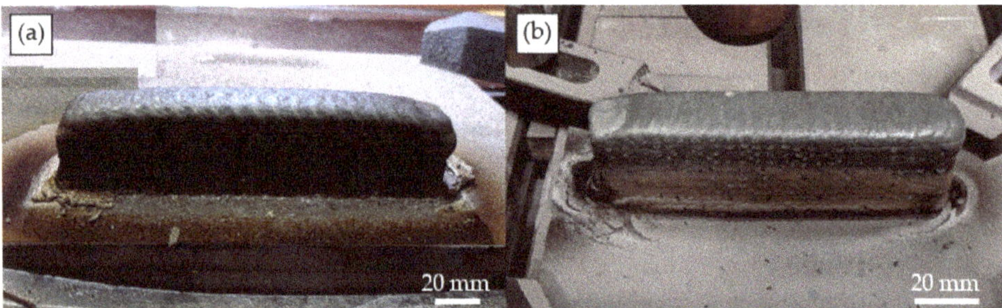

Figure 6. Walls in as-built conditions: (**a**) Produced under air atmosphere, (**b**) produced under argon atmosphere.

Next, the extraction of the specimens was conducted following a determinate distribution plan and marks of the location of the cut specimens from each wall. The cutting machine employed was an electrical discharge machine Mitsubishi FX-20 (Mitsubishi, Ratingen, Germany). The geometry of the tensile samples is shown in Figure 7, as well as the sample disposition on the wall. Three sets of samples were extracted.

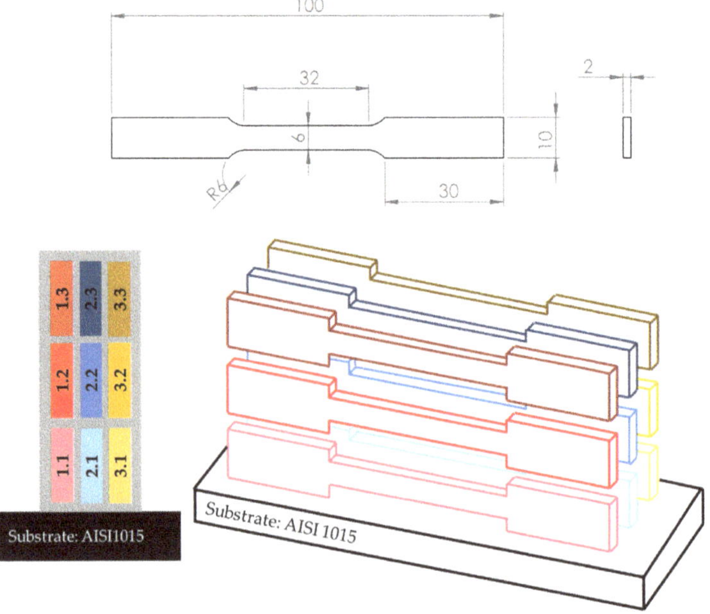

Figure 7. Tensile test sample geometry (in millimeters) and positioning of tensile test samples extracted from the produced sample walls.

The extracted specimens were studied under determined conditions: (1) As built, (2) after thermal treatment 1 (TT1), and (3) after thermal treatment 2 (TT2). Therefore, the heat treatment parameters were previously set up according to recent bibliography [25]. The cooling stage was the main difference between them. The two sets of samples were heat-treated at 1120 °C for 20 min in a high vacuum furnace under an argon atmosphere. On the one hand, the set named TT1 suffered a cooling down by rapid air cooling (RAC = TT1). On the other hand, the remaining treated set (named TT2) was cooled down by water quenching (WQ = TT2) after the heat treatment (Figure 8).

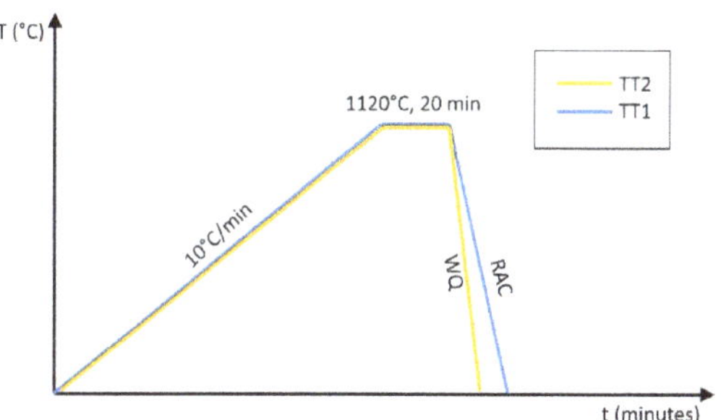

Figure 8. Thermal treatments applied to the specimens.

The study of the specimens was conducted through a detailed characterization. The microstructural study was performed by scanning electron microscopy (SEM) using a TENEO 6460LV microscope (FEI Teneo, Hillsboro, OR, USA), equipped with an energy dispersive X-ray spectrometry (EDS) system to carry out the phase analysis. X-ray diffraction (XRD) analysis was performed by a Bruker D8 Advance A25 (Bruker, Billerica, MA, USA) with Cu-K$_\alpha$ radiation. In a tester model, Struers-Duramin A300 (Struers, Ballerup, Denmark), the measurement of the mechanical properties was performed to ascertain the Vickers hardness (HV2). Room-temperature tensile tests were performed on a universal testing machine Instron 5505 (Instron, Norwood, MA, USA) with a strain rate of 0.5 mm·min^{-1}. Concluding the characterization, the tribological behavior of the samples was determined by a ball-on-disc tribometer (Microtest MT/30/NI, Madrid, Spain) using aluminum balls (6 mm in diameter) with a sliding speed of 200 rpm and a normal load of 5 N on the ball during 15 min on a circular path of 2 mm in radius. The surface morphology was studied by optical microscopy (OM) with a Leica Zeiss DMV6 (Leica Microsystems, Heerbrugg, Switzerland).

3. Results

3.1. Microstructural Study and XRD Analysis

The microstructural study revealed, in general, the apparition of precipitates that showed a distribution mainly oriented. Then, two typologies of phases were clearly differenced, a darker area named "matrix", and plenty of light grey precipitates. In Figure 9, the SEM images showed such precipitates, as well as their special orientation in the microstructure of a specimen fabricated under air and argon conditions during the PTA process without thermal treatment (named "as-built"), and placed at the position 1.2 (according to Figure 7) in both Hastelloy C-22 walls.

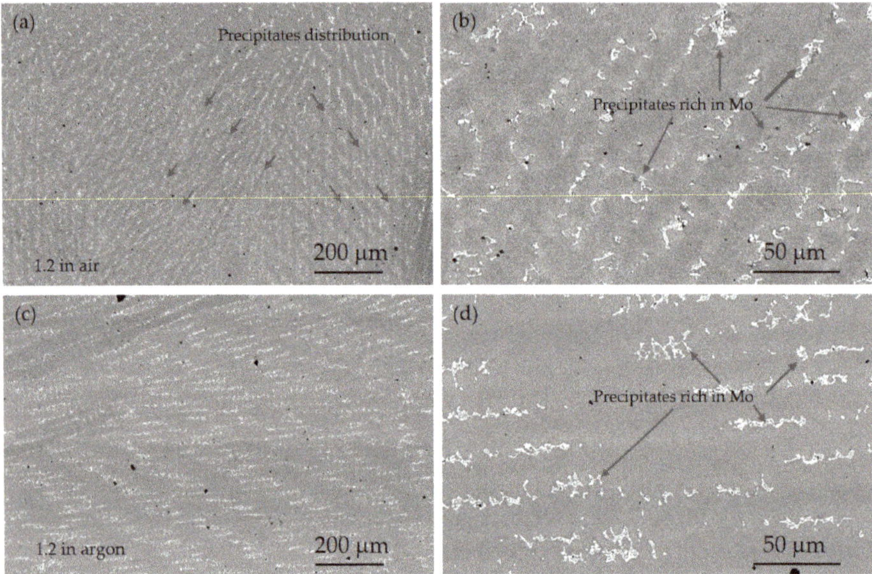

Figure 9. CBS-SEM images of Hastelloy C-22 specimens manufactured without thermal treatment (**a,b**), and in air (**c,d**) argon.

As can be seen in Figure 9a,c and Figure 10, at low magnifications (100×) the grain distribution could be clearly identified, regardless of the thermal treatment. Therefore, their origin and distribution could be attributed to possible segregation phenomena during the additive process. Residual porosity could be observed in the specimens, even after the heat treatments.

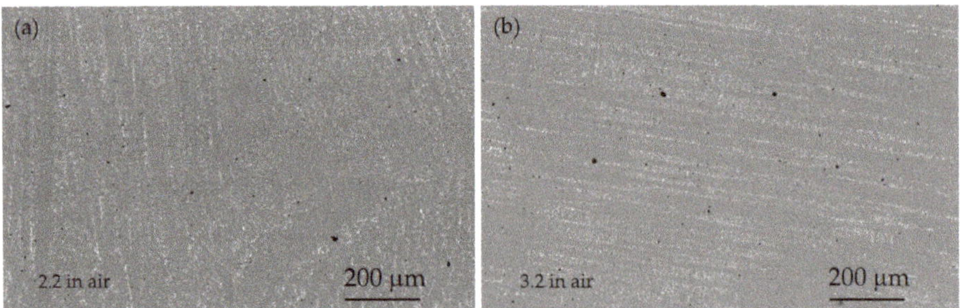

Figure 10. CBS-SEM images of Hastelloy C-22 specimen manufactured in air with thermal treatment, (**a**) thermal treatment 1 (TT1) (rapid air cooling (RAC)) and (**b**) TT2 water quenching (WQ).

Subsequent test by EDS analysis confirmed the compositional differences between the darker regions (matrix) and the light grey precipitates, observed in Figure 9. The representative spots checked, in addition to the results of the EDS analysis, are summarized in Figure 11 and Table 4. The elements analysis in light grey spots resulted in regions rich mainly in Mo, and in which the percentage in weight of Cr and Ni elements were lower than in the area considered as the matrix in the Hastelloy C-22. That suggested a possible diffusion of the Mo atoms, mainly during the layer depositions, as well as decrement of other characteristic elements of the Hastelloy C-22, as Ni and Cr. The mapping test (Figure 12) confirmed the elements' compositional variations occurred in the alloy, caused by the manufacturing process. Similar results were obtained in analysis carried out in other specimens processed in argon and thermal treated.

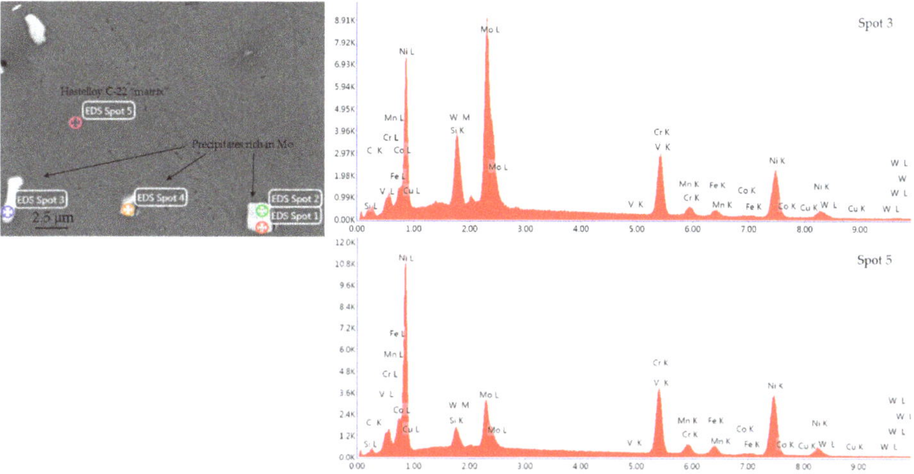

Figure 11. SEM image of the Hastelloy C-22 specimen manufactured under air without thermal treatment and EDS analysis (analyzed specimen 1.2).

Table 4. EDS composition for the Hastelloy C-22 specimen manufactured under air without thermal treatment.

Element	Spot 5 [%wt.]	Spot 3 in a Precipitate [%wt.]
C K	4.46	6.30
Si K	0.84	2.53
W M	1.81	5.63
Mo L	8.73	28.70
V K	0.11	0.07
Cr K	22.04	17.80
Mn K	1.43	0.79
Fe K	4.70	2.79
Co K	0.68	0.45
Ni K	54.41	34.59
Cu K	0.80	0.36

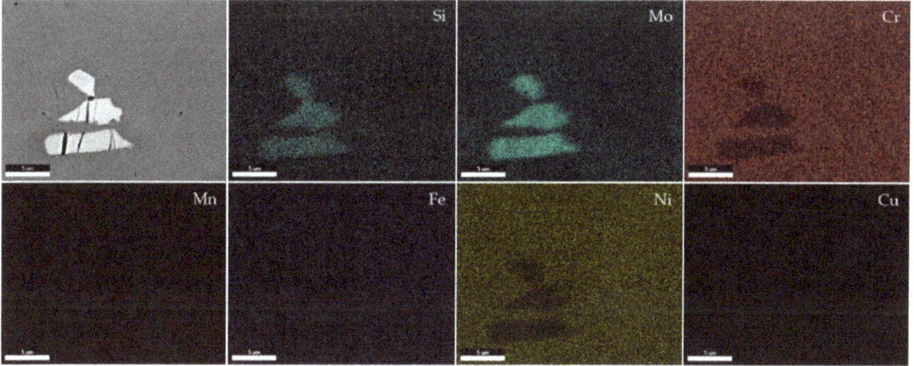

Figure 12. Mapping of specimen in air without thermal treatment (analyzed specimen 1.2): CBS-SEM (upper-left), Si, Mo, Cr, Mn, Fe, Ni, Cu.

The XRD analyses were performed, and the patterns of specimens produced under air are shown in Figure 13 and the ones manufactured in argon atmosphere are illustrated in Figure 14. The pattern of specimens without thermal treatment processing in air showed the highest peaks in comparison with the observed peaks in the pattern of the rest of the specimens under air or argon independently. This result suggested that the intensity of these peaks was related to the high crystallinity that this specimen could present. On the contrary, the alloy peaks belonging to the specimen processed under argon without any thermal treatment showed low intensity; note, in that regards, the processing conditions modified the characteristics on the crystallinity of the specimens. The pattern of specimens treated by cool down by water quenching (WQ = TT2) after the heat treatment showed almost the same tendency of the alloy peaks, independently of the processing conditions.

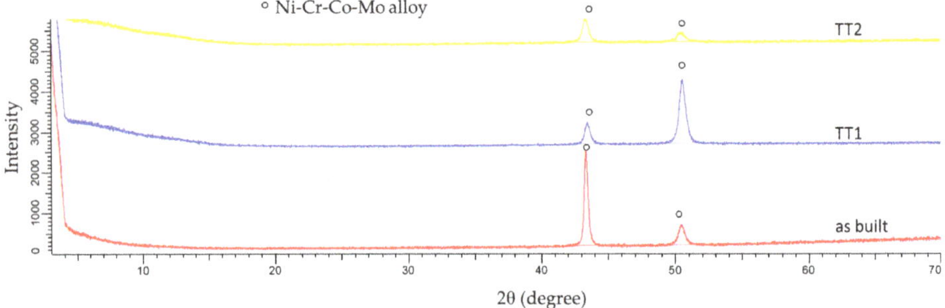

Figure 13. Pattern diffraction of specimens produced in air conditions.

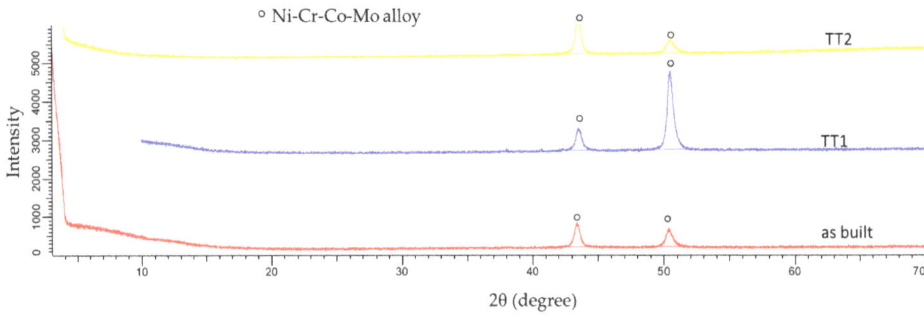

Figure 14. Pattern diffraction of specimens produced in argon conditions.

As it can easily appreciate, there were significant differences in the patterns of specimens whose thermal treatment finalized by rapid air cooling (TT1). This cooling process presented the lowest cooling rate, promoting in this stage singular variations in the crystallinity and phases presented in the specimens. Hence, in the framework of the thermal treatments, significant variations were found in the patterns of specimens where the crystallinity could be affected by the circumstances related to the cooling rate, independently of the air or argon fabrication conditions. Possible phases could be forming since peaks of these phases as FeNi and carbides could be matched to the main alloy peak. However, they were probably overlapped under the alloy peak. For that reason, the mentioned peaks were not marked in Figures 13 and 14.

3.2. Mechanical Properties and Wear Behavior

In this section, the mechanical properties obtained by tensile tests were studied and evaluated. The values of the most representative mechanical properties are summarized in Tables 5 and 6, being Young's modulus (E), yield strength (YS), ultimate tensile strength (UTS), and deformation. Three specimens were tested for each category, which means three specimens without treatment, three suffered a cooling down by rapid air cooling (TT1), and three after treatment by cool down by water quenching (TT2). When the empirical values of mechanical properties were compared to the theoretical ones, slight differences were observed [26]. The Hastelloy C-22 walls produced by the PTA technology, in average, (Tables 5 and 6) did not reach the mechanical properties that indicate the standard reference, the elongation properties, in particular, independently of the processing conditions and the thermal treatment [27]. This behavior is characteristic of the additive manufacturing process itself. While the YS values were closer to the minimum reference values (310 MPa) and UTS values were slightly lower than the reference ones (690 MPa), the elongation results were very low, not achieving even one-third or one-fourth of the standard value (reference value 45% of elongation). These results were similar with other ones obtained using additive manufacturing techniques [28]. Moreover, after performing the different thermal treatments (TT1 and TT2), the mechanical properties did not undergo some significant improvements. Although UTS properties in specimens fabricated in air and after TT1 and TT2 slightly improved on average, 660 MPa and 644 MPa, respectively, the elongation receded from the reference visibly below 11% and 9.2% in each case. Regarding the cooling rates referred to TT1 and TT2, the results suggested that the low cooling rate (TT1 = RAC) improved more the UTS than using a high cooling rate (TT2 = WQ) (see Table 5). In specimens processed in argon conditions, the mechanical properties showed almost the same trend, affecting the highest cooling rate (TT2 = WQ) to the elongation of the specimens, on average below 8.5%.

Table 5. Mechanical and physical properties of the specimens produced in air atmosphere condition.

Specimen	Treatment	E [GPa]	Yield Strength [MPa]	UTS [MPa]	Deformation [%]
1.1		209.30	352.52	611.84	16.59
1.2	As built	201.23	334.46	650.37	24.04
1.3		181.23	309.97	516.84	9.32
Average set 1		197.49	332.32	593.02	16.65
2.1		158.21	307.42	715.87	13.18
2.2	TT1	135.33	271.24	717.40	15.74
2.3		147.47	295.84	547.38	3.89
Average set 2		147.00	291.50	660.22	10.94
3.1		183.64	298.92	690.13	10.89
3.2	TT2	200.45	307.52	686.08	12.29
3.3		218.20	303.95	556.75	4.31
Average set 3		200.76	303.46	644.32	9.17

Table 6. Mechanical and physical properties of the specimens produced in argon atmosphere condition.

Specimen	Treatment	E [GPa]	Yield Strength [MPa]	UTS [MPa]	Deformation [%]
1.1		167.18	347.97	620.20	23.46
1.2	As built	137.53	336.75	516.90	11.51
1.3		130.11	329.71	513.90	12.26
Average set 1		144.94	338.14	550.33	15.74
2.1		131.23	348.52	711.72	15.22
2.2	TT1	130.92	350.08	646.63	9.39
2.3		155.37	366.25	625.52	7.96
Average set 2		139.18	354.95	661.29	10.86
3.1		161.62	365.35	628.71	7.04
3.2	TT2	152.37	349.25	604.39	6.86
3.3		135.85	337.09	625.98	10.50
Average set 3		149.95	350.56	619.69	8.13

Comparing the manufacturing atmosphere conditions, the alloy presented almost similar mechanical properties for both conditions in average values. While the wall built in air condition showed very low UTS values in specimens extracted on the top layer, the wall built under argon presented more homogeneity related to their mechanical properties, independently of the extracted location.

The obtained results after the tribological characterization were evaluated in order to determinate if there were correlations to the mechanical properties. In Figure 15, the friction coefficient of the tested specimens vs. distance is represented. This makes it possible to make a rough comparison of their friction behavior, according to the processing conditions and thermal treatment employed. Notwithstanding that the coefficient did not show significant differences among them, there was a slight improvement in the friction coefficient in specimens fabricated in air. Furthermore, the tendency of the friction coefficient in specimens cooled by RAC (TT1) suggested that this treatment produced a reduction of the friction coefficient with respect to the specimens tested as-built (without thermal treatment). Therefore, in some sense it might be said that this was a suitable treatment to decrease the friction coefficient. Regarding the effect of the TT2 in the friction coefficient, there were no major influences of such treatment in the friction coefficient, if specimens without treatment and specimens after TT1 were compared.

In the context of the mechanical properties and the friction coefficients, while the TT1 was considered a treatment that could cause embrittlement of the specimens, its effect promoted better friction behavior, in particular in specimens produced under air. In specimens produced under argon conditions, by contrast, a clear trend could not be appreciated easily.

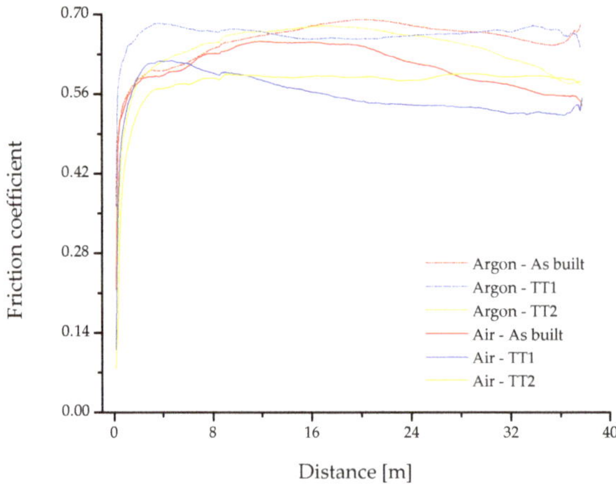

Figure 15. Friction coefficient vs. distance (m).

4. Discussion

In general, the processing atmosphere, air and argon, affected the final properties of the specimens; however, such influence was lower than expected. The findings revealed minus differences among specimens produced in air or argon conditions. This means that the processing in air could be feasible and it could involve easier manufacturing of specimens at a minor cost.

During the manufacturing process, segregation of precipitates occurred in air or argon conditions independently. The apparition of such precipitates rich in Mo could contribute to decreasing the mechanical properties, especially elongation behavior. Moreover, the origin of porosity could contribute to reducing the mechanical properties of the specimens. During the manufacturing, the crystallinity of the alloy was affected as shown by the different XRD patterns. This phenomenon may also be related to the friction coefficient, in addition to the brittleness of the specimens. The highest cooling rate promoted the highest brittleness of the specimens; however, the friction behavior was better.

Presumably, changes in the manufacturing process avoiding the segregation of the Mo rich precipitates could enhance the mechanical properties of the Hastelloy C-22 specimens, since the thermal treatments did not achieve this goal.

5. Conclusions

The findings obtained from this research were:

- Two walls from Hastelloy with large dimensions could be produced by plasma transferred arc in air or argon conditions.
- The atmosphere during the additive process presented lower impact on the final behavior than expected. The results of the thorough characterization of the specimens stated that their mechanical properties, as built, achieved values close to the standard ones, independently of the atmospheric conditions during the manufacturing.
- There was no effect of the thermal treatments in the final behavior of the specimens. The heat treatments were insufficient to modify and to improve the mechanical properties.
- In general, there were observed Mo-rich precipitates in all the specimens caused by processing; variations of the distribution of theses precipitates could be related to the mechanical properties measured.

Author Contributions: All the authors collaborated to obtain high-quality research work. E.M.P.-S. carried out the mechanical properties and the tribological tests, and performed the metallographic preparation. E.A. built the specimens with the PTA and selected the references. C.A. was responsible for the microstructure characterization for specimens: Optical and electron microscopy. I.M.-M. analyzed the data and the relationship between processing parameters and material properties, and designed the structure of the paper. M.K. controlled the fabrication process. E.N. optimized the equipment and applications. All authors read and agreed to the published version of the manuscript.

Funding: This research received funding from the European Union Horizon 2020 Programme (H2020) under grant agreement no. 768612.

Acknowledgments: The authors want to thank the Universidad de Sevilla for the use of experimental facilitates at CITIUS, Microscopy and X-Ray Laboratory Services (PPIT-2018-I.5 EVA M. PÉREZ SORIANO). The authors also wish to thank the technicians Jesús Pinto, Mercedes Sánchez, and Miguel Madrid for their experimental assistance.

Conflicts of Interest: The authors declare no conflict of interest.

References

1. Zhang, S.; Lane, B.; Whiting, J.; Chou, K. On thermal properties of metallic powder in laser powder bed fusion additive manufacturing. *J. Manuf. Process.* **2019**, *47*, 382–392. [CrossRef]
2. Fina, F.; Goyanes, A.; Gaisford, S.; Basil, A.W. Selective laser sintering (SLS) 3D printing of medicines. *Int. J. Pharmaceut.* **2017**, *529*, 285–293. [CrossRef] [PubMed]
3. Ríos, S.; Colegrove, P.A.; Williams, S.W. Metal transfer modes in plasma Wire + Arc additive manufacture. *J. Mater. Process. Technol.* **2019**, *264*, 45–54. [CrossRef]
4. Barros, R.; Silva, F.J.G.; Gouveia, R.M.; Saboori, A.; Marchese, G.; Biamino, S.; Salmi, A.; Atzeni, E. Laser powder bed fusion of Inconel 718: Residual stress analysis before and after heat treatment. *Metals* **2019**, *9*, 1290. [CrossRef]
5. Waqas, A.; Qin, X.; Xiong, J.; Zheng, C.; Wang, H. Analysis of ductile fracture obtained by charpy impact test of a steel structure created by robot-assisted GMAW-based additive manufacturing. *Metals* **2019**, *9*, 1208. [CrossRef]
6. Herderick, E. Additive manufacturing of metals: A review. *Mater. Sci. Technol. Conf. Exhib.* **2011**, *2*, 1413–1425.
7. Gibson, I. *Additive Manufacturing Technologies: 3D Printing, Rapid Prototyping, and Direct Digital Manufacturing*; Springer: Berlin, Germany, 2015.
8. Vály, L.; Grech, D.; Neubauer, E.; Kitzmantel, M.; Baca, L.; Stelzer, N. Preparation of titanium metal matrix composites using additive manufacturing. *Key Eng. Mater.* **2017**, *742*, 129–136. [CrossRef]

9.	Mercado Rojas, J.G.; Wolfe, T.; Fleck, B.A.; Qureshi, A.J. Plasma transferred arc additive manufacturing of Nickel metal matrix composites. *Manuf. Lett.* **2018**, *18*, 31–34. [CrossRef]

10.	Sharples, R.V. *The Plasma Transferred Arc (PTA) Weld. Surfacing Process*; Welding Institute: Cambridge, UK, 1985.

11.	Alberti, E.A.; Bueno, B.M.P.; D'Oliveira, A.S.C.M. Additive manufacturing using plasma transferred arc. *Int. J. Adv. Manuf. Technol.* **2016**, *83*, 1861–1871. [CrossRef]

12.	Dinovitzer, M.; Chen, X.; Laliberte, J.; Huang, X.; Frei, H. Effect of wire and arc additive manufacturing (WAAM) process parameters on bead geometry and microstructure. *Addit. Manuf.* **2019**, *26*, 138–146. [CrossRef]

13.	ISO/ASTM 52900:2015 (ASTM F2792). Additive manufacturing-General principles-Terminology. International Organization for Standardization. Available online: https://www.iso.org/standard/69669.html (accessed on 14 January 2020).

14.	Hoefer, K.; Mayr, P. Additive Manufacturing of titanium parts using 3D plasma metal deposition. *Mater. Sci. Forum* **2018**, *941*, 2137–2141. [CrossRef]

15.	Rodriguez, J.; Hoefer, K.; Haelsig, A.; Mayr, P. Functionally graded SS 316L to Ni-based structures produced by 3D plasma metal deposition. *Metals* **2019**, *9*, 620. [CrossRef]

16.	Tabernero, I.; Paskual, A.; Álvarez, P.; Suárez, A. Study on arc welding processes for high deposition rate additive manufacturing. *Procedia CIRP* **2018**, *68*, 358–362. [CrossRef]

17.	Ho, A.; Zhao, H.; Fellowes, J.W.; Martina, F.; Davis, A.E.; Prangnell, P.B. On the origin of microstructural banding in Ti-6Al4V wire-arc based high deposition rate additive manufacturing. *Acta Mater.* **2019**, *166*, 306–323. [CrossRef]

18.	Chen, L.; Bai, S.L.; Ge, Y.Y.; Wang, Q.Y. Erosion-corrosion behaviour and electrochemical performance of Hastelloy C22 coatings under impingement. *Appl. Surf. Sci.* **2018**, *456*, 985–998. [CrossRef]

19.	Chen, L.; Bai, S.L. The anti-corrosion behaviour under multi-factor impingement of Hastelloy C22 coating prepared by multilayer laser cladding. *Appl. Surf. Sci.* **2018**, *437*, 1–12. [CrossRef]

20.	Han, Q.; Mertens, R.; Montero-Sistiaga, M.L.; Yang, S.; Setchi, R.; Vanmeensel, K. Laser powder bed fusion of Hastelloy X: Effects of hot isostatic pressing and the hot cracking mechanism. *Mater. Sci. Eng. A Struct.* **2018**, *732*, 228–239. [CrossRef]

21.	Wang, Q.Y.; Bai, S.L.; Zhao, Y.H.; Liu, Z.D. Effect of mechanical polishing on corrosion behavior of Hastelloy C22 coating prepared by high power diode laser cladding. *Appl. Surf. Sci.* **2014**, *303*, 312–318. [CrossRef]

22.	Molin, S.; Dunst, K.J.; Karczewski, J.; Jasinski, P. High temperature corrosion evaluation of porous Hastelloy X alloy in air and humidified hydrogen atmospheres. *J. Electrochem. Soc.* **2016**, *163*, C296–C302. [CrossRef]

23.	Tu, R.; Goto, T. Corrosion behavior of Hastelloy-XR alloy in O_2 and SO_2 atmosphere. *Mater. Trans.* **2005**, *46*, 1882–1889. [CrossRef]

24.	Hussain, N.; Qureshi, A.H.; Shahid, K.A.; Chugtai, N.A.; Khalid, F.A. High-temperature oxidation Behavior of Hastelloy C-4 in steam. *Oxid. Met.* **2004**, *61*, 355–364. [CrossRef]

25.	Bal, K.S.; Majumdar, J.D.; Choudhury, A.R. Effect of post-weld heat treatment on the tensile strength of laser beam welded Hastelloy C-276 sheets at different heat inputs. *J. Manuf. Process.* **2019**, *37*, 578–594. [CrossRef]

26.	Wang, F. Mechanical property study on rapid additive layer manufacture Hastelloy X alloy by selective laser melting technology. *Int. J. Adv. Manuf. Technol.* **2012**, *58*, 545–551. [CrossRef]

27.	ASTM B574. *ASTM and SAE-AMS Standards and Specifications for Nickel Based Alloys*; ASTM: West Conshohocken, PA, USA, 2018.

28.	Montero-Sistiaga, M.L.; Pourbabak, S.; Van Humbeeck, J.; Schryvers, D.; Vanmeensel, K. Microstructure and mechanical properties of Hastelloy X produced by HP-SLM (high power selective laser melting). *Mater. Des.* **2019**, *165*, 107598. [CrossRef]

Article

Thermal Diffusivity Measurement of Laser-Deposited AISI H13 Tool Steel and Impact on Cooling Performance of Hot Stamping Tools

Jon Iñaki Arrizubieta [1,*], Magdalena Cortina [1], Arantza Mendioroz [2], Agustín Salazar [2] and Aitzol Lamikiz [1]

[1] Department of Mechanical Engineering, University of the Basque Country (UPV/EHU), Plaza Torres Quevedo 1, 48013 Bilbao, Spain; magdalena.cortina@ehu.eus (M.C.); aitzol.lamikiz@ehu.eus (A.L.)

[2] Department of Applied Physics I, University of the Basque Country (UPV/EHU), Plaza Torres Quevedo 1, 48013 Bilbao, Spain; arantza.mendioroz@ehu.eus (A.M.); agustin.salazar@ehu.eus (A.S.)

* Correspondence: joninaki.arrizubieta@ehu.eus; Tel.: +34-94-601-3932

Received: 19 December 2019; Accepted: 15 January 2020; Published: 20 January 2020

Abstract: Additive manufacturing is a technology that enables the repair and coating of high-added-value parts. In applications such as hot stamping, the thermal behavior of the material is essential to ensure the proper operation of the manufactured part. Therefore, the effective thermal diffusivity of the material needs to be evaluated. In the present work, the thermal diffusivity of laser-deposited AISI H13 is measured experimentally using flash and lock-in thermography. Because of the fast cooling rate that characterizes the additive process and the associated grain refinement, the effective thermal diffusivity of the laser-deposited AISI H13 is approximately 15% lower than the reference value of the cast AISI H13. Despite the directional nature of the process, the laser-deposited material's thermal diffusivity behavior is found to be isotropic. The paper also presents a case study that illustrates the impact of considering the effective thermal conductivity of the deposited material on the hot stamping process.

Keywords: DED; laser; additive manufacturing; thermal conductivity; thermal diffusivity; thermal modeling; hot stamping; AISI H13

---

1. Introduction

Metal additive manufacturing (AM) technologies are gaining increasing attention from both academia and industry because of the advantages they offer over conventional metal manufacturing techniques. These AM technologies enable the manufacture of near net shape complex structures and functionally graded components, which are impossible to fabricate through conventional methods.

In metal AM, an energy source (e.g., laser, electron beam, or plasma arc) irradiates and melts the surface of the substrate while filler material is added, building three-dimensional functional parts layer by layer. The advantages of using laser-based AM compared to conventional metallurgy and subtractive manufacturing were listed by Yan et al. [1], including a finer grain size, a small heat-affected zone, the possibility to process difficult-to-machine or refractory materials, as well as to combine materials, among others. Regarding the mechanical properties, Attar et al. [2] performed a comparative study of commercially pure titanium manufactured by laser-based AM processes and obtained comparable or even better mechanical properties than cast material. Much research has been devoted to deepening the understanding of laser-based metal AM by combining experimentation and modeling. In such works, the full characterization of the manufactured parts in terms of both mechanical and thermal properties is highly important.

Numerous investigations have focused on the mechanical properties of AM parts. For instance, Zhong et al. [3] presented an analysis of the mechanical properties and internal defects encountered in AM Inconel 718. Zhang et al. [4] studied the microstructures that developed in different zones of the substrate and deposited material, and Sun et al. [5] investigated the importance of stress-relieving treatments of AM AISI 4340 steel.

In addition, several works have modeled laser-based AM processes and studied their heat transfer characteristics. Roberts et al. [6] reported that the thermal conductivity value depends on factors such as the porosity of the material. Many models of laser-based powder bed fusion have been proposed to define the relationship between the thermal conductivities of powder-shaped and solid metals. All authors agree that the effective thermal conductivity of loose metallic powder is defined by the gas in the pores. Rombouts et al. [7] studied the effective thermal conductivity of the powder bed and concluded that its value is almost independent of the material composition and depends mainly on the size and morphology of the particles and the void fraction. Wei et al. [8] investigated five metal powders for powder bed additive manufacturing (Inconel 718, 17-4 stainless steel, Inconel 625, Ti-6Al-4V and AISI 316L) and concluded that the pressure and composition of the gas between the metallic particles have a significant effect on the thermal conductivity of the powder. Cernuschi et al. [9] calculated the thermal conductivity and density of the porous material using the Maxwell model.

Foteinopoulos et al. [10] reported an increase in the accuracy of the thermal model by assuming that the material's thermal properties, including thermal conductivity, are temperature dependent. In the same direction, Li et al. [11] found that, by neglecting the temperature dependence of the material properties, the size and volume of the melt pool were overpredicted. Although both Foteinopoulos et al. and Li et al. have considered the thermal conductivity reduction in the metallic powder (where the influence of the porosity is considered), once the material is melted, the tabulated value of the thermal conductivity of the cast material is used.

To the best of our knowledge, none of the previously published works considered that the thermal conductivity of AM parts may differ from that of the cast material. However, factors such as the existence of pores and micro-cracks affect the effective thermal conductivity of the manufactured part. Furthermore, the high cooling rate of laser processes (10^4–10^7 K·s^{-1}), in comparison with the much lower rate in casting (1–10 K·s^{-1}), influences the grain size that develops. In fact, Zhang et al. [12] reported that the grain size is much finer in parts produced by laser processes than that by casting, thus impacting the effective thermal conductivity of the AM part.

Therefore, this paper presents a study of the effective thermal conductivity of laser-deposited AISI H13 tool steel as a function of depth. In this work, the thermal diffusivity of the deposited material is measured experimentally. The accuracy of the methods employed to measure this thermal property enables the calculation of thermal conductivity from the diffusivity data. In addition, the impact of this issue on industrial applications is illustrated by means of a case study of the hot stamping process.

Hot stamping, also known as press hardening, is a process in which an ultra-high-strength steel blank is simultaneously formed and quenched. For this purpose, the blank is heated above the austenitic temperature, approximately 950 °C, and cooled at rates above 27 K·s^{-1} to ensure a complete martensitic phase transformation.

As stated by Shan et al. [13], the cooling of the blank consumes almost 30% of the total cycle time required to form and quench the material. According to Chen et al. [14], one approach to reducing the cycle time is to increase the thermal conductivity of the hot stamping tools, because it ensures quick heat transfer between the tools and the stamped part. In this direction, Directed Energy Deposition (DED) has arisen as an alternative to enhance the cooling performance of the tools. On the one hand, DED enables the manufacture of conformal cooling channels that can follow the surface of the tools, therefore avoiding hot spots. On the other hand, DED allows the combination of different materials to produce bimetallic tools. Materials with high thermal conductivity are used in the core of the tools, and high-resistance tool steel is used as a coating to withstand the high pressures and temperatures that are reached in the hot stamping process.

In the cooling stage of the hot stamping process, the thermal conductivity of the hot stamping tools is an important factor in the prediction of the temperature distribution within the blank and the cooling rate. This fact was highlighted by Karbasian and Tekkaya [15]. Different software programs have been developed to model the hot stamping process, such as LS-DYNA, Auto-Form, and PamStamp; however, all of these require the material data as inputs. Therefore, if DED-manufactured tools are to be used in hot stamping, their effective thermal properties must be known. Hence, defining the thermal conductivity of the DED-manufactured material is essential for determining the cooling capability of the tools. Many authors have developed diverse models of different additive manufacturing processes. Denlinger et al. [16] developed a thermomechanical model of electron beam deposition aimed at large parts, while Mukherjee et al. [17] studied the mitigation of thermal distortion during AM using a numerical heat transfer and fluid flow model. Peyre et al. [18] developed an analytical and numerical model of laser-aided DED, and Shi et al. [19] proposed a three-dimensional finite element model to investigate the effects of laser processing parameters on the thermal behavior and melting/solidification mechanism during selective laser melting. Regardless of the modeled process, all of them consider that properties such as thermal conductivity and specific heat are temperature dependent. However, all of them consider the thermal conductivity of only the cast material, neglecting possible variations caused by the manufacturing process to which the tool material has been subjected. This omission motivated the selection of the hot stamping process as the case study presented here.

2. Materials and Methods

2.1. DED Tests

The DED experiments were performed on a 5-axis laser-processing machine, with a work-piece size capacity of $700 \times 360 \times 380$ mm^3. A high-power Yb:YAG fiber laser, Rofin FL010 (ROFIN-SINAR Laser GmbH, Bergkirchen, Germany), with a maximum power output of 1 kW was employed. In addition, the powder was fed by means of a Sulzer Metco Twin 10 C powder feeder (Oerlikon Metco, Pfäffikon, Switzerland), and an in-house designed coaxial nozzle [20], using argon as both the drag and shielding gasses.

In the experimental tests, AISI 1045 (DIN 1.1191) and AISI H13 (DIN 1.2344) were used as the base and filler materials, respectively. AISI 1045 is a medium carbon steel commonly used in structural parts requiring high strength and hardness. AISI H13 is a Cr-Mo-V alloyed tool steel with a high level of resistance to thermal shock and fatigue and good temperature strength, which makes this material particularly valuable for tooling. The filler material was supplied by Flame Spray Technologies (Duiven, The Netherlands) and obtained via gas atomization, consisting of spherical particles with diameters of 53–150 μm. The chemical compositions of the employed materials are detailed in Table 1.

Table 1. Chemical compositions (wt %) of AISI 1045 [21] and AISI H13 [22].

Material	C	Si	Mn	Cr	Mo	V	Fe
AISI 1045	0.45	0.24	0.8	0.16	-	0.02	Balance
AISI H13	0.41	0.80	0.25	5.12	1.33	1.13	Balance

First, two specimens of $50 \times 50 \times 7$ mm^3 and $50 \times 50 \times 5$ mm^3, respectively, were manufactured by adding AISI H13 over an AISI 1045 substrate via DED, employing the process parameters detailed in Table 2. A zigzag pattern was used to deposit the filler material, alternating longitudinal and transversal directions for the deposition of successive layers, as shown in Figure 1a. This strategy reduces the anisotropic behavior inherent to the DED process and allows the manufacture of larger parts, which enables the transfer of the results obtained to real components. Figure 1b shows a photograph of the manufactured specimens.

Table 2. Process parameters employed for the deposition of AISI H13.

Process Parameters	Value
Continuous-wave laser power (W)	600
Feed rate (mm·min^{-1})	450
Track offset (mm)	1
Overlap (%)	50
Powder flow rate (g·min^{-1})	3.3
Shielding gas flow rate (L·min^{-1})	14

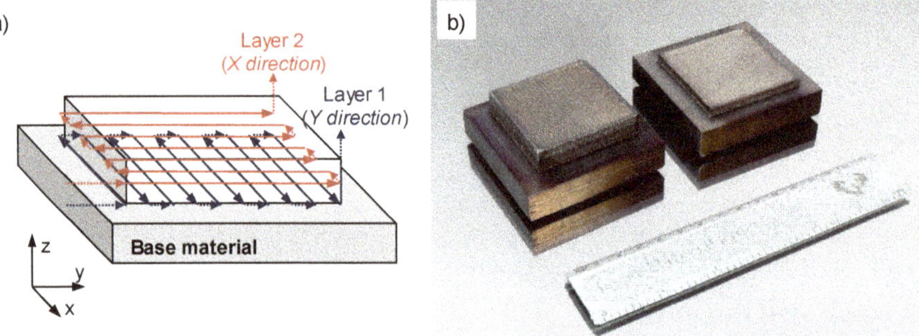

Figure 1. Schematic of the Directed Energy Deposition (DED) process (**a**) and (**b**) photograph of the manufactured AISI H13 specimens.

2.2. Thermal Diffusivity Measurement

To perform thermal diffusivity measurements, three slabs, each 2 mm thick, were extracted from the deposited material at different depths, as shown in Figure 2. From the 7-mm-thick specimen, two plates were cut: (a) the inner plate, Sample 1, contained the deepest and earliest deposition (0 to 2 mm from the substrate); (b) the outer plate, Sample 2, contained the outermost side of the coating (4 to 6 mm from the substrate). Sample 3 was extracted from the specimen with a 3.5 mm deposition thickness, and the sample spanned the interface between the filler and substrate, from −1 to 1 mm with respect to the interface, to evaluate the influence of the DED process on the substrate. Moreover, for comparison, a 2-mm-thick plate made of cast AISI H13 was also prepared. All samples were extracted by means of wire electrical discharge machining, and the white layer generated on the cut surfaces was ground to eliminate the heat-affected region.

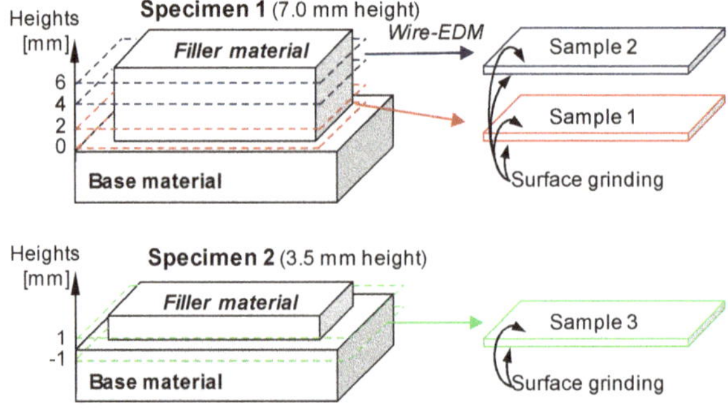

Figure 2. Sample extraction for thermal diffusivity measurements.

For each plate, the thermal diffusivities were measured at room temperature in two perpendicular directions: along the surface, the so-called in-plane thermal diffusivity ($\alpha_{\parallel}$), and in the direction perpendicular to the surface, the so-called through-thickness thermal diffusivity ($\alpha_{\perp}$).

To measure $\alpha_{\perp}$, a flash method was used, which was developed by Parker et al. [23]. In this technique, the front surface of the plate was illuminated homogeneously by the brief pulse of a flash lamp (3 kJ energy pulse, 3 ms duration) while the temperature evolution of the back-surface was recorded by a mid-infrared video camera (3–5 µm wavelength) operating at a rate of 950 frames·s^{-1}. The thermal diffusivity was obtained by measuring the time required to reach half of the maximum temperature rise ($t_{1/2}$), which was related to the thermal diffusivity through Equation (1), where L is the plate thickness:

$$t_{1/2} = 0.1388 \frac{L^2}{\alpha_{\perp}}. \tag{1}$$

In order to enhance both the absorption to the flashlight and the infrared emissivity, the sample surfaces were covered by a very thin graphite layer (≈ 3 µm thick). According to Maillet et al. [24], the influence of this layer on the accuracy of the thermal diffusivity values is less than 1% provided the sample is much thicker than the graphite layer (in the present case, 2 mm against 6 µm).

To measure $\alpha_{\parallel}$, a lock-in thermography setup with laser spot excitation was used, which was first used by Heath and Winfree [25] and enables measurements of the thermal diffusivities of the materials with high accuracy. This technology has been widely used for similar applications, for example, Nolte et al. [26] determined the thermal diffusivity of sheets of brass, stainless, and structural steel. The sample is illuminated by an intensity-modulated laser beam, tightly focused on the surface, and the oscillating component of the temperature rise is detected by an infrared video camera connected to a lock-in module. By analyzing the radial dependence of the temperature phase, the in-plane thermal diffusivity can be retrieved with ease, based on the linear relationship between the phase of the temperature and the lateral distance to the heating spot, the slope of which (m) is given by Equation (2), where f is the modulation frequency:

$$m = -\sqrt{\frac{\pi \times f}{\alpha_{\parallel}}}. \tag{2}$$

2.3. Thermal Modeling of the Tool Cooling

In order to quantify the influence of the effective thermal conductivity of the laser-deposited AISI H13 on a bimetallic hot stamping tool, two different cases were simulated using the same geometry, shown in Figure 3. The aim of the simulation was to quantify the impact of considering the real DED AISI H13 thermal conductivity or the data from the bibliography. Therefore, no optimization of the geometry of the cooling channels was performed and the cooling channels' position and geometry were maintained. The geometry has a $300 \times 170 \times 150$ mm^3 bounding box and the cooling channels have an 8 mm diameter and are positioned at a 12 mm distance from the contact face with the blank. The tool has an AISI 1045 core, which was coated with a 3-mm-thick DED AISI H13. In Case 1, the thermal conductivity value of the cast AISI H13 was used as a reference, whereas in Case 2, the effective thermal conductivity value of the deposited AISI H13 was considered. In both cases, the stamped blank was made of USIBOR 1500 steel (22MnB5), a boron alloyed steel that is well-suited for the entire range of automotive structural parts, which require high resistance to anti-intrusion during impact.

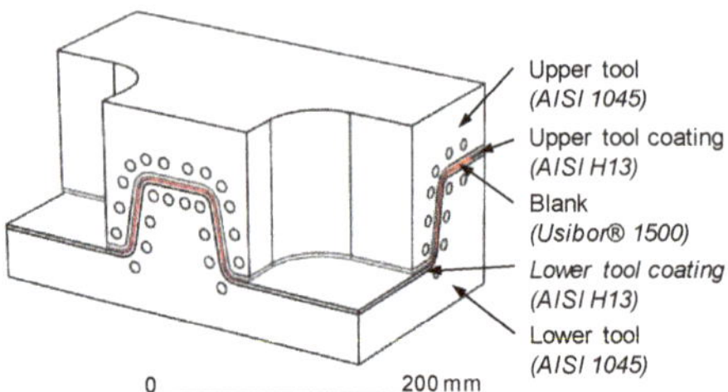

Figure 3. Simulated geometry of the bimetallic hot stamping tool.

The simulation was carried out using the thermal transient module of the FEM software ANSYS Workbench 19.2 (Ansys Inc., Canonsburg, PA, U.S.). The employed mesh consists of over 1 million first-order tetrahedral elements, with an average skewness of 0.246 and a maximum of 0.846. The initial temperature of the tools, as well as the reference temperature for the water-cooling convection, was set at 20 °C, whereas the temperature of the blank after the loading operation was 810 °C [27]. The blank was 1.85 mm thick, which is a typical thickness for an automotive sheet metal structural body part [28]. The geometric parameters of the tools are detailed in Table 3, and the thermal properties of the employed materials are shown in Table 4. The model simulated a 20 s cooling time, which is a typical value for hot stamping already used by other authors [27,29].

Table 3. Geometric parameters of the simulated tools.

Parameter	Cases 1 and 2
Diameter of the cooling ducts	8 mm
Length of the cooling ducts	170–280 mm
Distance between cooling ducts	15–20 mm
Distance from cooling duct center to surface	12 mm
Number of ducts in the upper/lower tools	12/10
Coating thickness	3 mm

Table 4. Thermal properties of AISI H13, AISI 1045, and USIBOR 1500, data obtained from [21,22,30].

Material	Thermal Properties	Temperature (°C)					
		20	200	400	600	800	1000
AISI H13	Specific heat ($J{\cdot}kg^{-1}{\cdot}K^{-1}$)	461	475	519	592	-	-
	Thermal conductivity ($W{\cdot}m^{-1}{\cdot}K^{-1}$)	24.9	27.4	29.1	28.5	-	-
AISI 1045	Specific heat ($J{\cdot}kg^{-1}{\cdot}K^{-1}$)	475	495	565	700	-	-
	Thermal conductivity ($W{\cdot}m^{-1}{\cdot}K^{-1}$)	47.6	40.4	36.2	32.0	-	-
USIBOR® 1500	Specific heat ($J{\cdot}kg^{-1}{\cdot}K^{-1}$)	444	520	561	581	590	603
	Thermal conductivity ($W{\cdot}m^{-1}{\cdot}K^{-1}$)	30.7	30.0	21.7	23.6	25.6	27.6

The tools are cooled by the convection of the water that is forced through the cooling channels, a parameter referred to as the convective heat transfer coefficient (CHTC). For cooling channels manufactured via drilling, Coldwell et al. measured the inner roughness between 0.14 and 0.48 μm [31]. Thus, an intermediate Ra value of 0.31 μm was considered in the present case. According to Arrizubieta et al. [32], for mechanically drilled 8-mm-diameter ducts with a 0.31 μm Ra value and a 20 °C cooling water, the CHTC is 4736.7 $W{\cdot}m^{-2}{\cdot}K^{-1}$.

The heat transfer between the hot blank and the tools needs to be established as an input parameter in the model. This parameter is referred to as the interfacial heat transfer coefficient (IHTC). In the present study, the correlation proposed by Hu et al. [33] was taken as a reference. Considering a 15 MPa contact pressure, a value which was already considered by Cortina et al. [34], the IHTC was estimated to be approximately 3000 $W{\cdot}m^{-2}{\cdot}K^{-1}$, based on the aforementioned approximation.

To compare the cooling performance of the tools using either the effective thermal conductivity of DED AISI H13 or the reference thermal conductivity, the time point at which the martensitic transformation was complete (280 °C) was calculated. In addition, the time at which the blank was cooled to below 70 °C was determined to define the total cycle time before the tools were opened.

3. Results and Discussion

3.1. Quality of the Deposited Material

Defects such as pores and cracks generate discontinuities within the material, lowering its density and possibly decreasing thermal conductivity. Therefore, the quality of the deposited material must be analyzed to evaluate defects and their impact on thermal properties. Therefore, three details of the cross-section of the DED AISI H13 were evaluated. The cross-sections were polished and etched using Murakami and Marble reagents. The samples are shown in Figure 4.

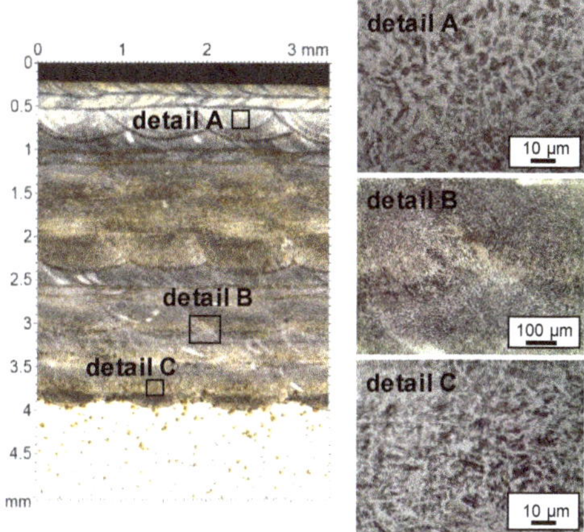

Figure 4. Cross-section of the DED AISI H13 and details of the microstructure.

As can be seen from the figure, clads free of cracks were attained, and this microstructure would be expected to ensure the continuity of the heat transfer within the deposited material. In addition, no defects were found at the interface between the DED AISI H13 and substrate AISI 1045 materials.

3.2. Effective Thermal Diffusivity and Conductivity Values

The results of the measurements of the through-thickness thermal diffusivity, $\alpha_\perp$, are summarized in the third column of Table 5. The statistical uncertainty was obtained by repeating each measurement five times and the uncertainty was thus found to be less than 3%.

Table 5. Thermal diffusivity results.

Sample	Distance from the Interface (mm)	$\alpha_\perp$ (mm^2·s^{-1})	$\alpha_\parallel$ (mm^2·s^{-1})	Material
1	0	5.72 ± 0.15	5.66 ± 0.16	DED AISI H13
1	2	5.72 ± 0.17	5.88 ± 0.17	DED AISI H13
2	4	6.03 ± 0.16	6.10 ± 0.18	DED AISI H13
2	6	6.03 ± 0.18	6.02 ± 0.17	DED AISI H13
3	−1	-	12.5 ± 0.4	Base AISI 1045
3	1	-	5.73 ± 0.16	DED AISI H13
Reference	-	6.75 ± 0.20	6.42 ± 0.19	Cast AISI H13

As shown in the table, the thermal diffusivity of the laser-deposited AISI H13 was always smaller than that of the cast sample. The thermal diffusivity of each plate was measured in two directions, i.e., from both the front (illuminated) surface and the rear (measured) surface, and the thermal diffusivities thus retrieved were the same. This homogeneity is obtained because the flash method measures the effective (or average) $\alpha_\perp$. Because Sample 3 included two different materials, $\alpha_\perp$ was not measured in that sample.

As for the in-plane thermal diffusivity measurements, $\alpha_\parallel$, Figure 5 shows the amplitude and phase thermograms of Sample 1, at the surface 0 mm from the substrate, with f = 7 Hz. For each specimen, the thermal diffusivity on both sides was measured. The round shape of the isophases and isotherms in Figure 5 is representative of all the cases analyzed and indicates in-plane thermal isotropy. The in-plane thermal diffusivity was obtained considering the vertical profiles of the phase thermograms (the white vertical line in Figure 5b) since they are free from diffraction effects, like those observed in the horizontal profile in Figure 5b. From the slope of the vertical phase profile, the in-plane thermal diffusivity was obtained, using Equation (2). In order to average local heterogeneities, the measurement was repeated at five different zones at the sample surface. The thermal diffusivities obtained using this method together with the uncertainty ($\approx$3%) are summarized in the fourth column of Table 5. The uncertainty takes into account the standard deviation in the slope of the phase profile and the standard deviation of the five repetitions.

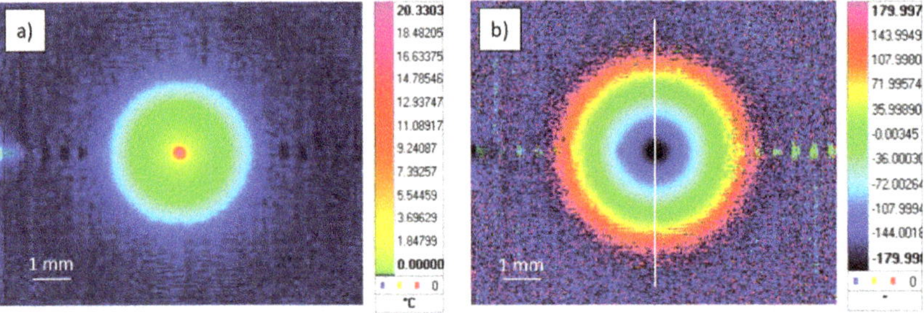

Figure 5. (**a**) Amplitude and (**b**) phase thermograms of Sample 1 at the surface 0 mm from the substrate, with a modulation frequency of 7 Hz. The white vertical line corresponds to the phase profile used for thermal diffusivity measurements. The scale of the amplitude is in °C and the phase is in degrees.

Although the thermal diffusivity of each surface increased with the surface's height above the substrate, at all heights, the thermal diffusivity of the sample remained below the diffusivity of the reference cast material. Comparing the values of $\alpha_\perp$ and $\alpha_\parallel$, the DED process did not introduce any thermal anisotropy.

Thermal diffusivity, α, and conductivity, k, are related to Equation (3), where ρ and c_p are the density and specific heat of the material, respectively:

$$\alpha = \frac{k}{\rho \times c_p}. \tag{3}$$

Equation (3) was used to calculate the thermal conductivity of DED AISI H13, cast AISI H13, and AISI 1045, and the results are shown in Table 6. The perpendicular thermal diffusivity values shown in Table 5 were used because the flash technique is generally acknowledged to be the most reliable method and is covered by standards, such as ASTM International [35], the British Standards Institution [36], and the Japanese Standards Association [37]. The density and specific heat were taken from the material specifications. In the case of DED AISI H13, the heat capacity was calculated using the rule of mixtures given in Equation (4) and considering this material as a mixture of AISI H13 and air.

$$\left(\rho \times c_p\right)_{\text{DED AISI H13}} = v_1 \times \left(\rho \times c_p\right)_{\text{cast AISI H13}} + v_2 \times \left(\rho \times c_p\right)_{\text{air}}. \tag{4}$$

Table 6. Thermal conductivities.

Material	Measured Thermal Diffusivity (mm^2·s^{-1})	Effective Thermal Conductivity (W·m^{-1}·K^{-1})	Reference Value from Bibliography (W·m^{-1}·K^{-1})
DED AISI H13 (Sample 1)	5.72	20.7	24.9 [22]
Cast AISI H13	6.75	24.4	
Base AISI 1045	12.5	46.7	47.6 [21]

In Equation (4), v_1 and v_2 are the volume fractions of cast AISI H13 and air, respectively. Because v_1 > 0.995, the same heat capacity was used for the cast reference and DED AISI H13. This result is consistent with the fact that the heat capacity, unlike the thermal transport properties (α and k), depends only on the composition of the sample, not the microstructure. Therefore, because the DED process does not affect the sample composition, the same heat capacity is expected for AISI H13 regardless of the production process.

In the cast AISI H13 and the base AISI 1045, the measured effective thermal diffusivities presented almost no differences compared to the reference values encountered in the literature [21,22], demonstrating the accuracy of the measurements acquired in this study. In the case of DED AISI H13, the effective thermal conductivity was 15.3% lower than that of cast AISI H13. This is because the thermal conductivity of alloys depends not only on the sample composition but also on the microstructure (grain size, micro-cracks, pores, etc.). Because the micrographs shown in Figure 4 do not indicate the presence of cracks or pores, the thermal conductivity reduction was attributed primarily to the smaller sizes of the grains produced by the fast cooling rate in DED in comparison with conventional manufacturing processes. According to Berman [38], the larger number of interfaces compared to the cast material reduces the electron mean free path and consequently the thermal conductivity. This effect is especially noticeable in the first layers, where the cooling rate is maximum and therefore, the microstructure is finer. As the number of deposited layers increases, the heat dissipation is slowed down, which leads to slower cooling rates and coarser grain sizes. Consequently, the thermal conductivity variation within the deposited material is attributed to the differences in the grain size, thus leading to lower values in the first deposited layers.

Focusing on the variation of thermal conductivity in relation to temperature, Zhang et al. [39] studied the thermal conductivity change of multi-stacked silicon steel sheets under different pressure and temperature conditions. Their results showed that although the thermal conductivity changed under different compressive stresses, the conductivity maintained the same rate of variation in response to temperature change. Therefore, in this study, the thermal conductivity reduction measured at 20 °C was assumed to affect the material in proportion to the temperature, and this result was extended to the whole temperature range, as shown in Table 7, and applied to the case study model described in the next section.

Table 7. Effective thermal conductivity of the DED AIS H13 considered in the thermal model.

Material	Thermal Properties	Temperature (°C)					
		20	200	400	600	800	1000
DED AISI H13	Specific heat (J·kg^{-1}·K^{-1})	461	475	519	592	592	592
	Thermal conductivity (W·m^{-1}·K^{-1})	20.7	22.8	24.2	23.7	23.7	23.7

3.3. Thermal Modeling and Cycle-Time Reduction

The influence of the thermal conductivity differences between the cast and DED AISI H13 tool steels was evaluated by means of thermal simulation of the upper part of an automotive structural body part with a B-pillar type geometry.

Based on the effective thermal conductivity of the laser-deposited AISI H13 tool steel, the cycle times required to lower the blank temperature below 280 and 70 °C were calculated, respectively.

Figure 6 shows the evolution of the maximum temperature of the blank. The lower thermal conductivity of the laser-deposited AISI H13 can be seen to reduce heat extraction from the blank, and this case thus requires a longer cooling time to achieve an equivalent thermal field. Table 8 presents the results of the simulated case study, as well as the error produced if the effective thermal conductivity of the laser-deposited AISI H13 is not considered in the model.

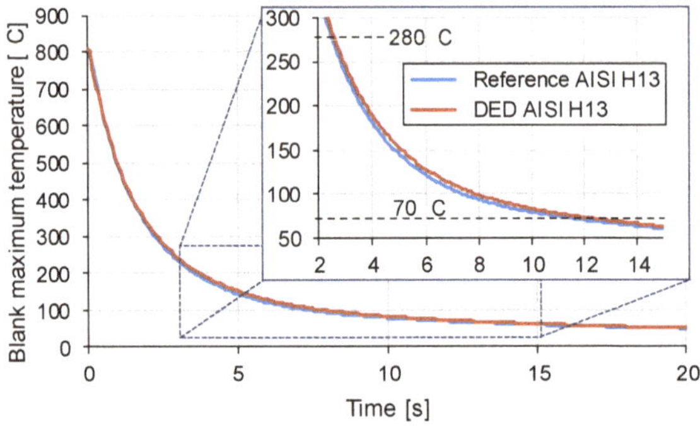

Figure 6. Blank maximum temperature evolution during the hot stamping process.

Table 8. Results of the simulated case study.

Blank Maximum Temperature	Time Point (s)		
	Reference AISI H13	DED AISI H13	Difference (%)
280 °C	5.50	5.59	1.64
70 °C	12.10	12.89	6.53

The errors generated when calculating the thermal fields were relatively low in comparison with the differences in thermal conductivity values. This is because the coating thickness was only 3 mm, and such low thickness values are commonly employed in bimetallic tools. Nevertheless, if fully DED-manufactured structures are employed, much higher errors could be generated in the simulation. Figure 7 shows the thermal field of the blank at 12.89 s in the case where the effective thermal conductivity of the DED AISI H13 coating is considered.

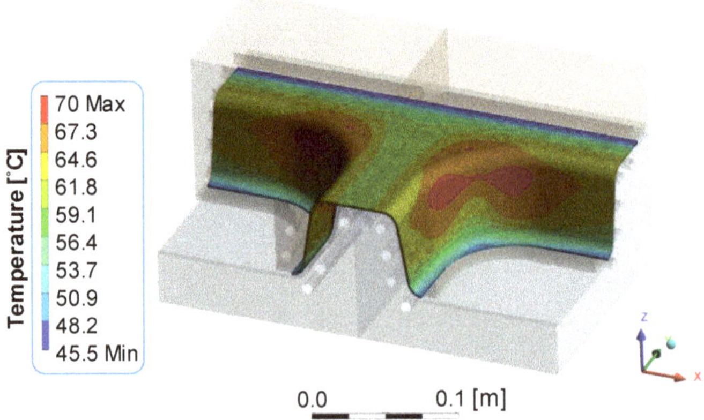

Figure 7. Temperature field of the blank at 12.89 s time instant, considering the effective thermal conductivity of the DED AISI H13 coating.

4. Conclusions

This work investigated the effective thermal diffusivity of laser-deposited AISI H13 tool steel, and the following conclusions were drawn:

(1) The effective thermal diffusivity of the laser-deposited AISI H13 tool steel is lower than that of the reference value of the cast material, which is critical for applications where the heat transfer is a key parameter.

(2) Despite the directional nature of the DED process, the resulting thermal properties presented no anisotropy and heat was conducted equally in all directions.

(3) Various cross-sections of the laser-deposited AISI H13 tool steel were studied and no porosity was found. Therefore, the thermal diffusivity reduction could not be attributed to the existence of internal voids. The reduction of the effective thermal diffusivity was due to the microstructure that developed during the fast cooling of the deposited material, in which the fine grain size reduced the heat transfer through the material.

(4) The microstructure that developed within the deposited material was directly related to the cooling rate, which was higher at the beginning of the DED process. This is why the effective thermal diffusivity of Sample 1 (situated at 0–2 mm from the interface) was lower than that of Sample 2 (at 4–6 mm from the interface).

(5) The effect of the DED process on the substrate was minimal and did not affect the thermal diffusivity of the base material.

(6) If the effective thermal conductivity is not considered the cooling capability of the DED-manufactured tools is overestimated.

This work extends current knowledge on the thermal properties of DED materials, and the thermal diffusivity and conductivity differences encountered may also affect other AM processes. Therefore, further research is necessary to fully characterize AM parts.

Author Contributions: Conceptualization, J.I.A. and M.C.; investigation, M.C., A.S. and A.M.; software, J.I.A.; writing—original draft preparation, J.I.A., M.C. and A.L.; supervision, A.L. All authors have read and agreed to the published version of the manuscript.

Funding: The authors gratefully acknowledge the financial support for this study from the European Union, through the H2020-FoF13-2016 PARADDISE project (contract number 723440) and from the Ministry of Economy and Competitiveness (grant number DPI2016-77719-R).

Conflicts of Interest: The authors declare no conflict of interest.

References

1. Yan, Z.; Liu, W.; Tang, Z.; Liu, X.; Zhang, N.; Li, M.; Zhang, H. Review on thermal analysis in laser-based additive manufacturing. *Opt. Laser Technol.* **2018**, *106*, 427–441. [CrossRef]
2. Attar, H.; Ehtemam-Haghighi, S.; Kent, D.; Wu, X.; Dargusch, M.S. Comparative study of commercially pure titanium produced by laser engineered net shaping, selective laser melting and casting processes. *Mater. Sci. Eng. A* **2017**, *705*, 385–393. [CrossRef]
3. Zhong, C.; Gasser, A.; Kittel, J.; Wissenbach, K.; Poprawe, R. Improvement of material performance of Inconel 718 formed by high deposition-rate laser metal deposition. *Mater. Des.* **2016**, *98*, 128–134. [CrossRef]
4. Zhang, Y.N.; Cao, X.; Wanjara, P. Microstructure and hardness of fiber laser deposited Inconel 718 using filler wire. *Int. J. Adv. Manuf. Technol.* **2013**, *69*, 2569–2581. [CrossRef]
5. Sun, G.; Zhou, R.; Lu, J.; Mazumder, J. Evaluation of defect density, microstructure, residual stress, elastic modulus, hardness and strength of laser-deposited AISI 4340 steel. *Acta Mater.* **2015**, *84*, 172–189. [CrossRef]
6. Roberts, I.A.; Wang, C.J.; Esterlein, R.; Stanford, M.; Mynors, D.J. A three-dimensional finite element analysis of the temperature field during laser melting of metal powders in additive layer manufacturing. *Int. J. Mach. Tools Manuf.* **2009**, *49*, 916–923. [CrossRef]
7. Rombouts, M.; Froyen, L.; Gusarov, A.V.; Bentefour, E.H.; Glorieux, C. Light extinction in metallic powder beds: Correlation with powder structure. *J. Appl. Phys.* **2005**, *98*, 13533. [CrossRef]

8.   Wei, L.C.; Ehrlich, L.E.; Powell-Palm, M.J.; Montgomery, C.; Beuth, J.; Malen, J.A. Thermal conductivity of metal powders for powder bed additive manufacturing. *Addit. Manuf.* **2018**, *21*, 201–208. [CrossRef]

9.   Cernuschi, F.; Ahmaniemi, S.; Vuoristo, P.; Mäntylä, T. Modelling of thermal conductivity of porous materials: Application to thick thermal barrier coatings. *J. Eur. Ceram. Soc.* **2004**, *24*, 2657–2667. [CrossRef]

10.   Foteinopoulos, P.; Papacharalampopoulos, A.; Stavropoulos, P. On thermal modeling of Additive Manufacturing processes. *CIRP J. Manuf. Sci. Technol.* **2018**, *20*, 66–83. [CrossRef]

11.   Li, Y.; Zhou, K.; Tor, S.B.; Chua, C.K.; Leong, K.F. Heat transfer and phase transition in the selective laser melting process. *Int. J. Heat Mass Transf.* **2017**, *108*, 2408–2416. [CrossRef]

12.   Zhang, D.; Feng, Z.; Wang, C.; Wang, W.; Liu, Z.; Niu, W. Comparison of microstructures and mechanical properties of Inconel 718 alloy processed by selective laser melting and casting. *Mater. Sci. Eng. A* **2018**, *724*, 357–367. [CrossRef]

13.   Shan, Z.D.; Ye, Y.S.; Zhang, M.L.; Wang, B.Y. Hot-stamping die-cooling system for vehicle door beams. *Int. J. Precis. Eng. Manuf.* **2013**, *14*, 1251–1255. [CrossRef]

14.   Chen, J.; Li, X.; Han, X. Hot stamping. *Compr. Mater. Process.* **2014**, *5*, 351–370. [CrossRef]

15.   Karbasian, H.; Tekkaya, A.E. A review on hot stamping. *J. Mater. Process. Technol.* **2010**, *210*, 2103–2118. [CrossRef]

16.   Denlinger, E.R.; Irwin, J.; Michaleris, P. Thermomechanical modeling of additive manufacturing large parts. *J. Manuf. Sci. Eng.* **2014**, *136*, 61007–61008. [CrossRef]

17.   Mukherjee, T.; Manvatkar, V.; De, A.; DebRoy, T. Mitigation of thermal distortion during additive manufacturing. *Scr. Mater.* **2017**, *127*, 79–83. [CrossRef]

18.   Peyre, P.; Aubry, P.; Fabbro, R.; Neveu, R.; Longuet, A. Analytical and numerical modelling of the direct metal deposition laser process. *J. Phys. D Appl. Phys.* **2008**, *41*, 25403. [CrossRef]

19.   Shi, Q.; Gu, D.; Xia, M.; Cao, S.; Rong, T. Effects of laser processing parameters on thermal behavior and melting/solidification mechanism during selective laser melting of TiC/Inconel 718 composites. *Opt. Laser Technol.* **2016**, *84*, 9–22. [CrossRef]

20.   Arrizubieta, J.I.; Tabernero, I.; Exequiel Ruiz, J.; Lamikiz, A.; Martinez, S.; Ukar, E. Continuous coaxial nozzle design for LMD based on numerical simulation. *Phys. Procedia* **2014**, *56*, 429–438. [CrossRef]

21.   Gao, K.; Qin, X.; Wang, Z.; Chen, H.; Zhu, S.; Liu, Y.; Song, Y. Numerical and experimental analysis of 3D spot induction hardening of AISI 1045 steel. *J. Mater. Process. Technol.* **2014**, *214*, 2425–2433. [CrossRef]

22.   Oh, S.; Ki, H. Deep learning model for predicting hardness distribution in laser heat treatment of AISI H13 tool steel. *Appl. Therm. Eng.* **2019**, *153*, 583–595. [CrossRef]

23.   Parker, W.J.; Jenkins, R.J.; Butler, C.P.; Abbott, G.L. Flash method of determining thermal diffusivity, heat capacity, and thermal conductivity. *J. Appl. Phys.* **1961**, *32*, 1679–1684. [CrossRef]

24.   Maillet, D.; Moyne, C.; Rémy, B. Effect of a thin layer on the measurement of the thermal diffusivity of a material by a flash method. *Int. J. Heat Mass Transf.* **2000**, *43*, 4057–4060. [CrossRef]

25.   Heath, D.M.; Winfree, W.P. Thermal diffusivity measurements in carbon-carbon composites. In *Review of Progress in Quantitative Nondestructive Evaluation*; Thompson, D.O., Chimenti, D.E., Eds.; Springer: Boston, MA, USA, 1989; Volume 8, pp. 1613–1619. [CrossRef]

26.   Nolte, P.W.; Malvisalo, T.; Wagner, F.; Schweizer, S. Thermal diffusivity of metals determined by lock-in thermography. *Quant. InfraRed Thermogr. J.* **2017**, *14*, 218–225. [CrossRef]

27.   Naganathan, A.; Penter, L. *Sheet Metal Forming—Processes and Applications Hot Stamping*; ASM International: Materials Park, OH, USA, 2012.

28.   Steels for Hot Stamping—Usibor and Ductibor. Available online: https://automotive.arcelormittal.com/products/flat/PHS/usibor_ductibor (accessed on 19 December 2019).

29.   Muvunzi, R.; Dimitrov, D.M.; Matope, S.; Harms, T.M. Development of a model for predicting cycle time in hot stamping. *Procedia Manuf.* **2018**, *21*, 84–91. [CrossRef]

30.   Shapiro, A.B. Using LS-Dyna for hot stamping. In Proceedings of the 7th European LS-DYNA Conference, Salzburg, Austria, 14–15 May 2009; p. 9. [CrossRef]

31.   Coldwell, H.; Woods, R.; Paul, M.; Koshy, P.; Dewes, R.; Aspinwall, D. Rapid machining of hardened AISI H13 and D2 moulds, dies and press tools. *J. Mater. Process. Technol.* **2003**, *135*, 301–311. [CrossRef]

32.   Arrizubieta, J.I.; Cortina, M.; Ostolaza, M.; Ruiz, J.E.; Lamikiz, A. Case Study: Modeling of the cycle time reduction in a B-Pillar hot stamping operation using conformal cooling. *Procedia Manuf.* **2019**, in press.

33. Hu, P.; Ying, L.; Li, Y.; Liao, Z. Effect of oxide scale on temperature-dependent interfacial heat transfer in hot stamping process. *J. Mater. Process. Technol.* **2013**, *213*, 1475–1483. [CrossRef]

34. Cortina, M.; Arrizubieta, J.I.; Calleja, A.; Ukar, E.; Alberdi, A. Case study to illustrate the potential of conformal cooling channels for hot stamping dies manufactured using hybrid process of laser metal deposition (LMD) and milling. *Metals* **2018**, *8*, 102. [CrossRef]

35. ASTM International. *ASTM Standard E1461–13. Standard Test Method for Thermal Diffusivity by the Flash Method*; ASTM International: West Conshohocken, PA, USA, 2013. [CrossRef]

36. British Standards Institution. *BS7134: Section 4.2: Method for the Determination of Thermal Diffusivity by the Laser Flash (or Heat Pulse) Method*; British Standards Institution: London, UK, 1990.

37. Japanese Standards Association. *JIS R 1611: Testing Methods of Thermal Diffusivity, Specific Heat Capacity and Thermal Conductivity for High Performance Ceramics by Laser Flash Method*; Japanese Standards Association: Tokyo, Japan, 1991.

38. Berman, R. *Thermal Conduction in Solids*; Clarendon Press: Oxford, UK, 1976.

39. Zhang, R.; Dou, R.; Wen, Z.; Liu, X. Effects of pressure and temperature on the effective thermal conductivity of oriented silicon steel iron core under atmospheric condition. *Int. J. Heat Mass Transf.* **2018**, *125*, 780–787. [CrossRef]

 *metals*

Article

Experimental Investigation of the Influence of Wire Arc Additive Manufacturing on the Machinability of Titanium Parts

Unai Alonso [1],*, Fernando Veiga [2], Alfredo Suárez [2] and Teresa Artaza [2]

[1]   Department of Mechanical Engineering, University of the Basque Country (UPV/EHU), 48013 Bilbao, Spain
[2]   TECNALIA, Parque Científico y Tecnológico de Gipuzkoa, 20009 Donostia-San Sebastián, Spain;
      fernando.veiga@tecnalia.com (F.V.); alfredo.suarez@tecnalia.com (A.S.); teresa.artaza@tecnalia.com (T.A.)
*   Correspondence: unai.alonso@ehu.eus; Tel.: +34-946-014-218

Received: 6 December 2019; Accepted: 20 December 2019; Published: 22 December 2019

Abstract: The manufacturing of titanium airframe parts involves significant machining and low buy-to-fly ratios. Production costs could be greatly reduced by the combination of an additive manufacturing (AM) process followed by a finishing machining operation. Among the different AM alternatives, wire arc additive manufacturing (WAAM) offers deposition rates of kg/h and could be the key for the production of parts of several meters economically. In this study, the influence of the manufacturing process of Ti6Al4V alloy on both its material properties and machinability is investigated. First, the mechanical properties of a workpiece obtained by WAAM were compared to those in a conventional laminated plate. Then, drilling tests were carried out in both materials. The results showed that WAAM leads to a higher hardness than laminated Ti6Al4V and satisfies the requirements of the standard in terms of mechanical properties. As a consequence, higher cutting forces, shorter chips, and lower burr height were observed for the workpieces produced by AM. Furthermore, a metallographic analysis of the chip cross-sectional area also showed that a serrated chip formation is also present during drilling of Ti6Al4V produced by WAAM. The gathered information can be used to improve the competitiveness of the manufacturing of aircraft structures in terms of production time and cost.

Keywords: WAAM; titanium; mechanical properties; drilling; chip geometry; cutting forces; hole quality

---

1. Introduction

The additive manufacturing market (AM) applied to the aerospace sector has experienced a big increase in the last decade. According to the report carried out by Wohlers Associates, the worldwide revenue from AM related services reached 6 billion dollars in 2016 and the forecast is to get to 21 billion by 2020 [1]. In fact, the first titanium 3D-printed part of the Airbus A350 XWB entered serial production after the approval of the European Union Aviation Safety Agency (EASA) in late 2017.

The most extended AM technologies (such as laser melting deposition (LMD) and selective laser melting (SLM)) are based on the fusion a metallic powder using a laser or an electron beam as a heating source. These processes can reach good dimensional accuracy, but the part size is limited by the low deposition rate (0.12–0.6 kg/h) [2]. To produce airframe parts of several meters economically, deposition rates of kilograms per hour are needed and this can be obtained by wire arc additive manufacturing (WAAM) [3].

WAAM uses a solid wire as the feedstock material and an electric arc as the heat source. Several research works have dealt with the fundamental aspects of the process such as the final material microstructure and its mechanical properties [4–7]. In this regard, the review recently presented by Rodrigues et al. [8] provides an overview of the different process variants and the materials for WAAM. Moreover, it also deals with the effect of the post-processing treatments in the final mechanical properties as well as with non-destructive testing.

Among the different workpiece materials, titanium-based alloys are being increasingly tested in WAAM owing to their adoption by the aerospace industry for airframe manufacturing. For example, Ti alloys represent 15% of the weight of the Boeing 787 [9]. The microstructure and mechanical properties of Ti6Al4V alloy parts made by WAAM were investigated by Wang et al. [6]. Microstructural banding was observed, and it was related to the repeated thermal cycles that occur during the layer by layer growth of the workpiece. Regarding mechanical properties, the yield and ultimate tensile strengths were found to be lower in the WAAM part than in that produced by forging. However, similar ductility values were observed, and the mean fatigue life was significantly higher in the workpiece produced by WAAM. In a recent study, Ho et al. [4] investigated in great detail the microstructure banding found in this alloy using both multiscale compositional and automated image mapping tools. This approach provided evidence of weak micro-segregation during solidification and explained the origin of such microstructure.

The main disadvantages of WAAM, when compared with other AM processes, are the reduced dimensional accuracy and poor surface finish [10,11]. Therefore, a machining operation is mandatory to achieve the functional requirements. During the assembly of an airframe, for example, a large number of holes are machined, and the control of the drilling operation becomes crucial.

Drilling of titanium is one of the most complex processes in the production of aircraft structures. Ti alloys have been reported as difficult to cut materials owing to their particular physical properties [12–14]. On the one hand, their low thermal conductivity (about 15 W/m K versus 270 W/m K for the steel CRS1018 at 700 °C) [15] leads to high temperatures in the cutting zone. However, they maintain their high mechanical strength and abrasiveness at such temperatures, resulting in high forces and fast tool wear. On the other hand, the chemical affinity with tool materials leads to a fast titanium adhesion to the cutting tool. The resulting built up edge (BUE) or chipping of the cutting edges often lead to poor surface finish of the inner surface of the hole (and, in the worst scenario, to premature failure of the cutting tool) [13].

The selection of the most appropriate cutting parameters (feed and cutting speed) is of vital importance for the success of the process. Previous studies have thoroughly analysed the effect of cutting parameter variation in the cutting forces and the resulting hole quality. Sharif et al. [14], for example, evaluated the performance of both uncoated and coated tungsten carbide drills at various cutting speeds. They observed that the average surface roughness value (Ra) decreased by approximately 20% by doubling the cutting speed. They also reported that uncoated drills are not suitable for cutting speeds over 25 m/min owing to their fast wear. In another study, Le Coz et al. [12] carried out a comprehensive study to test the effect of the variation of both the feed and cutting speed in the cutting forces. They observed an increase of a 60% in the cutting torque when the feed per revolution was increased in the same extent. This rise of cutting forces can be explained by the direct relationship between the feed and the cross-sectional area of the undeformed chip. The evolution of cutting forces was also investigated in the work of [14]. In this case, lower torque values were observed when increasing the cutting speed; this result was attributed to the reduction of the material hardness as the temperature increased.

The overall quality of the machined hole is also of great importance for the correct behaviour of the mechanical joint. Burr formation and poor geometrical quality, for example, can result in a reduction of fatigue life. Additional deburring or reaming operations are needed for the correction of such defects, and they generally imply a high cost [16]. Burr formation is dependent on various aspects such as tool geometry and material, workpiece material properties, and the selected machining parameters (among others) [17]. Apart from this defect, geometrical imperfection (such as irregular hole diameter) is also of critical concern. When drilling Ti alloys, the machined holes may shrink or expand depending on the cutting environment. Oversized holes are often obtained in dry cutting owing to the thermal expansion resulting from the high cutting temperatures. The influences of feed and cutting speed variation on hole diameter are complicated to study owing to their interacting effects [18]. Elastic recovery of compressive strain produced by thrust force may decrease the hole size. However, material expansion at high temperatures can lead to a hole larger than expected.

Despite extensive research work in Ti drilling, there are few works that have studied the machining of parts produced by AM. Priarone et al. [19] investigated the machinability of a gamma titanium aluminide obtained via the electron beam melting (EBM) process. In this work, drilling performance was measured in terms of surface roughness, dimensional errors, and tool wear. The experimental results showed a decrease of surface roughness when feed was decreased, as well as when the cutting speed increased. Hole quality also showed a dependence on the cutting parameters and tool wear level.

Even if WAAM could be a viable alternative to manufacturing aerostructure components, there is a lack of an experimental investigation on the drilling of titanium parts produced by this AM process. This work aims to reveal the influence of material properties by comparing the machining of commercial laminated plates to those produced by WAAM. In particular, the resulting chip morphology, cutting forces, and overall hole quality are analysed. For a deeper understanding of the cutting process, workpiece microstructure and mechanical properties are also studied. The novelty of this paper lies in the study of WAAM additive technology for the production of Ti6Al4V aeronautical titanium alloy parts and in experimentally addressing the drilling operation, as it is one of the most recurring machining operations in this type of applications.

2. Materials and Methods

2.1. WAAM Deposition Process and Material Analysis

Linear wall parts were produced using an Addilan V0.1 wire arc additive manufacturing machine (Durango, Spain) and plasma arc welding (PAW) as addition technology with compressed argon as a pilot gas to prevent oxidation and porosity. This machine has a closed loop control system and an inert chamber to work under controlled Argon atmosphere to protect the melt pool when using reactive materials like titanium or aluminum, and can reach deposition rates up to 6 kg/h.

The experiments were carried out on a Ti6Al4V annealed base plate (AMS 4911 according to the AMS4928 standard), which had a dimensions of 220 mm × 70 mm × 12 mm. Prior to welding, the substrate surface was cleaned with acetone to remove any organic contaminants. A 1.14 mm diameter Ti6Al4V solid wire electrode (AWS A5.16-13 ERTI-5) was used for deposition and the chemical compositions of both the substrate and the wire are shown in Table 1.

Table 1. Nominal composition of welding wire and substrate.

Material	Chemical Composition (wt %)								
	Al	V	Fe	O	C	N	H	Other	Ti
Wire	5.5–6.75	3.5–4.5	0.22	0.12–0.2	<0.05	<0.03	<0.015	<0.4	Bal.
Substrate	5.5–6.75	3.5–4.5	<0.3	<0.2	<0.08	<0.05	<0.015	<0.4	Bal.

The same deposition sequence was used for all the parts. First, the walls were obtained using a deposition rate of 2 kg/h and following the strategy shown Figure 1b. The pre-heating and inter-pass temperatures on the top layer were of 600 °C. Table 2 shows the welding parameters used for the sample processing. Finally, after the deposition was completed, the parts were stress relief treated by keeping them in a furnace at a temperature of 720 °C for 2.5 h.

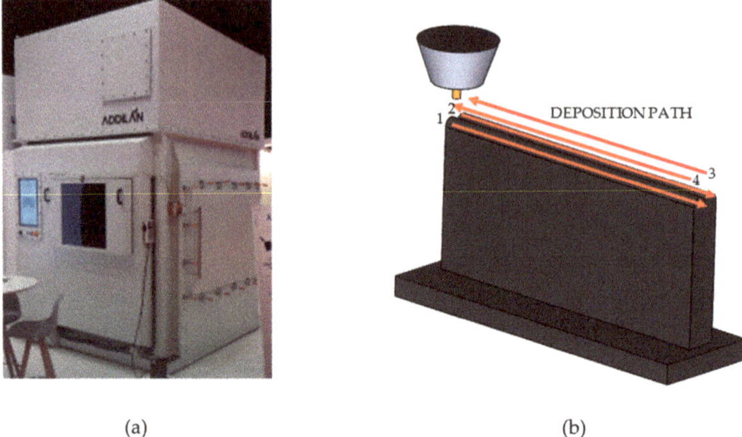

(a) (b)

Figure 1. (**a**) ADDILAN V0.1 wire arc additive manufacturing (WAAM) ([20]) and (**b**) schematic process sequence for wall manufacturing.

Table 2. Welding parameters used for sample deposition with wire arc additive manufacturing (WAAM).

Parameter	Unit	Value
Torch Velocity	mm/min	400
Wire feed	m/min	6.7
Arc Energy	J/mm	1.125
Gas Pilot	l/min	1.2
Shielding Gas	l/min	12
Inter-bead Temperature	°C	600
Nominal wall dimension (LxWXH)	mm	$210 \times 15 \times 105$

The determination of mechanical properties of the additively manufactured parts was obtained by tensile testing according to EN ISO 6892-1 standards. More specifically, the end start sections of the thin-walled structures were discarded and the samples for the tensile tests were prepared as shown in Figure 2. The tests were carried out at room temperature and with a speed of 1 mm/min, according ISO 6892-2, less than 5 Mpa/s in the elastic region. The chemical composition of the walls was also evaluated after the deposition process by infrared spectroscopy (IR spectrometry) for carbon C, oxygen O, and hydrogen H; thermal conductivity analysis for nitrate N; and optical emission spectroscopy (OES) for iron Fe, vanadium V, and aluminum Al.

Moreover, metallographic preparation was performed on samples sectioned along the XZ, XY, and YZ planes of the workpiece (see Figure 4), which were hot-mounted in a phenolic resin powder. Afterwards, they were subsequently polished using a series of abrasive grinding papers with decreasing coarseness from 1200 grit, followed by a final diamond suspension polishing. Finally, the samples were etched in Kroll's reagent and optical images of the microstructure were obtained. Micro-hardness tests were also conducted in two samples of the walls produced by the WAAM process and in the laminated material used as reference for the machining tests. In order to carry out the measurements, an EMCOTEST DuraScan G5 (Kuchl, Austria) microhardness tester with a 10-kgf (98 N) load was used according to ISO 6507 procedures.

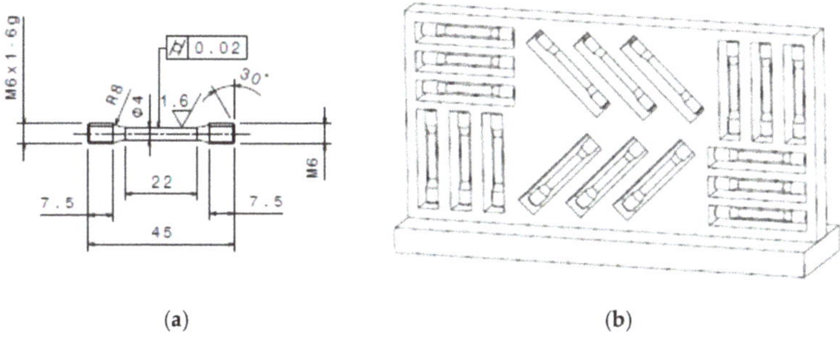

(a)   (b)

Figure 2. Dimensions of the samples for tensile tests in mm, Rz in µm, (**a**) and schematic representation of the location of the extracted samples (**b**).

2.2. Experimental Setup for Drilling Tests

The drilling trials were performed on a five-axis computer numerical control (CNC) machining center Ibarmia 5-axis ZV 65/U600 EXTREME (Azkoitia, Spain) and in collaboration with KENDU tools (Segura, Spain). WAAM specimens were rigidly clamped to the machine bed and the bottom and top faces were milled to produce a smooth surface for the final drilling operation. Figure 3 shows one of the parts after the WAAM process, during the milling operation and it final state before the drilling tests.

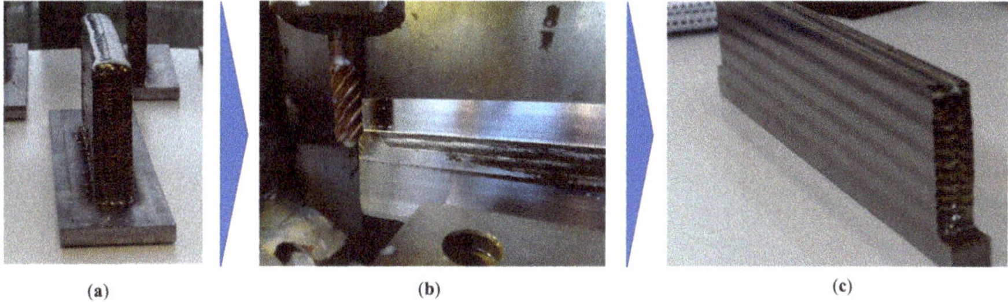

(a)   (b)   (c)

Figure 3. Additively manufactured part sample after the deposition process (**a**). Milling operation (**b**), part sample after milling process (**c**).

The drilling trials (Figure 4a) were performed using an uncoated ISO K30 KENDU tungsten carbide (WC) drill with diameter of 7.5 mm, a point angle of 140 mm, and a helix angle of 30°. An overall view of the tools is shown in Figure 4b. Two sets of experiments were carried out in order to investigate the influence of the feed and cutting speed, respectively. The levels of machining parameters selected were based on the recommendations of the tool supplier and are detailed in Table 3. A peck drilling strategy and minimum quantity lubrication (MQL) cooling were used to enhance chip evacuation and to reduce cutting temperature. In order to study the dispersion of the process, three repetitions were carried out with the same drilling parameters. Furthermore, the holes were arranged in a grid pattern and were randomly located to consider sample variability.

As previously mentioned, one objective of this work is to compare the performance of the drilling process of WAAM parts to those produced by conventional methods. Thus, these drilling tests were repeated in a conventional titanium plate.

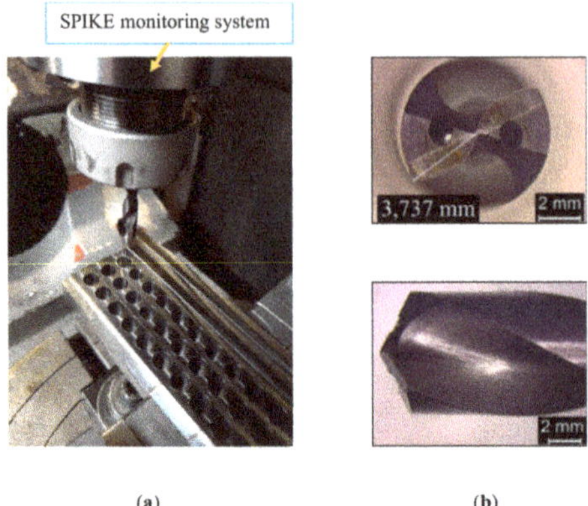

(a)   (b)

Figure 4. (**a**) General arrangement of the equipment used for the drilling tests. (**b**) Front and side view of the used tools.

Table 3. Cutting conditions for the drilling experiments.

Parameter	Level -1	Level 0	Level 1
Cutting speed (m/min)	10	20	30
Feed rate (mm/rev)	0.05	0.075	0.1

During the drilling tests, thrust force and cutting torque were registered using a sensory SPIKE® toolholder (Pro-micron GmbH, Kaufbeuren, Germany) wirelessly connected to a PROMICRON READ acquisition system. Afterwards, cutting force signals were treated using the software MATLAB v2018a (Matworks, Massachusetts, Estados Unidos). As an example, Figure 5 shows the evolution of the thrust force for one peck drilling cycle in which three different stages can be identified. The first one (a–b) is related to the que tool entrance in the workpiece. In the second one (b–c), the cutting edge is fully engaged in titanium leading to a steady drilling process. Finally, the exit of the drill is produced at the last stage (c–d). The mean value of the steady drilling process stage (b–c) was considered for the comparison of the different drilling conditions and workpiece materials.

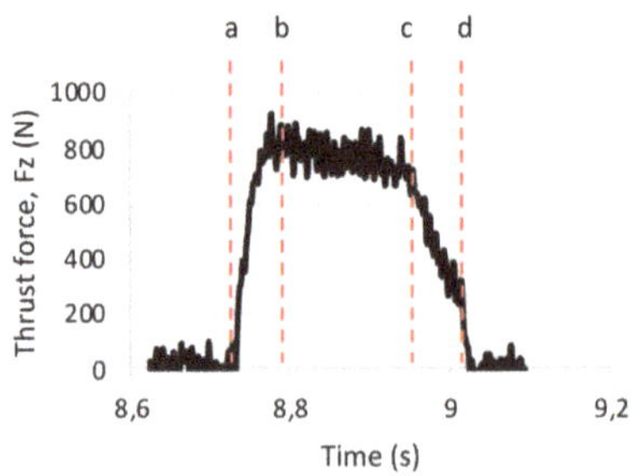

Figure 5. Thrust force evolution over time for a peck-drilling cycle.

After the drilling tests, hole quality was investigated by evaluating the diameter of the holes, their surface roughness, and the burr height at the hole exit. Hole diameter was measured at 6 mm from the tool entrance (i.e., the center of the inner surface) using a three-point internal digital micrometer (Mitutoyo Borematic® (Kawasaki, Japan) accuracy = 0.001 mm). Surface roughness was evaluated at the same hole depth using a Taylor Hobson Surtronik S100-series tester (Leicester, England) and the following filter values: λs = 2.5 µm and λc = 0.8 mm. Assessment of burr height was performed using a Leica DCM 3D confocal microscope (Wetzlar, Germany) at four equally-spaced positions around the hole periphery with an average calculation of the obtained data for each hole. Moreover, the overall chip geometry was also evaluated following the procedure proposed by Zhu et al. [21].

3. Results and Discussion

3.1. Mechanical Properties and Workpiece Microstructure Characterization

In this section, the mechanical properties of the added material are analyzed in terms of the following: titanium alloy metallographic structure, mechanical results derived from the tensile test, hardness, and chemical composition of the material.

Figure 6 top left shows the titanium wall manufactured by means of WAAM technology from which the specimens were extracted, and grains' structure was observed. Macrographic pictures clearly show the interleaves of the deposition process of the different beads as diffusion occurs between passes. The microstructure of the wall at different heights is shown, with the grain border and the dendritic structure derived from phases in which the temperature undergoes an abrupt decrease from the melting temperature necessary for the different welds production. Owing to the influences of thermal gradient and solidification rate, the acicular α interwoven with a basket-weave structure and martensite α' the β matrix were generated.

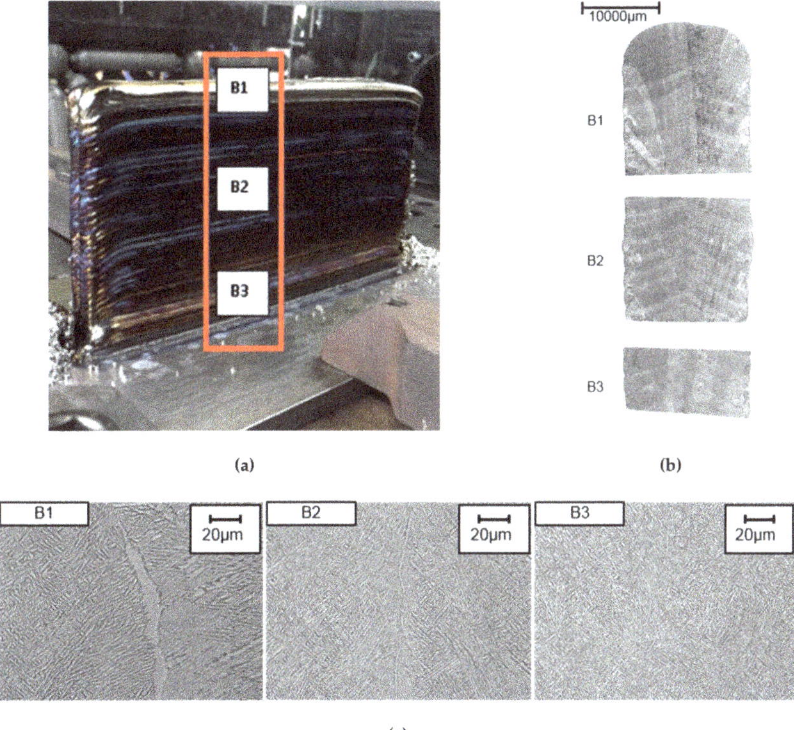

Figure 6. Grain structure of the wall manufactured by plasma arc welding (PAW)-WAAM (**a**) at different heights and (**b**) from the top to bottom, as well as (**c**) micrographic pictures at different heights at 500X magnification.

The results of the tensile tests performed for the characterization of the walls are displayed in Table 4. The material was tested in the different directions of the wall: vertical, horizontal, and oblique.

The results are consistent, and it can be observed that the ultimate tensile stress and yield strength are slightly higher in the horizontal direction than in the vertical direction, with the oblique direction being the one with the highest results. This behaviour is the opposite of that observed in the elongation before the fracture, being the most critical parameter compared with the standard required for these materials in their aeronautical application.

Table 4. Mechanical properties: ultimate tensile stress (UTS), yield strength (YS), and elongation at break (Elong). PAW-WAAM, plasma arc welding wire arc additive manufacturing.

Specimen	UTS (MPa)	YS (MPa)	Elong (%)
Horizontal PAW-WAAM	981 ± 36.3	917 ± 19.3	11 ± 0.9
Vertical PAW-WAAM	925 ± 18.2	864 ± 22.8	15 ± 1.3
Oblique PAW-WAAM	1094 ± 35.5	1020 ± 20.2	10 ± 0.5
Ti6Al4V	>931	>862	>10

Table 5 shows the surface hardness at different depths of the WAAM titanium alloy wall. The mean values of microhardness are similar from top to bottom of the wall. An average value of 301 HV in the microhardness measured is measured at the wall. No differences are observed between the different depths. Regarding the values of chemical composition of the material, Table 6 contains the results obtained from the analysis of the wall. The values meet the requirements set by the standard that characterizes this titanium alloy (Ti6Al4V), which makes this technology very advantageous. The selection of the use of WAAM technology, plasma arc welding (PAW), under a controlled atmosphere of argon allows to ensure the stability of the material so that it does not vary its composition significantly with respect to the thread used as raw material in the process.

Table 5. Microhardness at different heights.

Hardness (HV 10)		
Position	Average	St. Dev.
Top	294	±4
Center	304	±7
Bottom	304	±5
Mean	301	±8

Table 6. Chemical composition of WAAM and laminated.

Material	Chemical Composition (wt %)								
	Al	V	Fe	O	C	N	H	Other	Ti
PAW-WAAM	6.3	4	0.17	0.16	0.02	0.014	0.003	0.00	Bal.
Laminated	5.5–6.75	3.5–4.5	<0.3	<0.2	<0.08	<0.05	<0.015	<0.4	Bal.

3.2. Chip Morphology Analysis

Figure 7 shows the effect of cutting variables on chip morphology for the drilling tests of the specimen produced by additive manufacturing. As it can be noticed, the greater the feed, the smaller the chip length, while there is no big difference with the variation of cutting speed. This effect can be explained by the increase in the chip cross sectional area with the feed rate. A bigger cross-sectional area leads to a higher stiffness and, therefore, chips break more easily into segments. It is important to note that smaller chips are desirable in titanium drilling because a good chip extraction leads to lower thrust forces and cutting temperatures.

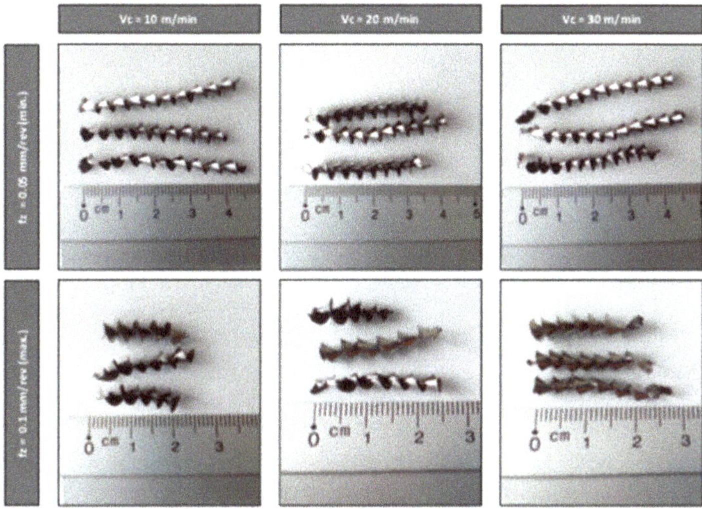

Figure 7. Effect of cutting speed and feed rate on chip morphology in drilling of WAAM specimen.

Moreover, as shown in Figure 8, it was also observed that, the higher the feed, and the smaller the cutting speed, and the greater the distance between two adjacent cones of the chip (geometrical parameter known as "pitch"). This effect also enhances chip breakability [21]. Similar results were obtained for the laminated specimen.

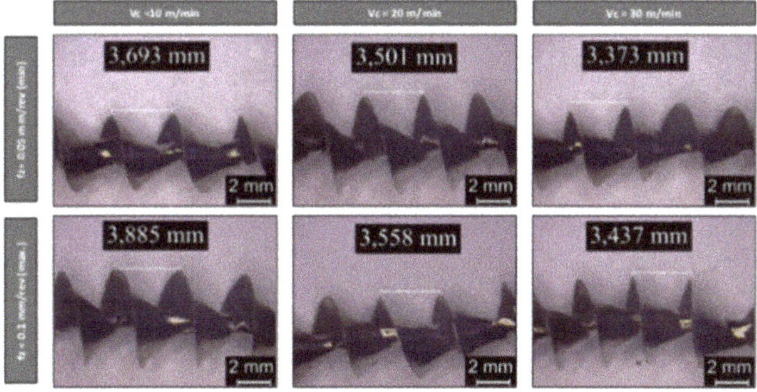

Figure 8. Effect of cutting speed and feed rate on the chip "pitch" (workpieces produced by WAAM).

To study in greater depth the chip formation process, a metallographic analysis of the chip cross section was carried out for some experiments. In particular, the part of the chip generated farther from the center of the tool was studied (see Figure 9). It should be noted that the cutting speed and rake angle are not the same in all the points of the cutting edge, leading to different material deformation and failure conditions.

Figure 9 shows the chip cross section for a workpiece produced by WAAM. A serrated chip formation process is noticeable with significant grain deformation in the primary shear zone. The appearance of such "shear bands" can be explained by a cycle of autocatalytic ruination [22]. During chip deformation, there is a local increase of the workpiece temperature produced by plastic work. During the short loading time, the temperature increases titanium's flow stress decreases and leads to an adiabatic shearing of the chip.

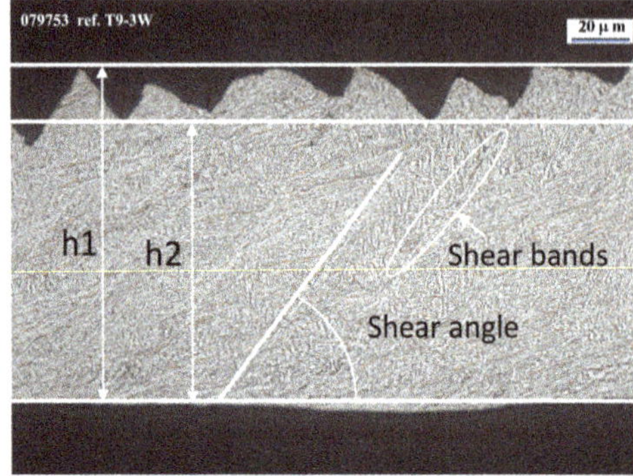

Figure 9. Metallographic analysis of the cross section of the chip for a workpiece produced by WAAM.

To compare the different machining conditions, two geometrical parameters were measured on the micrographs: the shear angle and the segment degree (see Figure 10). The segment degree of serrated chips is defined as G = (h1−h2)/h1 (h1 is the maximum teeth height and h2 is the minimum teeth height). For the drilling tests carried out in the workpieces produces by WAAM, a decrease in the shear angle with feed was observed for the tests with the same cutting speed. The bigger the feed, the higher would be the chip cross-sectional area and the plastic deformation level. As a consequence, workpiece temperature increases in shear zone and it becomes easier to slip. The segment degree shows an increase if a higher feed is used. For the studied machining parameter combinations, the results also showed that feed variation seems to have a bigger influence in the segment degree evolution than the change of the cutting speed.

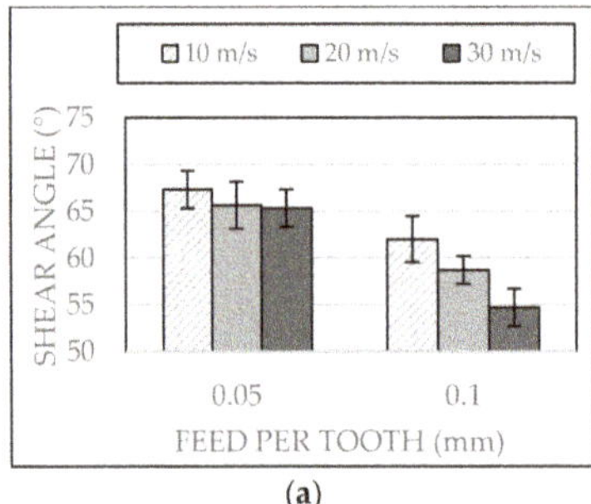

(a)

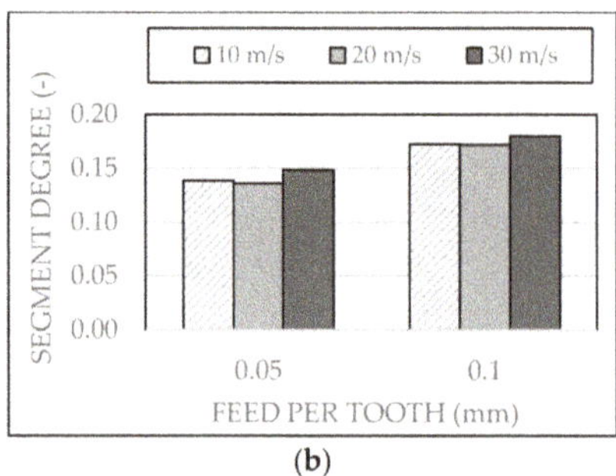

(b)

Figure 10. Effect of the cutting parameters on the (a) shear angle and (b)segment degree for a workpiece produced by WAAM.

3.3. Cutting Force Magnitudes

The knowledge of the evolution of cutting forces magnitudes with machining parameters is necessary for the process assessment. On the one hand, the machine-tool spindle should be able to generate enough torque for the cutting process. On the other hand, thrust force control is also important to minimize the vibrations of the drill along its axis as well as to reduce the bur height at the hole exit.

Figure 11 shows the effect of feed per tooth and cutting speed variation on the cutting torque value. An increase of cutting torque was observed when the feed per tooth was augmented. This result can be explained by the direct relationship between the cut chip thickness and the feed per tooth. Moreover, cutting torque decreased in both materials when the cutting speed was increased. The reduction of titanium's mechanical properties with temperature could explain this trend. These results are in accordance with the previous studies carried out in laminated titanium [13,15] and a trends was observed for thrust force (see Figure 12). Finally, it should be noted that higher force values were obtained for the tests carried out in the titanium produced by WAAM as compared with those in the laminated material. The higher hardness (and overall better mechanical resistance) of the titanium produced by WAAM could be the reason for such a difference.

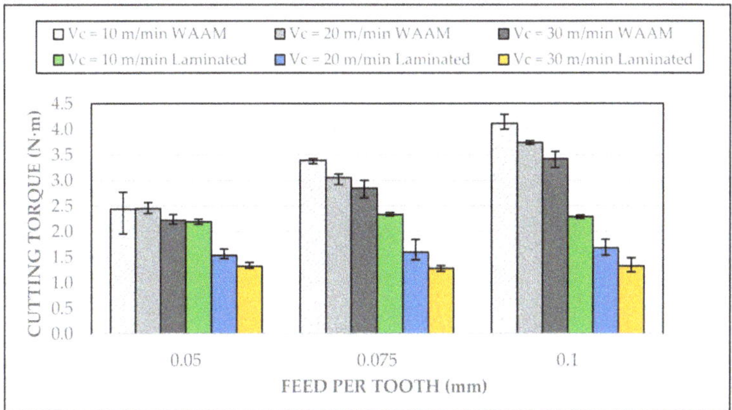

Figure 11. Effect of the cutting parameters and material manufacturing process on the cutting torque.

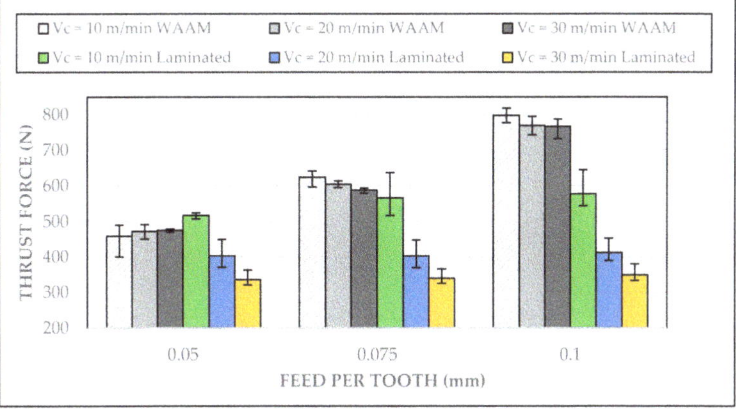

Figure 12. Effect of the cutting parameters and material manufacturing process on the thrust force.

3.4. Hole Quality Analysis

In this section, the quality of the hole drilled at different feed conditions per tooth and cutting speed will be analyzed, both in the laminated material and in the material manufactured by PAW-WAAM additive material in terms of the following: diameter, roughness, and burrs at the exit of the hole.

Regarding the diameter, the nominal values measured are slightly higher than the diameter of the drill, the difference being the one observed at low cutting speeds in the drilling of WAAM material, as can be seen in Figure 13.

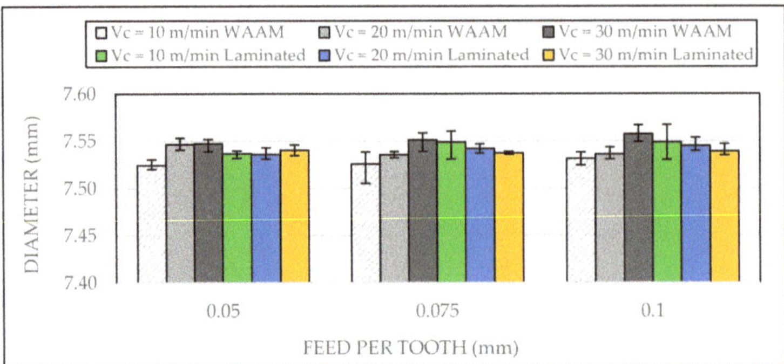

Figure 13. Effect of the cutting parameters and material manufacturing process on hole diameter.

The roughness is not affected by the set of material and conditions in the cases of rotting analyzed as can be seen in Figure 14. The observed variance is greater than the differences between the conditions, which does not allow to establish significant discrepancies between the values.

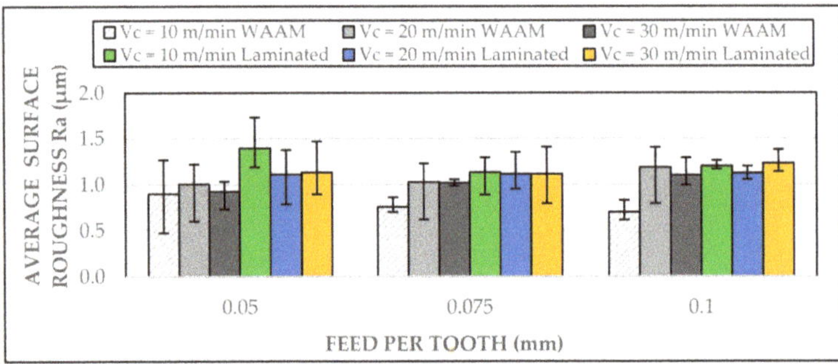

Figure 14. Effect of the cutting parameters and material manufacturing process on the surface roughness of the hole.

Figure 15 shows the average burr results measured at the exit of the hole for the drilling conditions tested, both in the case of the wall obtained by WAAM technology and in those obtained by conventional methods such as lamination. Cutting conditions, both cutting speed and feed per tooth, do not determine the size of the burr produced at the exit of the hole. The manufacturing process of the titanium alloy wall has a significant influence. Although the variance observed in the size of the burr is greater in the case of the WAAM wall, the hole quality in terms of average burr size is improved. The size of the burr is inversely correlated with the grain size and directly with the elongation before breakage.

The results show lower burr values in cases of drilling of material manufactured by WAAM technology owing to its higher hardness and mechanical resistance typical of the manufacturing process of the wall in which we find a heating–cooling cycle that increases the local hardness.

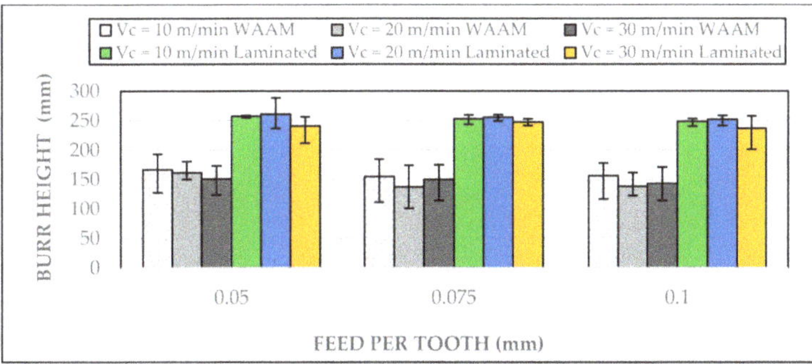

Figure 15. Effect of the cutting parameters and material manufacturing process on the burr height.

4. Conclusions

The titanium manufacturing process plays an important role in the mechanical properties and the workpiece machinability. An experimental study was performed to assess the influence of wire arc additive manufacturing on the material properties and on the resulting drilling performance. On the basis of the obtained results, the following conclusions can be drawn:

- It was shown that WAAM additive technology is a valid alternative for manufacturing Ti6Al4V titanium parts based on the mechanical results obtained. No typical welding defects such as pores or lack of fusion were observed. The microstructure is typical of rapid cooling normally associated with welding processes. The slight variations in hardness are because of the thermal cycles of the additive manufacturing layer by layer.
- The study also highlighted the influence of the workpiece manufacturing process on cutting forces. Higher cutting torque and thrust for values were obtained for the workpieces produced by WAAM.
- An increase of cutting forces with feed per tooth was observed for both materials. This result is in accordance with previous studied and can be linked to the higher cross-sectional area of the chip. Conversely, cutting forces were reduced when cutting speed was augmented. This result can be linked to the poorer mechanical resistance of titanium at higher temperatures.
- Regarding the chip geometry, a greater feed led to a smaller chip length, while no big differences were observed with the variation of cutting speed. Moreover, a metallographic analysis of the chip cross section showed that a serrated chip formation process was also present during drilling of titanium produced by WAAM.
- Hole quality was also found to be affected by the materials' manufacturing process. Burr height values were approximately 0.1 mm lower in the workpieces produced by WAAM. This can be explained by the higher hardness and mechanical properties of the additive-manufactured material. The hole diameter and surface roughness values were in the tolerance range required in the aeronautics industry for airframe parts.

Overall, from this study, it can be concluded that the workpieces produced by WAAM yield better performance in terms of mechanical properties and burr height, but higher cutting forces would be obtained under the same drilling conditions.

Future research works should consider the influence of the geometrical characteristics of the drill in the cutting process in order to obtain an optimized tool for this new workpiece material. Other lines of research to consider cover the study of different drilling parameters, considering the variance and repeatability with a large sample-size experiment analyzing the wear of the drill. Also of interest is the comparative study of the machinability of the material manufactured by this additive technology versus other material deposition technology such as selective laser melting (SLM) or laser metal deposition (LMD).

Author Contributions: Conceptualization, U.A., F.V., and A.S.; Data curation, U.A. and F.V.; Investigation, U.A. and F.V.; Methodology, U.A. and A.S.; Resources, A.S.; Supervision, A.S.; Validation, T.A.; Writing—original draft, U.A., F.V., and A.S.; Writing—review & editing, T.A. All authors have read and agreed to the published version of the manuscript.

Funding: This research was funded by the vice-counseling of technology, innovation, and competitiveness of the Basque Government grant agreement kk-2019/00004 (PROCODA project).

Acknowledgments: The authors thank the companies ADDILAN (Spain) and KENDU tools (Spain) for the technical and material support given to this study, as well as Paul Diego Martin for your collaboration.

Conflicts of Interest: The authors declare no conflict of interest.

Acronyms

Abbreviation	Definition
AM	Additive Manufacturing
WAAM	Wire Arc Additive Manufacturing
EASA	European Union Aviation Safety Agency
LMD	Laser Melting Deposition
SLM	Selective Laser Melting
EBM	Electron Beam Melting
PAW	Plasma Arc Welding
AMS	Aerospace Material Specification
IR	Infrared Spectroscopy
CNC	Computer Numerical Control
MQL	Minimum Quantity Lubrication
UTS	Ultimate Tensile Stress
YS	Yield Strength
Vc	Cutting Speed

References

1. Wohlers, T. *Additive Manufacturing and 3D Printing, State of the Industry*; Annual Worldwide Progress Report; Wohlers Associations: Colorado, USA, 2012.
2. Liberini, M.; Astarita, A.; Campatelli, G.; Scippa, A.; Montevecchi, F.; Venturini, G.; Durante, M.; Boccarusso, L.; Minutolo, F.M.C.; Squillace, A. Selection of optimal process parameters for wire arc additive manufacturing. *Procedia CIRP* **2017**, *62*, 470–474. [CrossRef]
3. Venturini, G.; Montevecchi, F.; Scippa, A.; Campatelli, G. Optimization of WAAM deposition patterns for T-crossing features. *Procedia CIRP* **2016**, *55*, 95–100. [CrossRef]
4. Ho, A.; Zhao, H.; Fellowes, J.W.; Martina, F.; Davis, A.E.; Prangnell, P.B. On the origin of microstructural banding in Ti-6Al4V wire-arc based high deposition rate additive manufacturing. *Acta Mater.* **2019**, *166*, 306–323. [CrossRef]
5. Wu, B.; Pan, Z.; Ding, D.; Cuiuri, D.; Li, H.; Fei, Z. The effects of forced interpass cooling on the material properties of wire arc additively manufactured Ti6Al4V alloy. *J. Mater. Process. Technol.* **2018**, *258*, 97–105. [CrossRef]
6. Wang, F.; Williams, S.; Colegrove, P.; Antonysamy, A.A. Microstructure and mechanical properties of wire and arc additive manufactured Ti-6Al-4V. *Metall. Mater. Trans. A Phys. Metall. Mater. Sci.* **2013**, *44*, 968–977. [CrossRef]
7. Artaza, T.; Alberdi, A.; Murua, M.; Gorrotxategi, J.; Frías, J.; Puertas, G.; Melchor, M.A.; Mugica, D.; Suárez, A. Design and integration of WAAM technology and in situ monitoring system in a gantry machine. *Procedia Manuf.* **2017**, *13*, 778–785. [CrossRef]
8. Rodrigues, T.A.; Duarte, V.; Miranda, R.M.; Santos, T.G.; Oliveira, J.P. Current Status and Perspectives on Wire and Arc Additive Manufacturing (WAAM). *Materials* **2019**, *12*, 1121. [CrossRef] [PubMed]
9. Lütjering, G.; Williams, J.C. *Titanium*, 2nd ed.; Springer: Berlin/Heidelberg, Germany, 2007.
10. Ding, D.; Pan, Z.; Cuiuri, D.; Li, H. Wire-feed additive manufacturing of metal components: Technologies, developments and future interests. *Int. J. Adv. Manuf. Technol.* **2015**, *81*, 465–481. [CrossRef]

11. Tabernero, I.; Paskual, A.; Álvarez, P.; Suárez, A. Study on Arc Welding Processes for High Deposition Rate Additive Manufacturing. *Procedia CIRP* **2018**, *68*, 358–362.
12. Le Coz, G.; Marinescu, M.; Devillez, A.; Dudzinski, D.; Velnom, L. Measuring temperature of rotating cutting tools: Application to MQL drilling and dry milling of aerospace alloys. *Appl. Therm. Eng.* **2012**, *36*, 434–441. [CrossRef]
13. Xu, J.; Mkaddem, A.; El Mansori, M. Recent advances in drilling hybrid FRP/Ti composite: A state-of-the-art review. *Compos. Struct.* **2016**, *135*, 316–338. [CrossRef]
14. Sharif, S.; Abd, E.; Sasahar, H. Machinability of Titanium Alloys in Drilling. In *Titanium Alloys: Towards Achieving Enhanced Properties for Diversified Applications*; Books on Demand: London, UK, 2012; pp. 117–138.
15. Nouari, M.; Makich, H. On the Physics of Machining Titanium Alloys: Interactions between Cutting Parameters, Microstructure and Tool Wear. *Metals* **2014**, *4*, 335–358. [CrossRef]
16. Abdelhafeez, A.M.; Soo, S.L.; Aspinwall, D.K.; Dowson, A.; Arnold, D. Burr formation and hole quality when drilling titanium and aluminium alloys. *Procedia CIRP* **2015**, *37*, 230–235. [CrossRef]
17. Dornfeld, D.A.; Kim, J.S.; Dechow, H.; Hewson, J.; Chen, L.J. Drilling burr formation in titanium alloy. *CIRP Ann.* **1999**, *48*, 73–76. [CrossRef]
18. Zhu, Z.; Sui, S.; Sun, J.; Li, J.; Li, Y. Investigation on performance characteristics in drilling of Ti6Al4V alloy. *Int. J. Adv. Manuf. Technol.* **2017**, *93*, 651–660. [CrossRef]
19. Priarone, P.C.; Rizzuti, S.; Ruffa, S.; Settineri, L. Drilling experiments on a gamma titanium aluminide obtained via electron beam melting. *Int. J. Adv. Manuf. Technol.* **2013**, *69*, 483–490. [CrossRef]
20. Available online: https://www.addilan.com/ (accessed on 29 November 2019).
21. Zhu, Z.; Guo, K.; Sun, J.; Li, J.; Liu, Y.; Chen, L.; Zheng, Y. Evolution of 3D chip morphology and phase transformation in dry drilling Ti6Al4V alloys. *J. Manuf. Process.* **2018**, *34*, 531–539. [CrossRef]
22. Daymi, A.; Boujelbene, M.; Salem, S.B.; Hadj Sassi, B.; Torbaty, S.; Sassi, B.H. Effect of the cutting speed on the chip morphology and the cutting forces. *Manuf. Process. Eng. Mater.* **2009**, *78*, 77–83.

Article

Erosion Resistance and Particle Erosion-Induced Tensile Embrittlement of 3D-Selective Laser Melting Inconel 718 Superalloy

Jun-Ren Zhao, Fei-Yi Hung * and Truan-Sheng Lui

Department of Materials Science and Engineering, National Cheng Kung University, Tainan 70101, Taiwan; a2x346yz03@gmail.com (J.-R.Z.); luits@mail.ncku.edu.tw (T.-S.L.)
* Correspondence: fyhung@mail.ncku.edu.tw; Tel.: +886-6-275-7575 (ext. 62950)

Received: 31 October 2019; Accepted: 20 December 2019; Published: 22 December 2019

Abstract: We used selective laser melting (SLM) Inconel 718 (coded AS) to carry out three heat treatment processes: (1) double aging (coded A), (2) solid solution + A (coded SA), (3) homogenization + SA (coded HSA) in order to investigate the effects of microstructure changes and tensile strength enhancement on erosion resistance. The as-SLM IN718 and three heat-treated specimens were subjected to clarify the effects of erosion-induced phase transformation on tensile mechanical properties. All heat-treated specimens showed better erosion resistance than as-SLM IN718 did at all impact angles. The as-SLM IN718 and the three heat-treated specimens produced new γ' phase or metal-oxide via particle erosion, which increased the surface hardness of the material. The thickness of the erosion affected zone is 200 μm, which is the main cause of tensile embrittlement.

Keywords: Inconel 718; selective laser melting (SLM); erosion; phase transformation; tensile; embrittlement

---

1. Introduction

Inconel 718 (IN718) is widely applied in aeronautics, astronautics, and energy industries because of its excellent corrosion resistance, oxidation resistance, weldability, and mechanical properties under circumstances like room temperature and elevated temperature. Its application area includes gas turbines, turbine blades, rocket engines, components for oil and gas extraction, and nuclear engineering [1–4]. IN718 is precipitation strengthened nickel-based superalloy which could be strengthened by precipitation of D022 γ'' (Ni3Nb), L12 γ' (Ni3(Al, Ti)), and D0a δ (Ni3Nb) in the γ matrix [5], thus could be imported in the engineering applications.

Additive manufacturing (AM) can reduce costs, save mold manufacturing costs, and create complex geometric shapes. It has been applied to various industries [6]. Selective laser melting (SLM) is classified as the family of AM technologies. SLM uses metal powder as raw material to be illuminated by a high-energy laser beam within a specified area. After the powder melts, it is rapidly solidified at high cooling rates [7–9].

Blades, turbines, and engines are often impacted by solid particle erosion wear, resulting in damage to pieces and equipment. Erosion wear is the phenomenon of gas or liquid-driven particles striking the surface of a material [10,11]. Few articles mention the particle erosion wear of nickel-based superalloys. Most of them concern their casting and forging [12–14]. We are interested in particle erosion resistance and the erosion induced phase transformation of 3D-printed IN718.

Particle erosion wear could induce phase transformation. This phenomenon has been confirmed by Al-Si-Mg alloy [15], austempered ductile iron (ADI) [16], 316 stainless steel [17], and even 3D-printed Ti 6Al 4V [18]. The heat generated by particle impact or strain helps dissolve a phase or form a new phase, further affecting the hardness and other mechanical properties.

Based on this, the purpose of this study is to investigate the microstructure and the mechanical properties of heat-treated SLM Inconel 718. To clarify the erosion-induced phase transformation mechanism and its effect on the material brittleness, we compared the particle erosion wear behavior of as-SLM Inconel 718 with those of three heat-treated specimens. These results have significant reference value for the aerospace industry.

2. Experimental Procedure

The purpose of the experiments was to compare the erosion resistance of as-SLM specimen (coded AS) with those of the three heat-treated specimens (A, SA, and HSA) under various impact angles. Table 1 shows the SLM process parameters used in this study. The raw material IN718 powder supplied by EOS GmbH (Electro-Optical Systems, Krailling, Germany) had an average particle size of 15 μm (Figure 1). The chemical compositions of IN718 included: Ni (50–55 wt. %), Cr (17–21 wt. %), Nb (4.7–5.5 wt. %), Ti (0.6–1.2 wt. %), Al (0.2–0.8 wt. %), Mo (2.8–3.3 wt. %), Co (≤1.0 wt. %), Cu (≤0.3 wt. %), C (≤0.08 wt. %), and Fe (bal.).

Table 1. Process parameters used for selective laser melting (SLM).

Laser Power	Scanning Velocity	Layer Thickness	Hatching Space	Idle Time	Preheat Temperature
230 W	760 mm/s	40 μm	110 μm	10 s	80 °C

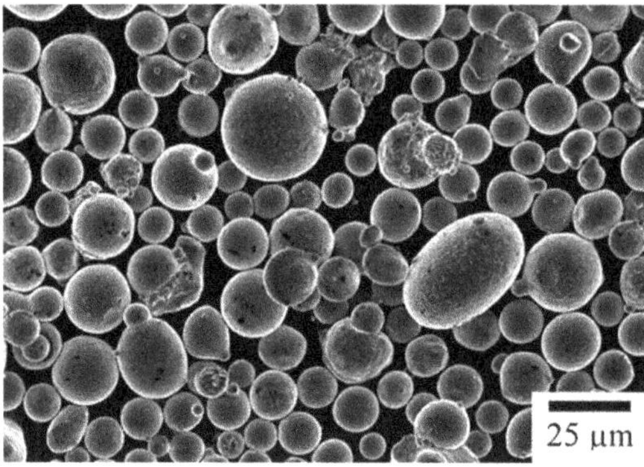

Figure 1. The morphology of Inconel 718 (IN718) powder.

The specimens were removed from the support by electrical discharge machining wire cutting. The as-SLM IN718 (shortly "AS") specimen was subjected to three heat treatment processes: (1) "A": double aging treatment (720 °C for 8 h, then furnace cooling at 55 °C/h to 620 °C, maintaining 620 °C for 8 h, finally air cooling); (2) "SA": solid solution treatment (980 °C for 1 h, then water cooling) followed by "A"; and (3) "HSA": homogenization treatment (1080 °C for 1.5 h, then air cooling) followed by "SA".

The equipment used for the erosion test is shown in Figure 2. We used Al_2O_3 particles as impact particle with an average size of 450 μm for the erosion tests. Their morphology is shown in Figure 3. The specimens were polished with SiC paper from #80 to #1000, then were soaked in acetone for ultrasonic cleaning before the erosion tests. Two hundred grams of the erosion particles were used under a compressed air flow of 3 kg/cm^2 (0.29 MPa). The impact angles were set to 15°, 30°, 45°, 60°, 75°, and 90°.

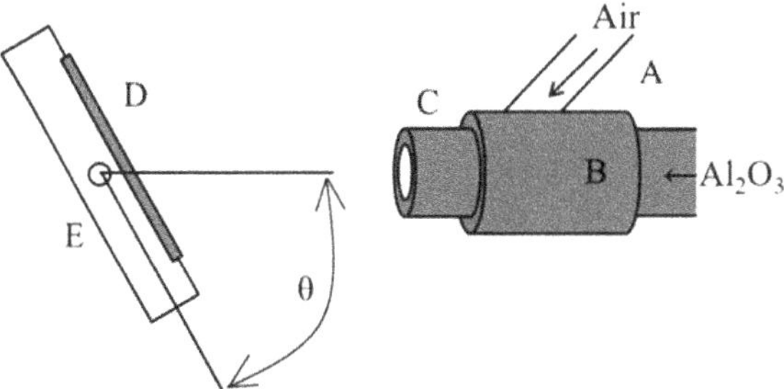

Figure 2. The equipment for particle erosion test. (A: compressed air flow; B: erosion particle (Al$_2$O$_3$) supplier; C: erodent nozzle; D: specimen; E: specimen holder; θ: impact angle.)

Figure 3. The morphology of the Al$_2$O$_3$ erosion particles.

After the erosion test, the specimens were polished with SiC paper (from #80 to #5000). Then they were soaked in Al$_2$O$_3$ aqueous solution (1 and 0.3 µm), and a 0.04 µm SiO$_2$ polishing solution. Finally, they were etched with a chemical solution consisting of 50% HCl, 10% HNO$_3$, 2% HF, and 38% of distilled water. Optical microscopy (OM, OLYMPUS BX41M-LED, Tokyo, Japan) and a scanning electron microscope (SEM, HITACHI SU-5000, HITACHI, Tokyo, Japan) were used to examine the surface and the subsurface of the erosion specimens.

The specimens eroded at the 90° impact angle were selected to analyze the phase composition pre and post erosion states by X-ray diffractometry (XRD, Bruker AXS GmbH, Karlsruhe, Germany). Transmission electron microscopy (TEM, Tecnai F20 G2, EFI, Hillsboro, OR, USA) was used to investigate the microstructure characteristics of the A and HSA pre and post erosion states since the difference in microstructure between these two is the most significant. Finally, the Vickers hardness test (Shimadzu HMV-2000L, Shimadzu, Kaohsiung, Taiwan) was used to analyze the microhardness distribution along the longitudinal direction after the erosion.

The dimensions of the SLM IN718 tensile specimen are shown in Figure 4. The parallel sections of the AS and the three heat-treated specimens were eroded by 200 g Al$_2$O$_3$ particles under a compressed air flow of 3 kg/cm^2 (0.29 MPa) at the 90° impact angle. The room temperature tensile tests of the pre and post erosion states were performed by a universal testing machine (HT-8336, Hung Ta, Taichung, Taiwan). The crosshead speed was chosen as 1 mm/min, corresponding to the initial strain rate of 8.33 $\times$ 10^{-4} s^{-1}. There were at least three specimens used for each tensile test, and the tensile results were taken as the mean values produced by those specimens.

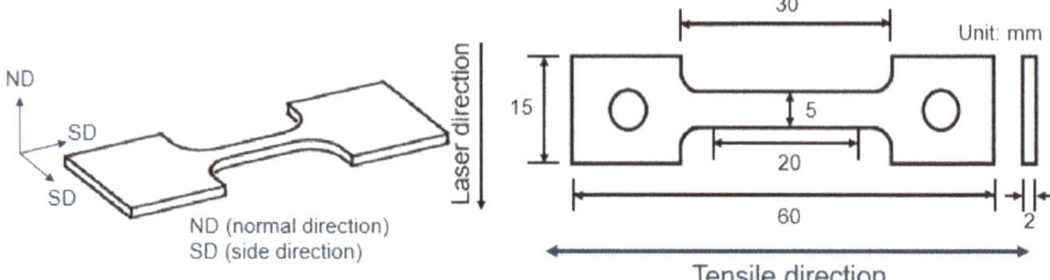

Figure 4. Schematic diagram of the selective laser melting (SLM) IN718 tensile specimen.

3. Results and Discussion

3.1. Particle Erosion Wear Characteristics and Mechanisms

Figure 5 shows the microstructure of the as-SLM Inconel 718 and the three strength-enhanced heat-treated specimens. The as-SLM (AS) specimen has a layered melt pool structure as shown in Figure 5a. Figure 5b shows the microstructure of the specimen "A". We can observe long grains penetrating the melt pools, while the melt pool traces are inconspicuous because of thermal diffusion. Figure 5c shows the microstructure of the specimen "SA". The melt pool traces and the dendritic structure are replaced by long grains parallel to the laser direction. This means that after the solid solution treatment, the dendrites decomposed, and the segregated solute atoms dissolved into the matrix to form long recrystallized grains. The microstructure of the specimen "HSA" is shown in Figure 5d. The HSA specimen exhibits passivated equiaxed recrystallized grains. These four microstructures are similar to those mentioned in the literature [19–22].

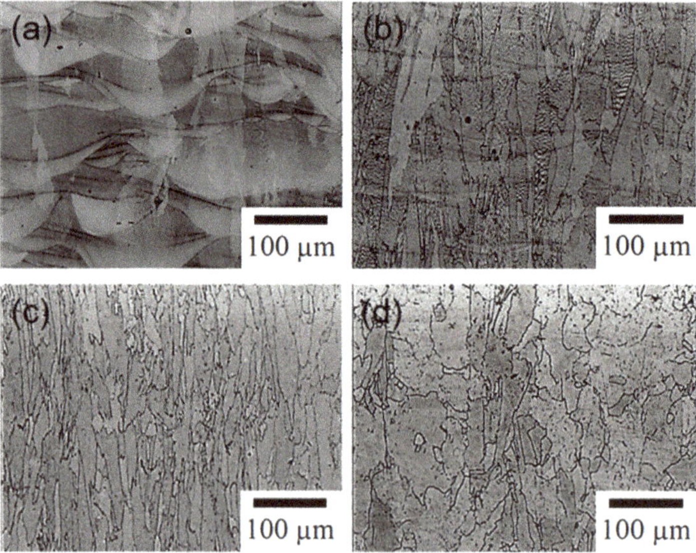

Figure 5. Microstructures of (**a**) AS, (**b**) A, (**c**) SA, and (**d**) HSA. (AS: as-SLM; A: double aging treatment (720 °C for 8 h, then furnace cooling at 55 °C/h to 620 °C, maintaining 620 °C for 8 h, finally air cooling); SA: solid solution treatment (980 °C for 1 h, then water cooling) followed by A; HSA: homogenization treatment (1080 °C for 1.5 h, then air cooling) followed by SA.).

According to previous literature [23,24], the maximal erosion rates of general ductile metals often take place at about 20–30° angles, but the maximal erosion wear rate of brittle materials (ceramics and glass) occurs at about a 90° angle. The erosion rates of the AS, A, SA, and HSA specimens are displayed in Figure 6. All specimens attain a maximal erosion rate at 30° angle, and a minimal erosion rate at 90° angle. The erosion rates decrease with impact angle, which means that ductile-cutting dominates the erosion behavior. At all impact angles, the erosion rate of AS was higher than those of A, SA, and HSA, indicating that heat treatment, especially double aging treatment, can significantly improve the erosion resistance of SLM IN718. In this system, the erosion resistance of the SLM specimen was better than the commercial rolling specimen.

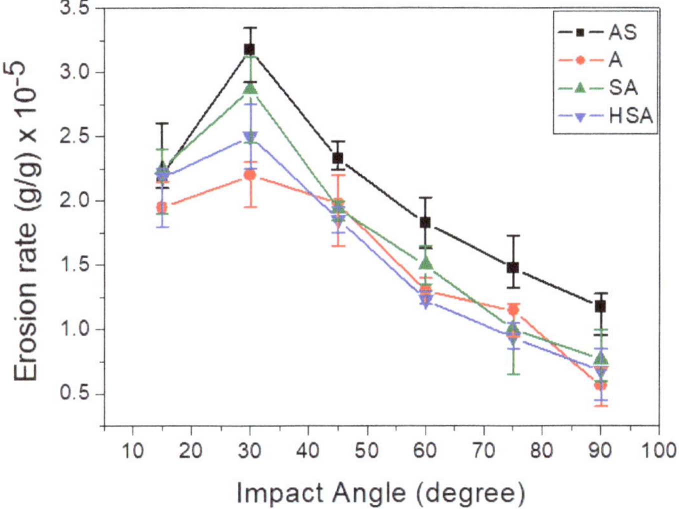

Figure 6. The erosion rates as a function of impact angle.

The difference in the erosion rates among the specimens culminated at 30°, and it was significant in the low-angle region (<45°), the domain of ductile domination. The erosion rates from high to low corresponded to AS, SA, HSA, and A. However, in the high-angle region (>45°), the domain of brittle domination, the erosion rates of A, SA, and HSA were similar to each other, and significantly lower than that of AS.

For an explanation, we turn to Figure 7 showing the surface morphology of specimens at 30°, 60°, and 90° angles. All specimens have lips and grooves on the erosion surface proportional to the erosion rate. The surface of AS has the most amount of lips and grooves, while A has the least. This means that below 45°, the erosion rate is positively correlated with the material being scraped off by erosion particles. At both 60° and 90°, AS exhibited a ductile scraped and squeezed morphology, different from those of the heat-treated specimens with a brittle pit morphology.

The subsurface morphology of all specimens at 30°, 60°, and 90° are shown in Figure 8. Specimen A had the lowest erosion rate with almost no lips, showing a smoother subsurface morphology. The subsurface morphology of A, SA, and HSA were more undulated. At 60°, AS and A exhibited similar subsurface morphology, with some lip features and oblique pits; SA and HSA were filled with oblique pits without lips. At 90°, narrow and deep pits were observed in AS, while A, SA, and HSA had wider and shallow pits. In addition, extrusion traces after the positively eroded particles were observed in SA and HSA, with the latter one being more obvious.

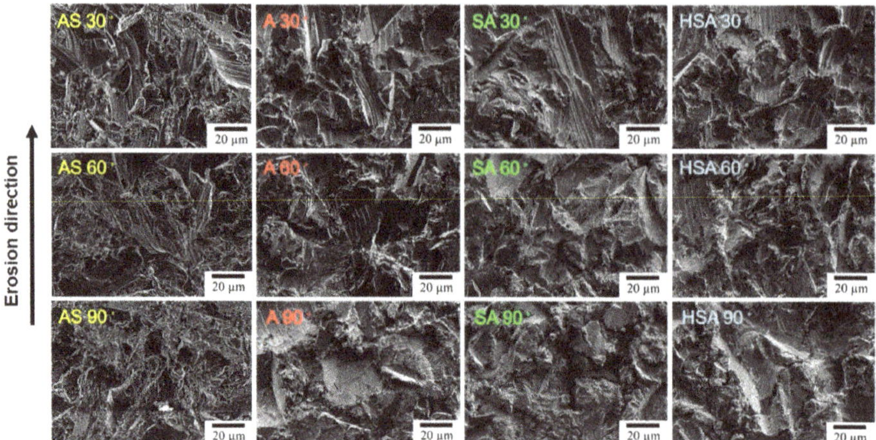

Figure 7. Surface morphology of AS, A, SA, and HSA at various impact angles.

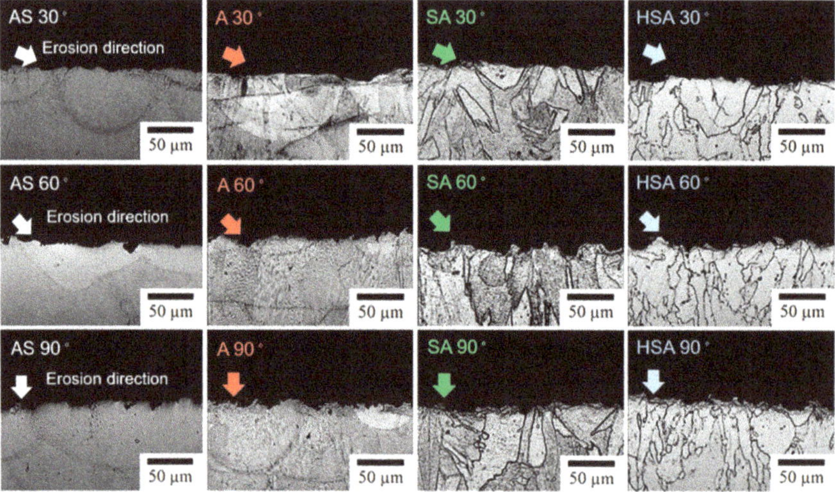

Figure 8. Subsurface morphology of AS, A, SA, and HSA at various impact angles.

3.2. Particle Erosion Induced Phase Transformation

Figure 9 displays the XRD diffraction analysis pre and post positive erosion endured at 90°. For post erosion labeling, we added an 'E' after the codes, i.e., ASE, AE, SAE, and HSAE. We can observe five peaks, namely (111), (200), (220), (311), and (222). As the peaks of γ, γ', and γ'' are almost overlapping, it is difficult to match peaks with phases. The literature [5,25] suggests, to match the (111) peak with the γ phase and the (222) peak with the γ' phase. The main peaks of A, SA, and HSA appear as (111) peaks, while the (222) peaks are more obvious than in the AS, indicating that the γ' phase was formed after heat treatments.

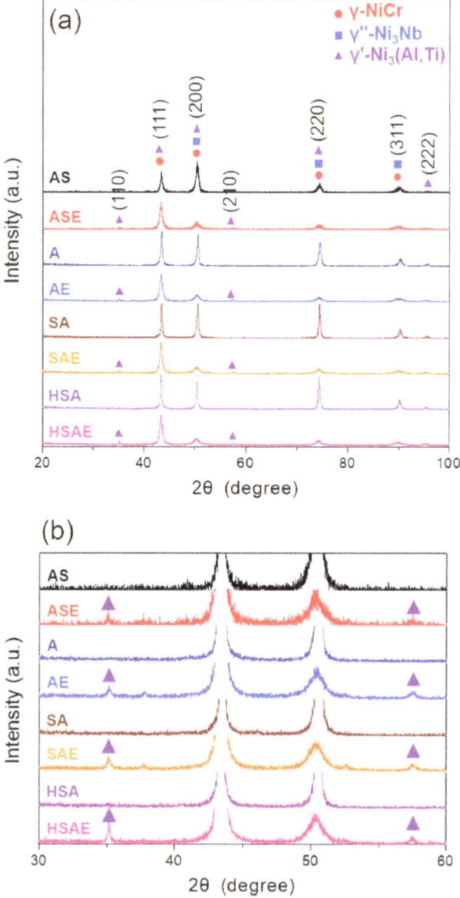

Figure 9. X-ray diffraction pattern of AS, A, SA, and HSA pre and post erosion: (**a**) 20–100°, and (**b**) 30–60°.

In addition, the orientation was changed, and several new peaks appeared after erosion. The new γ' phase and metal-oxide (MO) were generated. Meanwhile, these peaks are in agreement with Al_2O_3 corundum, meaning that there may be some residual Al_2O_3 phase left in the surface of the material.

We compared the changes of microstructures of A, AE, HSA, and HSAE between pre and post erosion (AE and HSAE) via TEM. Figure 10a is the bright field TEM image of the specimen A surface region. From the selected area electron diffraction (SAED) pattern shown in Figure 10b, it can be confirmed that the black and the white regions both represent γ phase matrices. Figure 10c displays the bright field TEM image of the surface region of the HSA specimen. The SAED pattern of the white area corresponds a γ phase matrix, shown in Figure 10d.

Figure 11a,b shows bright field TEM images of the surface region of AE and HSAE, respectively. The microstructures are completely different after the particle erosion. Figure 11c,d displays the well-defined SAED ring patterns of AE and HSAE, respectively. The fine grains within the region selected indicate that the surface regions of both A and HSA transformed post erosion.

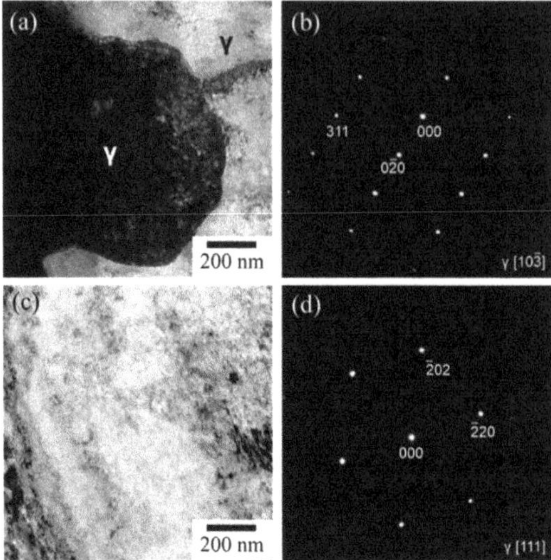

Figure 10. Transmission electron microscopy (TEM) images of specimens pre erosion: (**a**) A, (**b**) SAED (selected area electron diffraction) pattern of A, (**c**) HSA, and (**d**) SAED pattern of HSA.

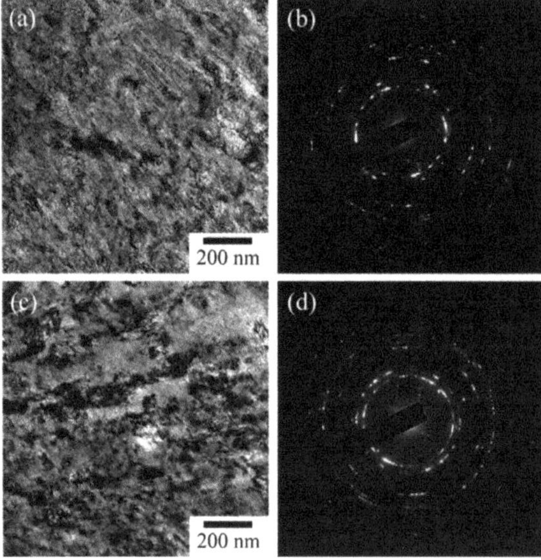

Figure 11. TEM images of specimens after erosion: (**a**) AE, (**b**) SAED pattern of AE, (**c**) HSAE, and (**d**) SAED pattern of HSAE.

In addition, when Al_2O_3 particles impact the surface of Ni-alloy specimen, it can induce impact-heat up to $\approx$500 °C and small sparks are also found during the erosion process. Furthermore, the aging precipitation temperature of the Ni alloy of this paper is about 600 °C. Based on our XRD and TEM: after erosion, the crystallization characteristics of the surface of the test specimen are different (including orientation changes, and metal-oxide formations).

The microhardness distribution along the longitudinal direction of the specimens post particle erosion is shown in Figure 12. After positive erosion, the specimens can be divided into two zones.

The erosion affected zone (EAZ) runs from the erosion surface toward the bottom, while the hardness stability zone (matrix zone) being located below the EAZ. At EAZ, microhardness gradually declines from the surface to a depth of 200 µm, where it reaches the matrix hardness stability zone.

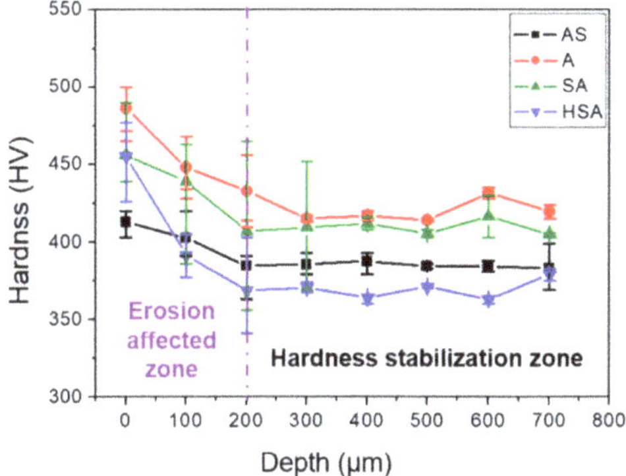

Figure 12. Microhardness distribution of ASE, AE, SAE, and HSAE in the longitudinal direction.

According to previous studies [15–18], temperature increase on the impacted specimen surface induces phase transformation. Wood [17] and Zhao [18] described that erosion would lead to high temperature, making the phase dissolve and the surface layer soft. However, this phenomenon did not occur during our observations. Instead, the surface hardness increased, which can be attributed to that of IN718 is precipitation hardened superalloy. The particle erosion resulted in heat-induced phase transformation, yielding a new γ' phase, which composite with work hardening owing to the eroded particles.

There is a difference between the surface hardness and the stable hardness of the specimens. This difference was maximal in the case of the HSA specimen. Our interpretation is that the HSA specimen had its surface squeezed down and stacked by erosion particles (see Figure 8). The structural evolution of the material after erosion has a significant effect on the tensile mechanical properties of the specimen.

3.3. Influence of Erosion Induced Phase Transformation on Mechanical Properties

Figure 13 shows the stress–strain curve of the four specimens pre and post positive erosion at a 90° impact angle. The tensile strength of A, SA, and HSA improved significantly, while the ductility reduced after heat treatments. After the positive erosion, the stress-strain curves of the four specimens shifted to the right slide, the ductility also reduced significantly.

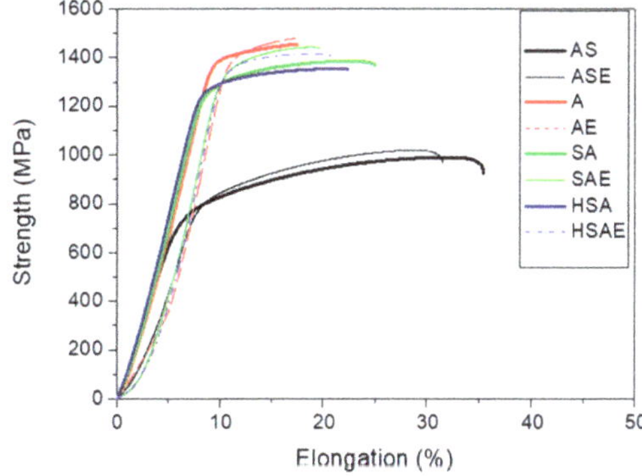

Figure 13. Stress–strain curve of specimen pre and post positive erosion.

Figure 14 shows the tensile mechanical proprieties of the four specimens pre and post positive erosion at a 90° impact angle. The AS specimen had better ductility (UE close to 30%), but the strength was lower than that in the ASM 5662 standard [26] for direct applicability. The strength obviously increased, and the ductility significantly reduced after three heat treatments. The strength of ASE, AE, SAE, and HSAE were similar to those pre erosions, but ductility is significantly reduced, and embrittlement occurred.

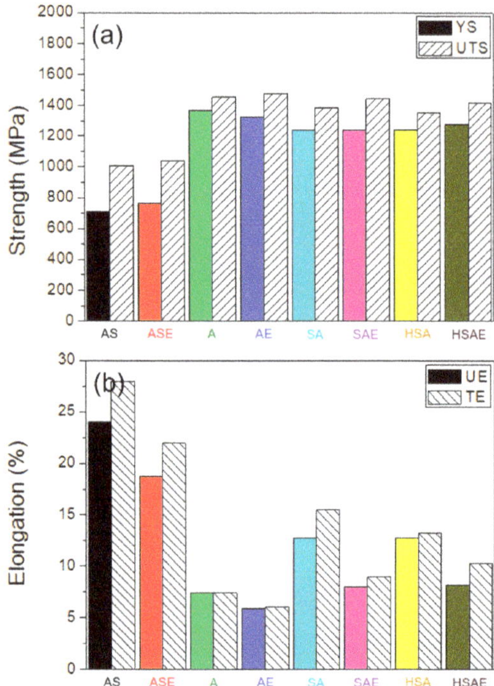

Figure 14. Tensile properties of specimen pre and post positive erosion: (**a**) strength, (**b**) ductility.

Figure 15 displays the SLM-IN718 phase transformation after erosion. According to previous research [27–29], deformation at a moderate temperature promotes the precipitated of strengthening phases. The heat generated by particle impact during erosion increases the surface temperature. The deformation caused by the extrusion lets the γ' phase precipitate into the γ matrix, thereby increasing the hardness of EAZ. Figure 12 suggests that the erosion induced phase transformation precipitates the hard γ' phase, generates MO in the surface erosion zone and enhances the hardness of EAZ.

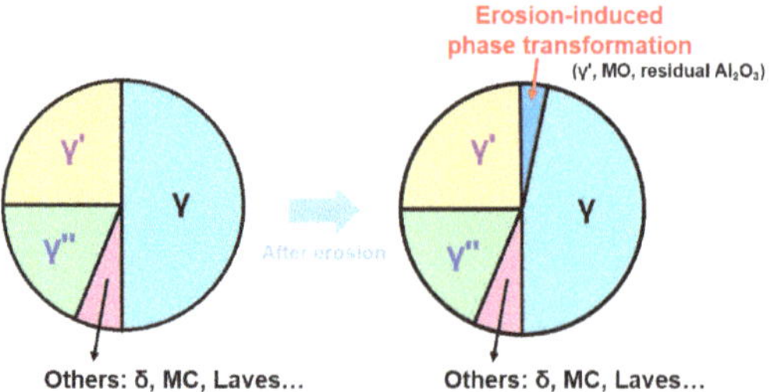

Figure 15. Schematic diagram of the IN718 phase transformation between pre and post erosion.

The precipitation behavior of the double aging heat treatment causes a large decrease in ductility (Figure 14). This effect is similar to the ductility reducing mechanism of the erosion induced precipitation of the γ' phase. Based on this, it can be confirmed that the phase transformation occurs on the IN718 specimen surface post particle erosion, and increases the amount of erosion (see Figure 16), resulting in the tensile ductility decreases and the erosion embrittlement effect occurs.

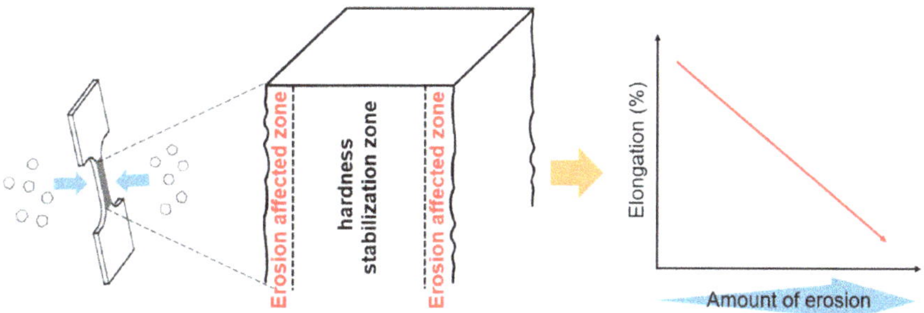

Figure 16. Decrease in ductility caused by the impact of erosion particles.

4. Conclusions

1. The maximal corrosion rate of AS, A, SA, and HSA occurred at a 30° angle, while the lowest one occurred at 90°. This corresponds to the dominant corrosion behavior of ductile cutting. Heat treatment, especially double aging heat treatment, could increase erosion resistance. In the ductile domination region (<45°), the difference of increasing was remarkable, which positively correlated with the ductility of each specimen. In the brittle domination region (>45°), the erosion rates of A, SA, and HSA remained similar. In this system, the erosion resistance of the SLM specimen was better than the commercial rolling specimen.

2. After particle erosion, orientation changes and phase transformation occurs to generate γ' phase or MO. With the work-hardening effect induced by the impact of Al_2O_3 particles, the specimens obtained their highest hardness at the surface region.

3. After heat treatment, the strength of SLM IN718 increased, while ductility decreased. The double aging heat treatment precipitated γ' to enhance the strength and reduce the ductility. After erosion, orientation change and phase transformation to generate γ' phase or MO on both sides of EAZ, the tensile strength merely varied a little, yet tensile embrittlement occurred.

Author Contributions: Methodology, J.-R.Z.; investigation, J.-R.Z.; data curation, J.-R.Z.; writing—original draft preparation, J.-R.Z.; writing—review and editing, F.-Y.H. and T.-S.L.; supervision, F.-Y.H. and T.-S.L. All authors have read and agreed to the published version of the manuscript.

Funding: This research received no external funding.

Acknowledgments: The authors are grateful to The Instrument Center of National Cheng Kung University and the Ministry of Science and Technology of Taiwan (Grant No. MOST 107-2221-E-006-012-MY2) for their financial support for this research.

References

1. Strondl, A.; Fischer, R.; Frommeyer, G.; Schneider, A. Investigations of MX and γ'/γ'' precipitates in the nickel-based superalloy 718 produced by electron beam melting. *Mater. Sci. Eng. A* **2008**, *480*, 138–147. [CrossRef]

2. Hong, C.; Gu, D.; Dai, D.; Gasser, A.; Weisheit, A.; Kelbassa, I.; Zhong, M.; Poprawe, R. Laser metal deposition of TiC/Inconel 718 composites with tailored interfacial microstructures. *Opt. Laser Technol.* **2013**, *54*, 98–109. [CrossRef]

3. Murr, L.E.; Martinez, E.; Gaytan, S.M.; Ramirez, D.A.; Machado, B.I.; Shindo, P.W.; Martinez, J.L.; Medina, F.; Wooten, J.; Ciscel, D.; et al. Microstructural Architecture, Microstructures, and Mechanical Properties for a Nickel-Base Superalloy Fabricated by Electron Beam Melting. *Metall. Mater. Trans. A* **2011**, *42*, 3491–3508. [CrossRef]

4. Kistler, N.A. Characterization of Inconel 718 Fabricated through Powder Bed Fusion Additive Manufacturing. Bachelor's Thesis, The Pennsylvania State University, University Park, PA, USA, 2015.

5. Strößner, J.; Terock, M.; Glatzel, U. Mechanical and Microstructural Investigation of Nickel-Based Superalloy IN718 Manufactured by Selective Laser Melting (SLM). *Adv. Eng. Mater.* **2015**, *17*, 1099–1105. [CrossRef]

6. Sun, J.F.; Yang, Y.Q.; Wang, D. Parametric optimization of selective laser melting for forming Ti6Al4V samples by Taguchi method. *Opt. Laser Technol.* **2013**, *49*, 118–124. [CrossRef]

7. Shiomi, M.; Osakada, K.; Nakamura, K.; Yamashita, T.; Abe, F. Residual Stress within Metallic Model Made by Selective Laser Melting Process. *CIRP Ann.* **2004**, *53*, 195–198. [CrossRef]

8. Mercelis, P.; Kruth, J.P. Residual stresses in selective laser sintering and selective laser melting. *Rapid Prototyp. J.* **2006**, *12*, 254–265. [CrossRef]

9. Liu, F.; Lin, X.; Yang, G.; Song, M.; Chen, J.; Huang, W. Microstructure and residual stress of laser rapid formed Inconel 718 nickel-base superalloy. *Opt. Laser Technol.* **2011**, *43*, 208–213. [CrossRef]

10. Finnie, I. Some observations on the erosion of ductile metals. *Wear* **1973**, *23*, 87–96. [CrossRef]

11. Dai, W.S.; Chen, L.H.; Lui, T.S. A study on SiO_2 particle erosion of flake graphite and spheroidal graphite cast irons. *Wear* **2000**, *239*, 143–152. [CrossRef]

12. Shanov, V.; Tabakoff, W. Erosion resistance of coatings for metal protection at elevated temperatures. *Surf. Coat. Technol.* **1996**, *86–87*, 88–93. [CrossRef]

13. Mishra, S.B.; Prakash, S.; Chandra, K. Studies on erosion behavior of plasma sprayed coatings on a Ni-based superalloy. *Wear* **2006**, *260*, 422–432. [CrossRef]

14. Bircan, B.; Fidan, S.; Çimenoğlu, H. Solid particle erosion behavior of Inconel 718 super alloys under elevated temperatures. *Jamme.* **2014**, *66*, 5–12.

15. Liou, J.W.; Lui, T.S.; Chen, L.H. SiO_2 particle erosion of A356.2 aluminum alloy and the related microstructural changes. *Wear* **1997**, *211*, 169–176. [CrossRef]

16. Hung, F.Y.; Chen, L.H.; Lui, T.S. Phase transformation of an austempered ductile iron during an erosion process. *Mater. Trans.* **2004**, *45*, 2981–2986. [CrossRef]

17. Wood, R.J.K.; Walker, J.C.; Harvey, T.J.; Wang, S.; Rajahram, S.S. Influence of microstructure on the erosion and erosion–corrosion characteristics of 316 stainless steel. *Wear* **2013**, *306*, 254–262. [CrossRef]

18. Zhao, J.R.; Hung, F.Y.; Lui, T.S. Particle erosion induced phase transformation of different matrix microstructures of powder bed fusion Ti-6Al-4V alloy flakes. *Metals* **2019**, *9*, 730. [CrossRef]

19. Chlebus, E.; Gruber, K.; Kuźnicka, B.; Kurzac, J.; Kurzynowski, T. Effect of heat treatment on the microstructure and mechanical properties of Inconel 718 processed by selective laser melting. *Mater. Sci. Eng. A* **2015**, *639*, 647–655. [CrossRef]

20. Mostafa, A.; Rubio, I.P.; Brailovski, V.; Jahazi, M.; Medraj, M. Structure, texture and phases in 3D printed IN718 alloy subjected to homogenization and HIP treatments. *Metals* **2017**, *7*, 196. [CrossRef]

21. Zhang, S.; Zhao, D. *Aerospace Materials Handbook*, 1st ed.; CRC Press: Boca Raton, FL, USA, 2013.

22. Lewandowski, M.S.; Sahai, V.; Wilcox, R.C.; Matlock, C.A.; Overfelt, R.A. High temperature deformation of Inconel 718 castings. In *Superalloys 718, 625, 706 and Various Dérivatives*; Loria, E.A., Ed.; TMS-AIME: Warrendale, PA, USA, 1994; pp. 345–354.

23. Lindsley, B.A.; Marder, A.R. The Effect of velocity on the solid particle erosion rate alloys. *Wear* **1999**, *225–229*, 510–516. [CrossRef]

24. Winter, R.E.; Hutchings, I.M. Solid particle erosion srudies using single angular particles. *Wear* **1974**, *29*, 181–194. [CrossRef]

25. Liu, F.; Lin, X.; Jeng, H.; Cao, J.; Liu, Q.; Huang, C.; Huang, W. Microstructural changes in a laser solid forming Inconel 718 superalloy thin wall in the deposition direction. *Opt. Laser Technol.* **2013**, *45*, 330–335. [CrossRef]

26. SAE Aerospace. *Aerospace Material Specification*; AMS 5662; SAE International: Warrendale, PA, USA, 2009.

27. Nalawade, S.A.; Sundararaman, M.; Singh, J.B.; Verma, A.; Kishore, R. Precipitation of γ' phase in δ-precipitated alloy 718 during deformation at elevated temperatures. *Mater. Sci. Eng. A* **2010**, *527*, 2906–2909. [CrossRef]

28. Kuo, C.M.; Yang, Y.T.; Bor, H.Y.; Wei, C.N.; Tai, C.C. Aging effects on the microstructure and creep behavior of Inconel 718 superalloy. *Mater. Sci. Eng. A* **2009**, *510–511*, 289–294. [CrossRef]
29. Chamanfar, A.; Sarrat, L.; Jahazi, M.; Asadi, M.; Weck, A.; Koul, A.K. Microstructural characteristics of forged and heat treated Inconel-718 disks. *Mater. Design.* **2013**, *52*, 791–800. [CrossRef]

Article

The Effect of Stress Relief on the Mechanical and Fatigue Properties of Additively Manufactured AlSi10Mg Parts

Busisiwe J. Mfusi [1,2,*], Ntombizodwa R. Mathe [2,*], Lerato C. Tshabalala [2] and Patricia AI. Popoola [1]

[1] Department of Chemical and Metallurgical Engineering, Tshwane University of Technology, Staatsartillerie Rd, Pretoria West, Pretoria 0183, South Africa; PopoolaAPI@tut.ac.za
[2] National Laser Centre, Council for Scientific and Industrial Research, Meiring Naudé Road, Brummeria, Pretoria 0185, South Africa; ltshabalala1@csir.co.za
[*] Correspondence: mfusibusisiwe08@gmail.com (B.J.M.); nmathe@csir.co.za (N.R.M.); Tel.: +012-841-4667 (B.J.M.); +012-841-4667 (N.R.M.)

Received: 25 July 2019; Accepted: 14 October 2019; Published: 12 November 2019

Abstract: The heating and cooling profiles experienced during laser additive manufacturing results in residual stresses build up in the component. Therefore, it is necessary to perform post build stress relieving towards the retention and improvement of the mechanical properties. However the thermal treatments for conventional manufacturing do not seem to completely accommodate these rapid heating and cooling cycles of laser processing techniques such as powder bed fusion. Characterizations such as density measurements on the samples were performed employing the Archimedes principle; hardness testing was performed on the Zwick micro/macro (Hv) hardness tester, SEM and Electron backscatter diffraction (EBSD). Fracture toughness and crack growth was conducted on a fatigue crack machine. All characterization was done after stress relieving of Selective Laser Melting (SLM) produced samples at 300 °C for 2 hrs was performed in a furnace. The mechanical properties appear to be rather compromised instead of being enhanced desirably. As-built SLM produced tensile specimens built in different directions exhibited significantly favorable mechanical properties. However, post stress relieve thermal treatment technique deteriorated the strength while increasing the ductility significantly. Nonetheless, fatigue crack growth and fracture toughness illustrated positive outcome in terms of fatigue life on SLM produced AlSi10Mg components in application.

Keywords: selective laser melting (SLM); Alsi10Mg; stress relieve; additive manufacturing

---

1. Introduction

Additive manufacturing (AM) is a new technology that has made its mark as an innovative and flexible manufacturing technology [1]. Continuously, there is a need for the development of this technique, with the main objective being to reach 100% component density [2,3]. The additive manufacturing of aluminum alloys in particular finds recent applications in the aerospace, rail and automotive industries for structural and non-structural parts, which are usually die casted. In particular, the powder bed fusion processing of aluminum alloys has gained interest in the aerospace, rail and automotive industries due to the versatile nature of the process. The shapes of the components are generally attained from rigidity-focused strategies. Usually low stress requirements are often the objective for these components relative to stressful loading circumstances where finite-life fatigue resistance should be considered as distinctively possible such as circumstances where load or vibrations counteracting on component could be extreme [4]. Fatigue resistance crack propagation relative to the direction of the load is inherently dependent to anisotropy of the process of manufacturing. It is common knowledge that defects that are found on the surface are the most dangerous of them all since crack propagation is most likely to be initiated on the surface. In AM applications, majority of failures observed are those that generally instigated from the surface where there are defects [4].

Silicon based aluminum alloys such as AlSi10Mg, AlSi12, etc. are currently used for laser AM and these are characterized by good castability, low shrinkage and moderately low melting temperature and AlSi10Mg is one of the most common alloys characterized with a hypoeutectic composition [5,6]. Prashanth et al. [7] studied the heat treatment of hypereutectic Al–Si alloys, which are said to have a wide application in the automobile and aerospace sectors due to their high wear and corrosion resistance.

During laser AM processing, the rapid heating and cooling of the laser processing results in the residual stresses build up. These residual stresses affect the properties of the components such as the ultimate tensile strength and the fatigue life [8]. It is said to be common knowledge that AM material has comparatively lower fatigue resistance than traditionally manufactured materials in the as-built condition, the reason being that fatigue life is affected directly by impurities and the inhomogeneity of the microstructure [9]. In this case, post heat treatment is necessary in order to relieve the residual stresses while maintaining the desired mechanical properties.

For instance, Cabrini et al. [10] performed various heat treatment techniques on the AlSi10Mg samples with various direct metal laser sintering (DMLS). These were stress relieving, annealing at high temperature and water quenching. Their results determined that annealing resulted in intensification to the matrix of aluminum phase with precipitation of rounded silicon on the surface also. Fiocchi et al. [11] studied the low temperature annealing of Selective Laser Melting (SLM) produced AlSi10Mg. After stress relieving at 263 °C on the as built samples, minor microstructural differences were observed. However, for heat treatment at 294 °C the silicon network appeared to be disconnected. Other work by Fousova et al. [12], investigated the modifications in the microstructure and mechanical properties of additively manufactured AlSi10Mg alloy after exposure to temperatures of 120–180 °C. The current, heat treatment profiles, mainly die casting used for AM parts, are adapted from conventional profiles, which have been shown to affect the AM parts negatively in some instances [13].

Therefore in order to improve the fatigue performance of AlSi10Mg, T6 heat treatment, processes such as solution treatment at 520 °C, water quenching and artificial ageing at 160 °C, which are also known as peak-hardening have been used [14]. Microstructural coarsening and material softening during annealing of SLM produced AlSi12 were reported and revealed the same results as conventionally cast material [15].

This paper will focus on the effect of stress relieving on the microstructure, porosity, mechanical properties and fatigue life of SLM produced AlSi10Mg samples with a focus on the build direction effect. It is a continuation of the previously published work conducted by the authors investigating the effect of build direction on the as-built SLM samples using AlSi10Mg [16]. This is because although the effect of the build direction of powder bed fusion manufactured Ti6Al4V has been extensively studied, there is little information on the AlSi10Mg alloy.

2. Materials and Methods

2.1. Powder Bed Fusion Processing

AlSi10Mg powder with a spherical morphology and particle size distribution of 30–65 μm was purchased from TLS Technik GmbH, Germany, and it was used as received (see reference [6]). The study was carried out on specimens produced by the SLM Solution M280 GmbH from Lubeck in Germany using a laser power of 150 W, 1000 mm/s scan speed, 50 μm hatch spacing and 50 μm powder layer thickness fixed processing parameters (the image of the samples on the base late is presented in previous work [16]). The samples were built in the XY, 45° and Z orientations. Tensile, fatigue and crack growth samples (three of each) were prepared according to the images in Scheme 1. Post build stress relieving on the specimens was carried out in a rotary furnace using a temperature of 300 °C and hold time of 2 h then furnace cooled to room temperature.

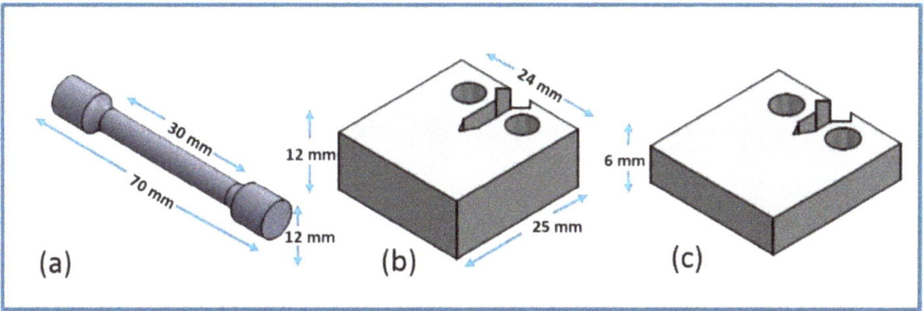

Scheme 1. An illustration of the Selective Laser Melting (SLM)produced AlSi10Mg mechanical samples; (a) tensile, (b) fracture toughness and (c) fatigue crack growth. Note: The length and width of (b) and (c) are the same.

2.2. Characterization Techniques

Density measurements on the samples were performed using the Ohaus densitometer employing the Archimedes principle with ethanol as a liquid medium. For testing the mechanical properties (ASTME8), the 20kN Zwick/Roell Tensile/Compression tester GmbH from Radeberg in Germany was used, while hardness testing was performed on the Zwick micro/macro (Hv) hardness tester GmbH from Radeberg in Germany. The microstructures of the polished samples after etching with Keller's reagent were viewed on the optical microscope. The samples were also analyzed for phase and grain information using a Zeiss LEO 1530-FESEM fitted with the Oxford Energy Dispersive Spectroscopy (EDS) and Hardware Lab Kit (HKL) Electron Backscatter Diffraction (EBSD) detector GmbH from Radeberg in Germany. The fractured tensile samples also underwent fracture analysis on the JEOL JEM-210 SEM from Peabody in USA. A 1342 Instron 30 kN fatigue crack machine from Norwood in US was used to conduct fracture toughness and fatigue crack growth tests according to ASTM399. The dimensions of the samples were as seen in Scheme 1, with a notch length of 7.37 mm for both fatigue crack growth and fracture toughness samples. The load used on the samples was 156 MPa, cycling at a rate of 2.1 mm/min and frequency of 15 Hz.

3. Results and Discussion

3.1. Density and Porosity Measurements

The density results of the AlSi10Mg as-built samples were investigated and presented by Mfusi et al. [16], where the average density was 2.68 g/cm^3, with relative densities above 99% for the samples in different build directions. However, after stress relieving at 300 °C the density values dropped drastically to 2.58–2.61 g/cm^3, which were 0.07–0.1 g/cm^3 lower than as-built, presented in Table 1. It was also observed that the porosity values were higher at 2.67–3.81%, which was above the accepted levels for applications.

Table 1. Showing the orientation density and porosity levels after stress relieve.

Orientation	Density (g/cm^3)	Porosity (%)	Relative Density (%)
A (0°)	2.61 ± 0.025	2.67 ± 0.92	97.33 ± 0.92
B (45°)	2.60 ± 0.023	3.19 ± 0.86	96.81 ± 0.86
C (90°)	2.58 ± 0.027	3.82 ± 1.00	96.18 ± 1.00

This behavior was also observed by Calignano [17], where they also determined that stress relief led to a decline the density and mechanical properties of the specimen. Ahmed [5] determined that the greatest challenge in producing aluminum alloys parts by SLM technique was to minimize porosity, which is the major effect of the relative density as aluminum alloys are easily subjects to oxidation during processing, stimulating pore formation. Therefore, more research is needed to address the impact of heat treatment temperatures on the properties of the SLM produced samples in order to optimize the properties for intended applications [18].

3.2. Hardness Measurements

The Vickers hardness method was used to measure the hardness of the samples in different orientations after stress relieving and is presented in Table 2.

Table 2. Build orientation hardness results after stress relief.

Orientation	HV
A (0°)	43.91 ± 2.44
B (45°)	47.32 ± 3.35
C (90°)	47.14 ± 4.16

After stress relief the samples suffered an enormous drop in hardness from 126–128 HV for the as-built samples [16], to 46–49 HV as shown in Table 2. This illustrated the uncertainty of the stress relieving method undertaken towards the improvement of the mechanical properties of SLM produced AlSi10Mg, irrespective of the build direction. However, the largest drop in hardness was observed for batch A that was built in the 0° direction. This decrease in hardness after stress relief in the SLM produced aluminum alloy study was also observed by Fiocchi et al. [11] and Aboulkhair et al. [16,19], where a negative response was seen after heat treating the AlSi10Mg alloy using the T6 treatment and stress relief at 300 °C. In their work, stress relief showed a significant decrease in hardness of up to 66%, compared to the decrease of up to 62% obtained in this investigation. They attributed this behavior to the material softening that was seen to expose the additively manufactured component's weakness for application that requires high hardness properties. Trevisan et al. investigated the effect of stress relieving on SLM produced AlSi10Mg parts and also reported negative results as well [17], which might not render the temperature profiles viable for the improvement of properties, however the process is sometimes necessary in order to relieve the residual stresses obtained in SLM produced parts.

3.3. Microstructure Analysis

The evolution of the microstructures as a function of heat treatment is an important aspect of metallurgical evaluation as it is related to the mechanical properties of the samples. Therefore, the microstructures of the SLM processed AlSi10Mg samples after stress relieving were taken and these are presented in Figure 1a–c, demonstrating the three build directions at 50× magnifications.

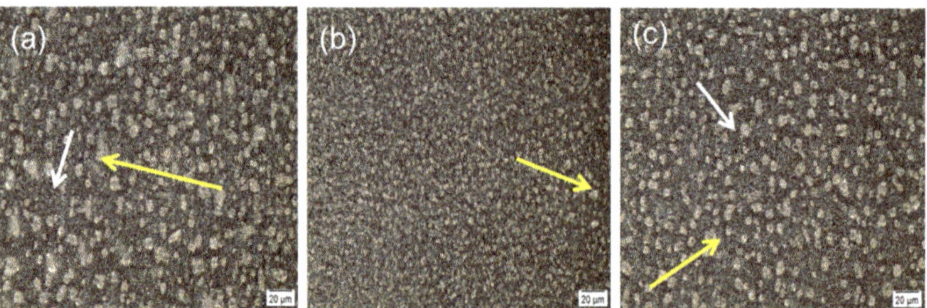

Figure 1. Optical Microscopy (OM) microstructure of AlSi10Mg (orientation 0°, 45° and 90°) samples: (**a–c**) at 50× magnification post stress relief.

Figure 1a–c illustrates that post stress relieving, the grain boundaries disappeared completely and a different homogeneous "sand like" structure was observed. The microstructural observation acquired a big change in the morphology compared to the as-built micrographs, which had scale-like morphologies [16], as presented by Mfusi et al. From the observation, this structure exhibited a rough surface, which was speculated to be a result of the silicon (depicted by yellow arrow) and aluminum (depicted by white arrow) being uniformly distributed. Brandl et al. [15] also observed the same change in the microstructure after performing peak hardening heat treatment for SLM processed AlSi10Mg. Finer silicon particles were observed in 45° orientation relative to 0° and 90° orientation, which were anticipated to be attributed by the defects contained in the specimen. Built direction played a role as it was observed also that 0° orientation had more courser particles than the rest. Tang determined that there were three stages that silicon undergoes during stress relief heat treatment where silicon and Mg_2Si precipitates, spheroidization of precipitates and silicon particles coarsening [20].

Fousova et al. [12], explained the effect as the comparative infringement of the silicon network causing the coarsening of distinct Si particles, which are observed in all the images. Zhang et al. [21] also stated that after stress relieving, there is a diffusion of silicon dendrites into a discontinuous state.

3.4. Electron Backscatter Diffraction (EBSD)

EBSD was performed on the as-built and stress relieved samples to determine the effect that stress relieving has on the phases and grains produced by SLM processing. Table 3 presents the phase fractions of the main elements of AlSi10Mg alloy for the as built samples, as well as for samples after stress relieve heat treatment respectively, for the three build directions.

Table 3. Showing as built and stress relieved Electron Backscatter Diffraction (EBSD) phase fraction.

	AS-BUILT			
Orientation	**EBSD Phase Fraction (%)**			
	Al	Si	Mg2Si	Zero solution
A (0°)	2.02	8.52	1.04	88.42
B (45°)	8.80	54.17	1.19	35.84
C (90°)	11.48	37.23	3.47	47.82
	STRESS RELIEVED			
Orientation	**EBSD Phase Fraction (%)**			
	Al	Si	Mg2Si	Zero solution
A (0°)	8.13	29.13	3.20	59.54
B (45°)	3.09	10.15	1.67	85.09
C (90°)	11.56	37.49	2.57	48.38

In Table 3, 0° orientation samples, before stress relief had a minimal amount of the main elements and a high percentage of zero solution. After stress relieving an increase in aluminum and especially silicon was observed on the surface of the sample, this supports the observation made in Figure 1 OM images. The 45° orientation showed a huge fraction of silicon element and a significant amount of aluminum before stress relieve. After stress, a huge rise to zero solution is seen. On the other hand 90° orientation showed to have an insignificant change before and after stress relieve with a drop in the Mg_2Si as compared to other orientations. During the SLM process, Mg_2Si appeared upon final solidification while appearing beneath solidus temperature below equilibrium circumstances. Mg generally melted into the aluminum matrix considering that it was not present as any intermetallic phases during solidification at equilibrium as stated by Tang et al. [20] then the increase of Mg_2Si and silicon after heat treatment was caused by the precipitation onto the surface.

Figure 2 shows the pictures of the electron backscatter diffraction results as-built and after stress relief.

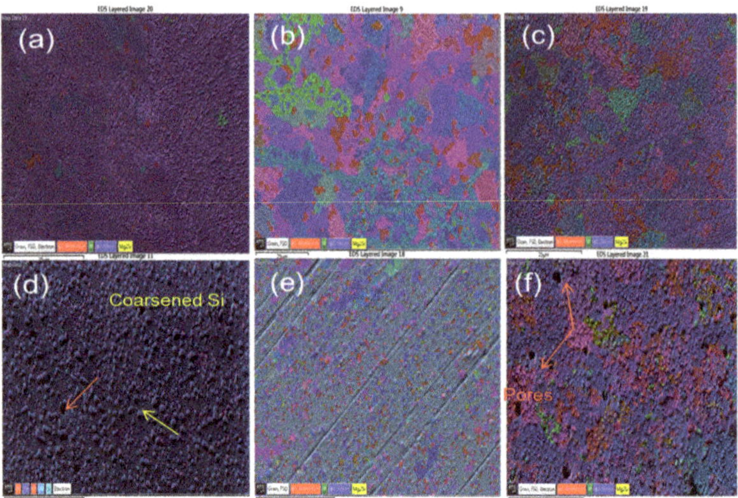

Figure 2. EBSD of AlSi10Mg samples at the Energy Dispersive Spectroscopy (EDS layer image, orientation A (**a**) and (**d**), orientation B (**b**) and (**e**) and orientation C (**c**) and (**f**), as-built and post stress relief respectively).

Figure 2a showed cellular and dendritic growth, which was also observed in the optical microstructure. Wu et al. [22] attributed these cellular growths as long cells that are formed as a result of cooling conditions in place of dendrites. The dark region seen in Figure 2b,c was drastically eradicated in Figure 2e,f, where more silicon and aluminum were visible. These dark regions, according to Wu et al. [22] and Mathe et al. [6], are aluminum grains that also contain Al–Si eutectic, which is difficult to distinguish from cell boundaries because of the disappearance of diffraction data. Figure 2d has developed roughness on its surface that looks "pimple" like as compared to Figure 2a before stress relief. The same roughness was observed in the optical microstructures.

Heat treatments at high temperature can encourage the combination of second phases as well as change in distribution of those phases [23]. This was observed in Figure 2c after stress relief. In these samples, the columnar and equiaxed grains were observed before stress relief. These have been proposed to be formed directly by solidification that takes places as cellular dendrites as a consequence of rapid cooling with little information for the mechanism of formation [22,24]. In the optical microstructure, it was stated as silicon segregation to the grain boundaries as a result of rapid cooling [19]. Aboulkhair et al. [25] and Li et al. [23] proposed that these silicon rich boundaries isolated by the aluminum grains and the cellular structure are the fine eutectic composed of aluminum grains with silicon particles. Longer columnar grain sizes after stress relief are seen even though they were covered by the black region. In Figure 2f some Mg_2Si precipitation was also observed and the silicon phase that was more visible on the as-built samples, while after stress relief it seemed to fade.

Wu et al. [22] stated that the reason SLM processed AlSi10Mg is optimum in strength is because the larger aluminum regions contains silicon particles, which are surrounded by the thick eutectic boundaries that prevents dislocation movement inside the aluminum grains. In these pictures, it is observed that the dark lines seen before stress relieved are removed. These dark lines were also observed by Mathe et al. [6] for SLM produced AlSi10Mg, which were attributed as shear bands that are caused by the shear strain from the manufacturing process. Grain refinement was also observed on the samples after stress relieving.

3.5. Tensile Strength after Stress Relieving

To determine the effect that build directions have on the mechanical properties of post stress relief tensile specimen tensile measurements were performed. The average values of triplicate measurements are presented in Figures 3 and 4. In the previous study of the as-built AlSi10Mg SLM samples [16], orientation 90° had the highest ultimate tensile strength compared to the other orientations. After stress relieving, the Ultimate Tensile Strength (UTS) values dropped drastically from 420–470 MPa as-built, [16], to 110–160 MPa. The same was observed also for the modulus values in Figure 3b.

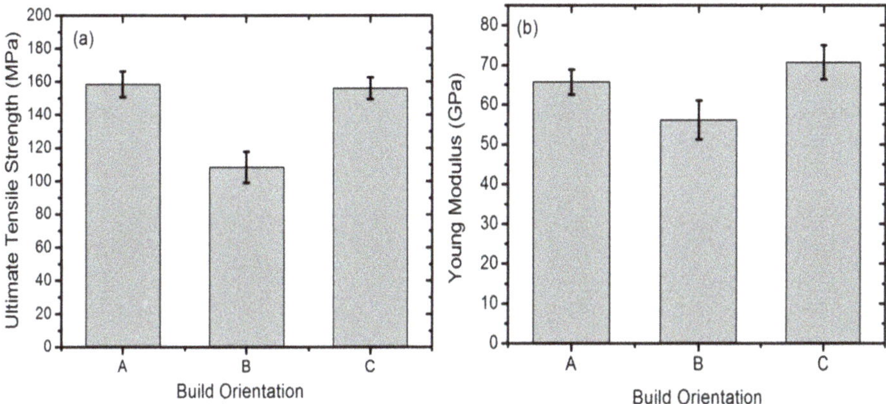

Figure 3. The illustration of (**a**) Ultimate Tensile Strength UTS variation and (**b**) modulus of elasticity for the Selective Laser Melting (SLM) processed AlSi10Mg samples as a function of build direction.

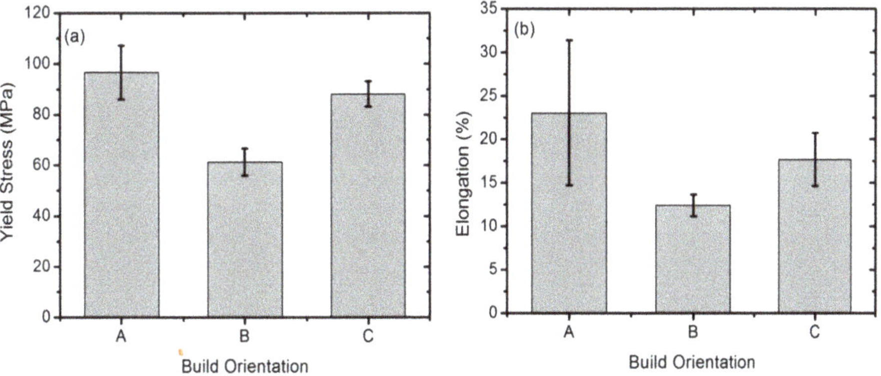

Figure 4. The illustration of (**a**) yield strength and (**b**) elongation for the SLM processed AlSi10Mg samples as a function of build direction.

In Figure 3a, 0° and 90° orientation exhibited the higher UTS values compared to 45° orientation. This is in contrast with the as-built samples where the 0° orientation had the lowest UTS value and the 90° orientation still had the highest value. The drastic decrease in the mechanical properties of the SLM produced has also been observed by Brandl et al. [15,19], for the same material, which they attributed to the changes in the grain structures and phases of the samples present when heat treatment occurs above the eutectic temperature.

The yield strength results in Figure 4a showed higher values for the 0° and 90° orientation, with the values ranging from 61–96 MPa. The 90° orientation was ductile but could only endure elastic deformation up to just below 88.1 MPa, this might be because of the number of pores suffered by the material for the as-built samples [16]. The yield strength of the 45° orientation was radically dropped by a magnitude of approximately 4.3 times to 108 MPa. Figure 4b showed the elongation results of the samples after stress relieving, where for all the build directions the elongation increased significantly from 6.25–7.25 mm as-built [16] to 12.5–23 mm post stress relief. This means with stress relief the ductility of the material is drastically improved. The effect of post build stress relieving in this case had both a negative and positive impact with the most improvement seen for the elongation.

Represented graphically in Figure 5a are the stress–strain curves of the tensile tests after stress relieving. All the samples after stress relieve experienced Lueder's bands during tensile testing (see Figure 5b), which was a result of high total elongation before fracture. This phenomenon occurs when a specimen that cannot yield to the given load, yields, which is also known as discontinuous yielding [3,7], the behavior is typically observed in the area where an increase in strain occurs without an increase in stress. It was observed in the comparison of mechanical properties that before stress relieving the samples had more strength and less ductility [16], however, after thermal treatment, in the attempt to acquire ductility, strength was compromised.

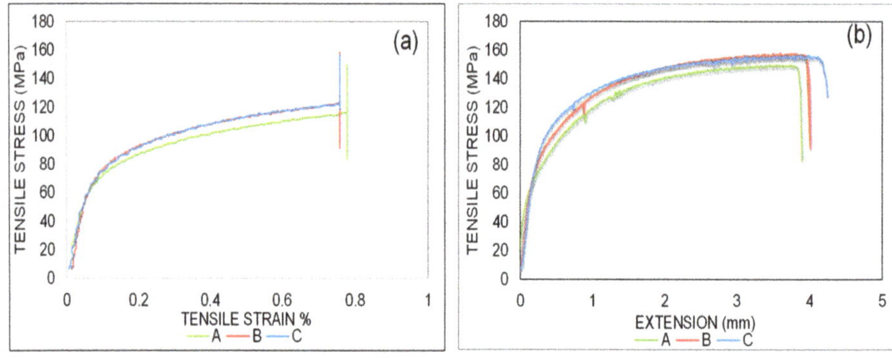

Figure 5. Showing Batch A, B and C (0°; 45° and 90°) samples, (**a**) tensile strain and (**b**) extension.

Zhang et al. [26] determined that stress relieve reduces the strengths and fatigue properties of the alloy significantly due to reduced solid solubility and the precipitation of the fine silicon as well as the demolition of the fine sub-structures within the aluminum grains. Li et al. [23,27] stated that even though heat treatment is required for quality improvement of a part by microstructural refinement, it has been found to decrease the UTS while increasing the ductility. This is as a result of the silicon trapped in the aluminum matric that precipitates to the grain boundaries, reducing the solid solution strength [12].

Rosenthal et al. [28], also reported a decline in the ductility that is inversely proportional to the strain rate, which suggests the effects of confined strain rate hardening had not taken place, and that the existence of the silicon phase could be predominantly the reason for this conduct. Brandao et al. [29] used stress relief to prevent the residual stress from distorting the components. The effect was the same as in this work, as it led to a drastic decrease in the static yield strength.

3.6. Fractography Analysis

Figure 6 illustrates the SEM morphology of the fractured surfaces from the tensile specimens before and after stress relieving. Figure 6a showed dense and smooth surface for the 0° orientation as-built samples, with fracture defects observed. Fracture observed to begin from the one end of the structure propagating to the other end.

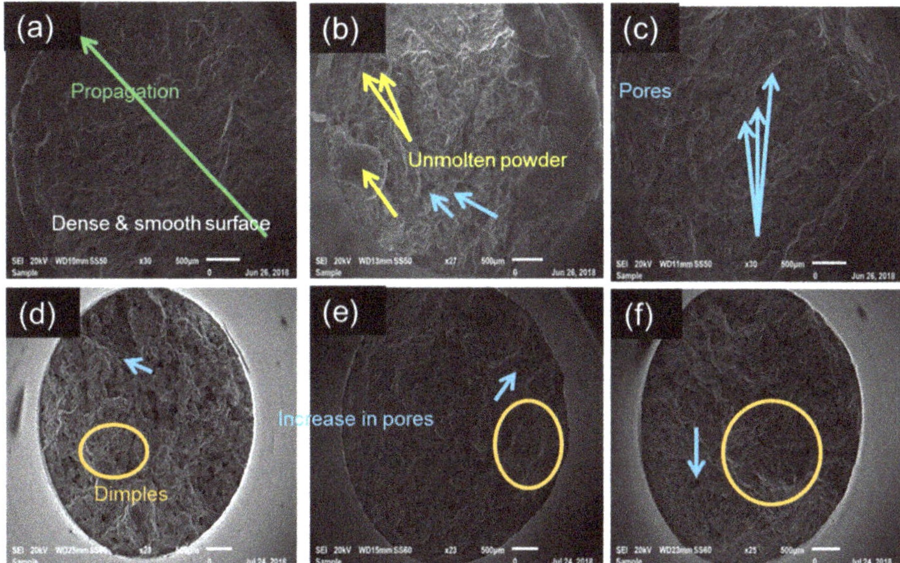

Figure 6. SEM morphology of SLM produced AlSi10Mg samples, orientation 0° (**a**) and (**d**), orientation 45° (**b**) and (**e**) and orientation 90° (**c**) and (**f**) as-built and post stress relief respectively.

The stress relieved samples, (Figure 6c–f), showed dimples, which illustrated the ductility that was desirable for the manufacturing of industrial parts, but in this case the tensile testing strength was proven to have been compromised. There was also an increase in the number of surface pores observed on the stress relieved samples compared to the as-built samples, which was in support of the Archimedes density and porosity results presented in Table 1.

Various defects were observed for the as-built samples, for instance in Figure 6b, ballus defects were observed and Dheyaa et al. [30] explains these ballus defects, marked with yellow arrows, to be unmolten or partly molten powder. According to Read et al. [13,31], these unmolten powder particles are a result of thick oxides layer existing on the particles, which did not allow full consolidation during processing. Defects such as unmolten powder and pores (marked in red), impacts the fatigue life of a component severely by reducing the effective load bearing area thus causing stress concentrations that consequently form static and dynamic strength reduction [15]. In Figure 6e, post stress relieving for the 45° orientation samples, the unmelted powder disappeared but there seemed to not be an apparent improvement in the mechanical properties. This was observed in the tensile results for orientation B, which had the highest elongation post stress relieve.

Figure 6c (90° orientation) specimen showed a ductile fracture with dimples, which suggests forced fracture. It was determined that there were micro-cavities formed where there were defects visible in Figure 6d–f post stress relieve [15,25,31,32]. These micro-cavities lead to more micro-cavities that join together to cause a fast growing tear in the structure. The tear spreads laterally to the interface between the melt pool core and the boundary, constantly along the fracture sides. This is due to the fact that the melt pool boundary is weaker than the melt pool core, containing a coarser microstructure and minimal content of silicon for the reduced grain boundary, which were also observed on the OM and EBSD microstructures. Overall post processing by stress relieve has shown the microstructure of the fracture surfaces of all three different orientations were similar. Ductility dimples are virtually the same with voids that develop and merge together. Zaretsky et al. [33] believe that the homogeneous spreading of sites is suitable for void nucleation all over the sample and these are the characteristics of SLM processing.

3.7. Fracture Toughness and Fatigue Crack Growth Rate Analysis

The stress relieved samples for the 90° orientation were further analyzed for fracture toughness and crack growth in order to determine the effect that the heat treatment process has on the properties. One orientation was chosen because all samples after stress relieve exhibited more or less the same mechanical properties. The samples were machined to dimensions specified in Scheme 1 and ASTM399, before undergoing the tests. According to Rosenthal [34], as-built AlSi10Mg relative to traditionally produced AlSi10Mg exhibited inferior fracture toughness properties so the samples were stress relieved before testing.

Figure 7 was plotted based on the Paris equation: $da/dN = C(\Delta K)^m$ and the three states were marked in the graph. The K_q obtained were 29.51, 30.47 and 29.99 MPa.m$^{1/2}$ for samples 1 to 3 respectively according ASTM399 standard for fracture toughness. Results show correspondence with those of Rosenthal [34], which were 30.4 MPa.m$^{1/2}$ after stress relieving. These results show that even though the stress relieving profile use in this case resulted in a decline in strength, it increased the fatigue life of the samples as presented in Figure 8. The samples have dimple fractures, which Kobayashi [35] equated to nucleation–growth–coalescence of voids. According to Brandl [15], additively manufactured AlSi10Mg demonstrated an increase in fatigue life after thermal treatment compared to cast AlSi10Mg.

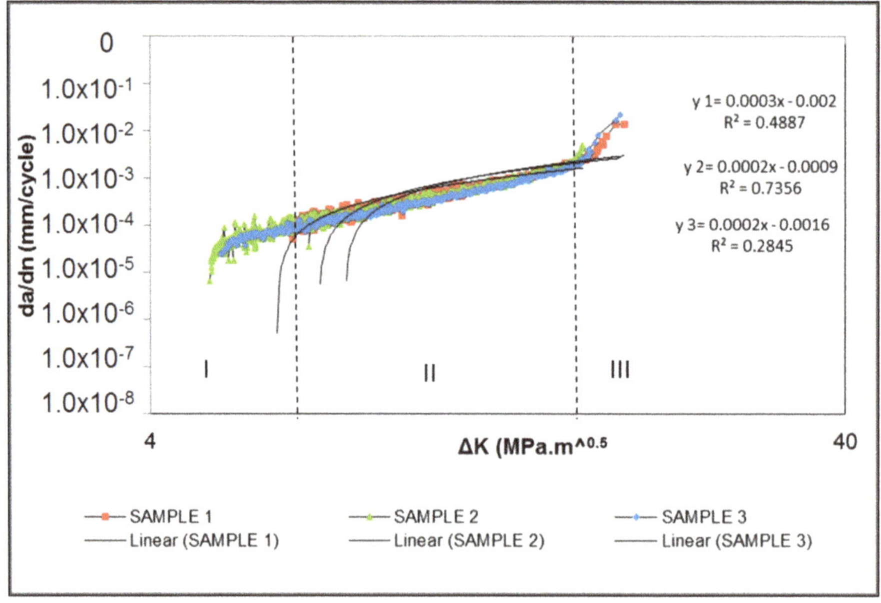

Figure 7. Fracture toughness results of the SLM produced AlSi10Mg samples.

A representative sample was chosen from the three samples and is presented in Figure 8; it shows the propagation until the edge of the sample. The cleavage fracture was observed in these samples as "particle like" brittle phases where there were holes. These are called Griffith-like-microcrack as studied by Ruggieri and Dodds Jr. [36]. This is the microcrack that is nucleated then promptly spreads into the inner grain boundary to cause a fracture when it is not blocked by any obstructions in the grain boundary. It was observed in the gentleness of the steep of the graph in Figure 7 that the fracture was gradual. Therefore these samples failed due to nucleation overpowering various consecutive obstructions in the grains structures [37].

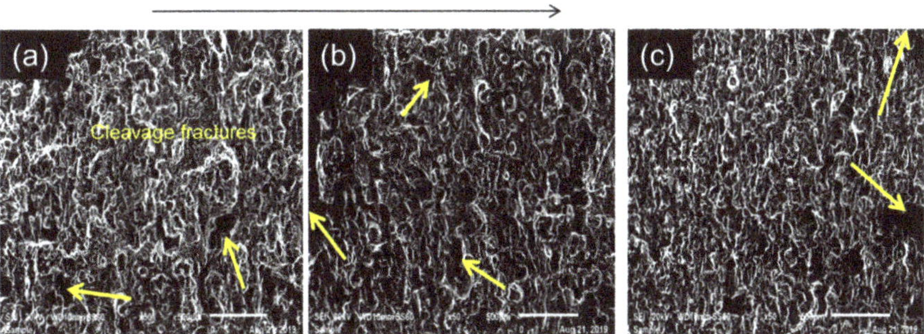

Figure 8. Illustration of the three fracture toughness samples (**a**), (**b**) and (**c**), SEM images of the SLM produced AlSi10Mg built in 90°.

The fatigue crack growths results show that the samples were able to sustain the load for up to 75,000 cycles. The samples exhibited brittle propagation then dimples where it seemed to have tried to endure further before yielding to cracking. Aluminum alloys exhibited ductile fracture that formed dimples even though formation of dimples is dependent on shape, properties, volume fraction and condition of the particle matrix [35].

Figure 9 presents fatigue sample 1 (a,b,c), sample 2 (d,e,f) and sample 3 (g,h,i) of the fatigue crack growth sample. It was observed in sample 1 numerous pores that were anticipated to have assisted the crack to grow faster relative to sample 2 and 3.

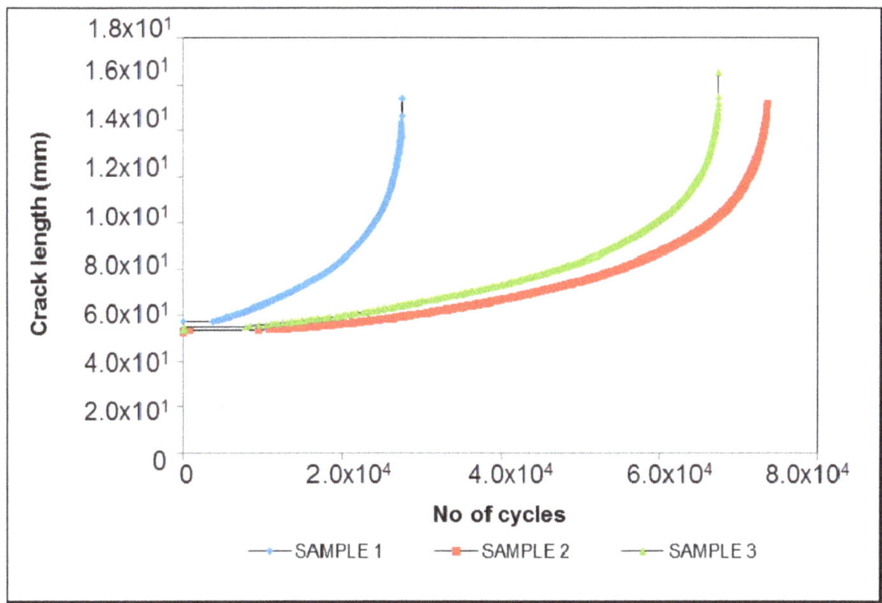

Figure 9. Crack growth rate results of the SLM produced AlSi10Mg samples.

There seem to have been an increase of propagation possibilities even after a certain number of obstructions in the grain structure that arose from the pores [35,36] resulting to rapid failure. Samples 3 displayed a smoother propagation before fracture relative to the other two samples as also observed in Figure 10. Defects and pores at most were mainly the cause of rapid failure on components since generally cracks were initiated where there were defects of some sort in the structure such as oxides in the case of SLM produced AlSI10Mg [37].

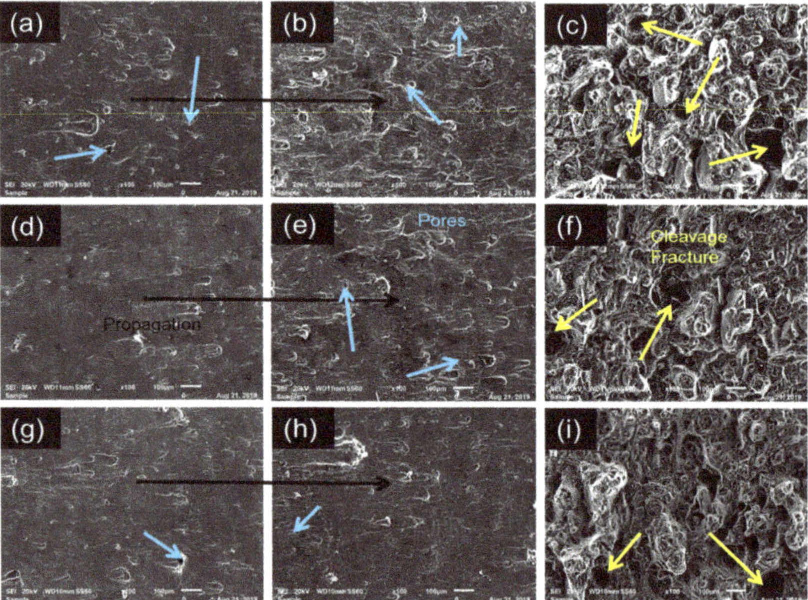

Figure 10. Crack growth rate SEM results of the SLM produced AlSi10Mg samples where ((**a–c**) is sample 1, (**d–f**) is sample 2 and (**g–i**) is sample 3).

4. Conclusions

The results of this work substantiated the anisotropy, which is the demonstration of different mechanical properties in different axis, in SLM produced tensile samples even post stress relief. There was an increase in the porosity post stress relief due to the softening of the microstructure subsequent to the precipitation of silicon and Mg_2Si during the heat treatment process. Stress relief, however, proved to deteriorate all the mechanical properties accomplished by the SLM process except for the improvement in the ductility/elongation and fatigue life. However, fatigue crack growth and fracture toughness showed that these mechanical properties could be suited for certain applications even though there is a need for the development of the stress relief thermal treatment suitable for additively manufactured materials. SLM produced AlSi10Mg showed to have fatigue resistance after stress relief due to its ductility that could endure pressure from the loading. It is suggested that post SLM thermal treatment for AlSi10Mg be further explored, since seemingly most of these thermal treatments as explored by other researchers' as well demonstrated poor results. Perhaps, looking into developing post SLM rules for heat treatment as compared to using the conventional method rules for SLM produced samples.

Author Contributions: Conceptualization, N.R.M.; writing—original draft preparation, B.J.M.; writing—review and editing, B.J.M. and N.R.M.; supervision, L.C.T. and P.AI.P.

Funding: The project was funded by the South African Council of Scientific and Industrial Research Parliamentary Grant and the National Research Foundation Grant#114675.

Acknowledgments: Council for Scientific and Industrial Research, Tshwane University of Technology, and Metal Heart Additive Manufacturing, are gratefully acknowledged. The authors would also like to thank National Laser Center (NLC) Metallurgical Laboratory and Material Science and Manufacturing Mechanical Testing Laboratory for sample preparation and characterization. The National Research Foundation Grant No. 114675 is acknowledged for funding. Acknowledgement also goes to the following individuals for their support, Thabo Lengopeng, Khoro Malabi, Nana Arthur and Chwayita Madikizela.

Conflicts of Interest: The authors declare no conflict of interest.

References

1. NSAI Standards. *Additive Manufacturing–General Principles–Terminology*; NSAI Standards: Dublin, Ireland, 2015; p. 2.
2. Manakari, V.; Parande, G.; Gupta, M. Selective laser melting of magnesium and magnesium alloy powders: A review. *Metals* **2016**, *7*, 2. [CrossRef]
3. Danilov, V.; Gorbatenko, V.; Zuev, L.; Orlova, D.; Danilova, L. Luders deformation of low-carbon steel. *Steel Trans.* **2017**, *47*, 662–668. [CrossRef]
4. Romano, S.; Brückner-Foit, A.; Brandão, A.; Gumpinger, J.; Ghidini, T.; Beretta, S. Fatigue properties of AlSi10Mg obtained by additive manufacturing: Defect-based modelling and prediction of fatigue strength. *Eng. Fract. Mech.* **2018**, *187*, 165–189. [CrossRef]
5. Ahmed, A.; Wahab, M.; Raus, A.; Kamarudin, K.; Bakhsh, Q.; Ali, D. Effects of Selective Laser Melting Parameters on Relative Density of AlSi10Mg. *Int. J. Eng. Technol.* **2016**, *8*, 2552–2557. [CrossRef]
6. Mathe, N.R.; Tshabalala, L.C. The validation of the microstructural evolution of selective laser-melted AlSi10Mg on the in-house built machine: Energy density studies. *Prog. Addit. Manuf.* **2019**. [CrossRef]
7. Ma, P.; Prashanth, K.; Scudino, S.; Jia, Y.; Wang, H.; Zou, C.; Wei, Z.; Eckert, J. Influence of annealing on mechanical properties of Al-20Si processed by selective laser melting. *Metals* **2014**, *4*, 28–36. [CrossRef]
8. Yadollahi, A.; Mahtabi, M.; Khalili, A.; Doude, H.; Newman, J., Jr. Fatigue life prediction of additively manufactured material: Effects of surface roughness, defect size, and shape. *Fatigue Fract. Eng. Mater. Struct.* **2018**, *41*, 1602–1614. [CrossRef]
9. Cabrini, M.; Lorenzi, S.; Pastore, T.; Pellegrini, S.; Ambrosio, E.P.; Calignano, F.; Manfredi, D.; Pavese, M.; Fino, P. Effect of heat treatment on corrosion resistance of DMLS AlSi10Mg alloy. *Electrochim. Acta* **2016**, *206*, 346–355. [CrossRef]
10. Fiocchi, J.; Tuissi, A.; Bassani, P.; Biffi, C. Low temperature annealing dedicated to AlSi10Mg selective laser melting products. *J. Alloys Compd.* **2017**, *695*, 3402–3409. [CrossRef]
11. Fousová, M.; Dvorský, D.; Michalcová, A.; Vojtěch, D. Changes in the microstructure and mechanical properties of additively manufactured AlSi10Mg alloy after exposure to elevated temperatures. *Mater. Charact.* **2018**, *137*, 119–126. [CrossRef]
12. Read, N.; Wang, W.; Essa, K.; Attallah, M.M. Selective laser melting of AlSi10Mg alloy: Process optimisation and mechanical properties development. *Mater. Des.* **2015**, *65*, 417–424. [CrossRef]
13. Brandl, E.; Heckenberger, U.; Holzinger, V.; Buchbinder, D. Additive manufactured AlSi10Mg samples using Selective Laser Melting (SLM): Microstructure, high cycle fatigue, and fracture behavior. *Mater. Des.* **2012**, *34*, 159–169. [CrossRef]
14. Mfusi, B.; Tshabalala, L.C.; Mathe, N.R. The effect of selective laser melting build orientation on the mechanical properties of AlSi10Mg parts. *IOP Conf. Ser. Mater. Sci. Eng.* **2018**, *430*, 012028. [CrossRef]
15. Trevisan, F.; Calignano, F.; Lorusso, M.; Pakkanen, J.; Ambrosio, E.; Lombardi, M.; Pavese, M.; Manfredi, D.; Fino, P. Effects of heat treatments on A357 alloy produced by selective laser melting. In Proceedings of the World Powder Metallurgy Congress 2016, Hamburg, Germany, 9–13 October 2016; pp. 1–6.

16. Williams, J.C.; Starke, E.A., Jr. Progress in structural materials for aerospace systems. *Acta Mater.* **2009**, *51*, 5775–5799. [CrossRef]

17. Aboulkhair, N.T.; Maskery, I.; Tuck, C.; Ashcroft, I.; Everitt, N.M. Improving the fatigue behaviour of a selectively laser melted aluminium alloy: Influence of heat treatment and surface quality. *Mater. Des.* **2016**, *104*, 174–182. [CrossRef]

18. Tang, M. Inclusions, Porosity, and Fatigue of Alsi 10mg Parts Produced by Selective Laser Melting. Ph.D. Thesis, Carnegie Mellon University, Pittsburgh, PA, USA, 2017.

19. Zhao, Y.-H.; Liao, X.-Z.; Cheng, S.; Ma, E.; Zhu, Y.T. Simultaneously increasing the ductility and strength of nanostructured alloys. *Adv. Mater.* **2006**, *18*, 2280–2283. [CrossRef]

20. Wu, J.; Wang, X.; Wang, W.; Attallah, M.; Loretto, M. Microstructure and strength of selectively laser melted AlSi10Mg. *Acta Mater.* **2016**, *117*, 311–320. [CrossRef]

21. Li, W.; Li, S.; Liu, J.; Zhang, A.; Zhou, Y.; Wei, Q.; Yan, C.; Shi, Y. Effect of heat treatment on AlSi10Mg alloy fabricated by selective laser melting: Microstructure evolution, mechanical properties and fracture mechanism. *Mater. Sci. Eng. A* **2016**, *663*, 116–125. [CrossRef]

22. Javidani, M.; Arreguin-Zavala, J.; Danovitch, J.; Tian, Y.; Brochu, M. Additive manufacturing of AlSi10Mg alloy using direct energy deposition: Microstructure and hardness characterization. *J. Ther. Spray Technol.* **2017**, *26*, 587–597. [CrossRef]

23. Aboulkhair, N.T.; Maskery, I.; Tuck, C.; Ashcroft, I.; Everitt, N.M. The microstructure and mechanical properties of selectively laser melted AlSi10Mg: The effect of a conventional T6-like heat treatment. *Mater. Sci. Eng. A* **2016**, *667*, 139–146. [CrossRef]

24. Zhang, C.; Zhu, H.; Liao, H.; Cheng, Y.; Hu, Z.; Zeng, X. Effect of heat treatments on fatigue property of selective laser melting AlSi10Mg. *Int. J. Fatigue* **2018**, *116*, 513–522. [CrossRef]

25. Anwar, A.B.; Pham, Q.-C. Selective laser melting of AlSi10Mg: Effects of scan direction, part placement and inert gas flow velocity on tensile strength. *J. Mater. Proc. Technol.* **2017**, *240*, 388–396. [CrossRef]

26. Rosenthal, I.; Tiferet, E.; Ganor, M.; Stern, A. Post-processing of AM-SLM AlSi10Mg specimens: Mechanical properties and fracture behavior. *Ann. Dunarea Jos Univ. Galati Fascicle XII Weld. Equip. Technol.* **2015**, *26*, 33–38.

27. Brandão, A.D.; Gumpinger, J.; Gschweitl, M.; Seyfert, C.; Hofbauer, P.; Ghidini, T. Fatigue properties of additively manufactured AlSi10Mg–surface treatment effect. *Procedia Struct. Integr.* **2017**, *7*, 58–66. [CrossRef]

28. Al-Saedi, D.S.; Masood, S.; Faizan-Ur-Rab, M.; Alomarah, A.; Ponnusamy, P. Mechanical properties and energy absorption capability of functionally graded F2BCC lattice fabricated by SLM. *Mater. Des.* **2018**, *144*, 32–44. [CrossRef]

29. Wang, L.; Jue, J.; Xia, M.; Guo, L.; Yan, B.; Gu, D. Effect of the thermodynamic behavior of selective laser melting on the formation of in situ oxide dispersion-strengthened aluminum-based composites. *Metals* **2016**, *6*, 286. [CrossRef]

30. Kempen, K.; Thijs, L.; Yasa, E.; Badrossamay, M.; Kruth, J.P. Process optimization and microstructural analysis for selective laser melting of AlSi10Mg. In Proceedings of the Solid Free Form Fabrication Symposium, Texas, TX, USA, 8–10 August 2011; pp. 484–495.

31. Zaretsky, E.; Stern, A.; Frage, N. Dynamic response of AlSi10Mg alloy fabricated by selective laser melting. *Mater. Sci. Eng. A* **2017**, *688*, 364–370. [CrossRef]

32. Rosenthal, I.; Stern, A.; Frage, N. Strain rate sensitivity and fracture mechanism of AlSi10Mg parts produced by selective laser melting. *Mater. Sci. Eng. A* **2017**, *682*, 509–517. [CrossRef]

33. Kobayashi, T. Strength and fracture of aluminum alloys. *Mater. Sci. Eng. A* **2000**, *280*, 8–16. [CrossRef]

34. Ruggieri, C.; Dodds, R.H., Jr. A local approach to cleavage fracture modeling: An overview of progress and challenges for engineering applications. *Eng. Fract. Mech.* **2018**, *187*, 381–403. [CrossRef]

35. Scibetta, M. Application of a new cleavage fracture framework to ferritic steels. *Procedia Struct. Integr.* **2016**, *2*, 1610–1618. [CrossRef]

36. Shlyannikov, V. Creep–fatigue crack growth rate prediction based on fracture damage zone models. *Eng. Fract. Mech.* **2019**, *214*, 449–463. [CrossRef]
37. Tang, M.; Pistorius, P.C. Oxides, porosity and fatigue performance of AlSi10Mg parts produced by selective laser melting. *Int. J. Fatigue* **2017**, *94*, 192–201. [CrossRef]

Article

The Microstructure, Mechanical Properties, and Corrosion Resistance of UNS S32707 Hyper-Duplex Stainless Steel Processed by Selective Laser Melting

Feng Shang [1,2], Xiaoqiu Chen [2], Zhiyong Wang [2], Zuchun Ji [3], Fei Ming [1], Shubin Ren [1,*] and Xuanhui Qu [1,*]

[1] Beijing Advanced Innovation Center for Materials Genome Engineering, University of Science and Technology Beijing, Beijing 100083, China; shangfeng@hhit.edu.cn (F.S.); fryming@163.com (F.M.)
[2] School of Mechanical and Ocean Engineering, Jiangsu Ocean University, Lianyungang 222005, China; cxq950808@163.com (X.C.); w19851821109@163.com (Z.W.)
[3] Mingyang Technology (Suzhou) Co., Ltd., Suzhou 215000, China; jzc110@126.com
* Correspondence: sbren@ustb.edu.cn (S.R.); quxh@ustb.edu.cn (X.Q.); Tel.: +86-10-82377286 (S.R.); +86-10-62332700 (X.Q.)

Received: 15 August 2019; Accepted: 12 September 2019; Published: 17 September 2019

Abstract: UNS S32707 hyper-duplex stainless steel (HDSS) parts with complex shapes for ocean engineering were prepared by selective laser melting (SLM) process. In the process of SLM, the balance between austenite and ferrite was undermined due to the high melting temperature and rapid cooling rate, resulting in poor ductility and toughness. The solution annealing was carried out with various temperatures (1050–1200 °C) for one hour at a time. The evolution of microstructures, mechanical properties, and corrosion resistance of UNS S32707 samples prepared by SLM was comprehensively investigated. The results indicate that a decrease in nitrogen content during the SLM process reduced the content of austenite, and a nearly balanced microstructure was obtained after appropriate solution annealing. The ratio between ferrite and austenite was approximately 59.5:40.5. The samples with solution treated at 1150 °C and 1100 °C exhibited better comprehensive mechanical properties and pitting resistance, respectively.

Keywords: selective laser melting; hyper-duplex stainless steel; microstructure; mechanical property; corrosion resistance

---

1. Introduction

The UNS S32707 hyper-duplex stainless steel (HDSS) has excellent mechanical properties and corrosion resistance. The pitting resistance equivalent number (PREN = mass% Cr + 3.3 mass% Mo + 16 mass% N) is 49, and the critical pitting temperature (CPT) is 90 °C. This material has been widely used in marine engineering equipment [1,2]. The hot working performance of the HDSS is poor due to the high content of Cr, Mo, and N, leading to crack during hot rolling. Nitrogen escapes easily during welding, resulting in the performance of weld metal deteriorates [3,4]. In general, it is difficult to manufacture a duplex stainless steel, while the selective laser melting (SLM) is an effective technique to produce HDSS with the complex shapes. In detail, a higher relative density of duplex stainless steel can be obtained [5], and the precipitation of the σ phase and other harmful phases can be inhibited due to a rapid cooling rate [6].

In recent years, the austenitic stainless steel [7], precipitation hardening stainless steel [8], martensitic stainless steel [9], high manganese steel [10], super-duplex stainless steel (SDSS) [11–14] prepared using SLM were studied. Although SLM technology has a certain basis for the preparation of SDSS, SLM of HDSS with higher nitrogen content has rarely been reported. Shang et al. [15] have shown that the microstructure of UNS S32707 HDSS prepared by SLM from the powder prepared by plasma rotating electrode process (PREP) is mainly ferrite. Although it has high strength and hardness, the ductility and toughness of duplex stainless steel are poor due to the precipitation of nitride at ferrite grain boundary. In this study, the microstructure, mechanical properties and corrosion resistance at different solution annealing temperatures were examined. The objective is to obtain an UNS S32707 HDSS prepared by SLM with balanced austenite-ferrite arrangement, ideal tensile strength, ductility, toughness, and corrosion resistance.

2. Materials and Methods

The PREP process of UNS S32707 HDSS powder, the SLM parameters, and the samples size can be found in Shang et al. [15]. The chemical composition of the powder is shown in Table 1, and the median diameter of the prepared metal powder was 45 µm. The samples were water-quenched using an atmosphere-protected quenching furnace (OTF-1500X-80-VTQ, Hefei Kejing Materials Technology Co., Ltd., Hefei, China) following solution annealing at 1050, 1100, 1150, and 1200 °C, respectively. The sample numbers of different processing technologies are shown in Table 2.

Table 1. Chemical composition of UNS S32707 hyper-duplex stainless steel (HDSS) powder.

Element	Cr	Ni	Mo	N	Si	Mn	Co	Cu	Al	C	O	Fe
mass%	27.19	6.48	5.00	0.36	0.58	1.5	1.03	0.98	0.02	0.02	0.018	Bal.

Table 2. The sample numbers of different processing technologies.

Processing Technology	Sample Number
SLM	S1
SLM+ solution annealing (1050 °C × 1 h) + water quenching	S2
SLM+ solution annealing (1100 °C × 1 h) + water quenching	S3
SLM+ solution annealing (1150 °C × 1 h) + water quenching	S4
SLM+ solution annealing (1200 °C × 1 h) + water quenching	S5
Casting bar UNS S32707 for PREP+ solution annealing (1150 °C × 1 h) + water quenching	S6
Welded and seamless UNS S32707stainless steel pipe (ASTM A790)	S7

The mechanical properties were measured by a WDW-10E microcomputer controlled electronic universal testing machine. The tensile specimens were built with a nominal gauge section of 25 mm length × 5 mm width × 2 mm thickness. Rectangular samples with the size of 55 × 10 × 10 mm were machined into V-notches for the impact test. The microhardness of the samples was measured by an HV-30 hardness tester. The nitrogen content in the powder and selective laser melted parts was measured by a LECO ONH 836 Oxygen Nitrogen Hydrogen Analyzer (LECO Corporation, St. Joseph, MI, USA).

The microstructure of the selective laser melted sample was observed by a Quanta 450 FEG scanning electron microscope (SEM) (FEI Company, Hillsboro, OR, USA). Prior to the SEM observations, the samples were electro etched in 10% oxalic acid at an operating voltage of 7 V for 25 s. An HKL Channel5 electron backscatter diffraction (EBSD) analysis system (Oxford Instruments, Oxford, UK) equipped in an FEI Quanta 650F field emission scanning electron microscope (SEM) (FEI Company, Hillsboro, OR, USA) was used to obtain inverse pole figure mappings and phase mappings with a step size of 0.3 µm and 0.5 µm at an acceleration voltage of 20 kV with a working distance of 13 mm. The EBSD samples were polished by Leica EMRES 102 argon ion polishing instrument (Leica Microsystems, Wetzlar, Germany). The samples were first polished at a voltage of 5 kV for 40 min, and then polished at a voltage of 4.5 kV for 50 min. The angle of the ion gun was 15 degrees (2 ion guns). Finally, the voltage was reduced to 3.5 kV, and the polishing time was 60 min at low voltage with the same angle of the ion gun. The microstructure was observed by an FEI Tecnai G2 F20 transmission electron microscope (TEM) (FEI Company, Hillsboro, OR, USA). TEM samples were first mechanically polished to a thickness of about 50 μm, followed by ion-beam milling in a Gatan 691 precision ion polishing system (PIPS) at 5 kV with a final polishing step at 1 kV of ion energies. An FEI Super-energy dispersive spectrometer (EDS) system equipped in an FEI Titan Themis spherical aberration electron microscope (FEI Company, Hillsboro, OR, USA) was used for a high-resolution energy surface scan test. The effect of nitrogen content on the austenite content was simulated by JMatPro software (JMatPro 7.0, Sente Software Ltd., Guildford, UK). The simulated solution annealing temperature was 1150 °C.

A Gill AC Bi-STAT electrochemical workstation was used to measure the potentiodynamic polarization curve at room temperature in 3.5% NaCl solution. The classical three-electrode system is as follows: Saturated calomel electrode as the reference electrode, platinum electrode as the auxiliary electrode, and sample as the working electrode. The tests were recorded at a scan rate of 20 mV/min and ranged between −150 and +2000 mV. Prior to the corrosion testing, the samples were electro polished in 10% oxalic acid at an operating voltage of 12 V for 25 s. The composition of the passive film was analyzed by a Thermo Scientific ESCALAB *250Xi* X-ray photoelectron spectroscopy (XPS) (Thermo Fisher Scientific Inc., Waltham, MA, USA), the source of X-ray excitation was Al K_α, and the vacuum of analysis chamber was 10^{-9} mbar. The high resolution spectrum of XPS was resolved by PeakFit software (PeakFit 4.12, Seasolve Software Inc., Framingham, MA, USA).

3. Results and Discussion

3.1. Microstructure

The XRD diagrams of samples processed by different technologies are shown in Figure 1. It can be seen that the samples prepared by SLM are mainly composed of ferrite phase, whereas the solution treated samples at 1050 °C were mainly composed of austenite and the σ phase, and mainly composed of ferrite and austenite phases for samples treated at 1100–1200 °C, respectively. The TEM bright-field image in Figure 2a shows that dislocations were found in the S1 sample, suggesting that the high stress caused by rapid cooling led to the appearance of dislocations. Dislocations provide a driving force for recrystallization during solution annealing [14]. The scanning transmission electron microscope-energy dispersive spectrometer (STEM-EDS) mapping of Figure 2b,c shows that the nanoscale Cr_2N were found at the ferrite grain boundary.

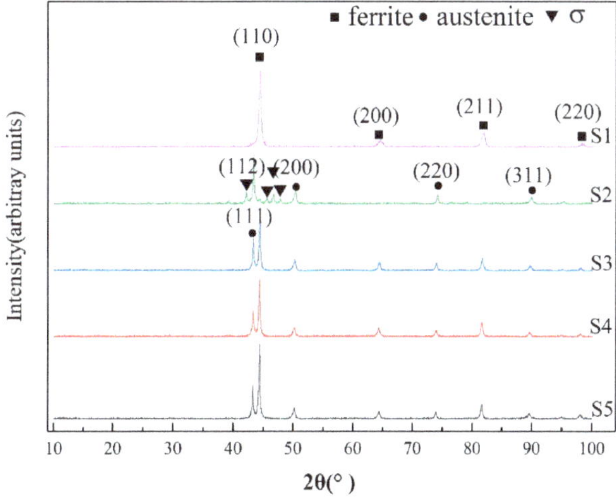

Figure 1. The XRD diagrams of samples processed by different technologies.

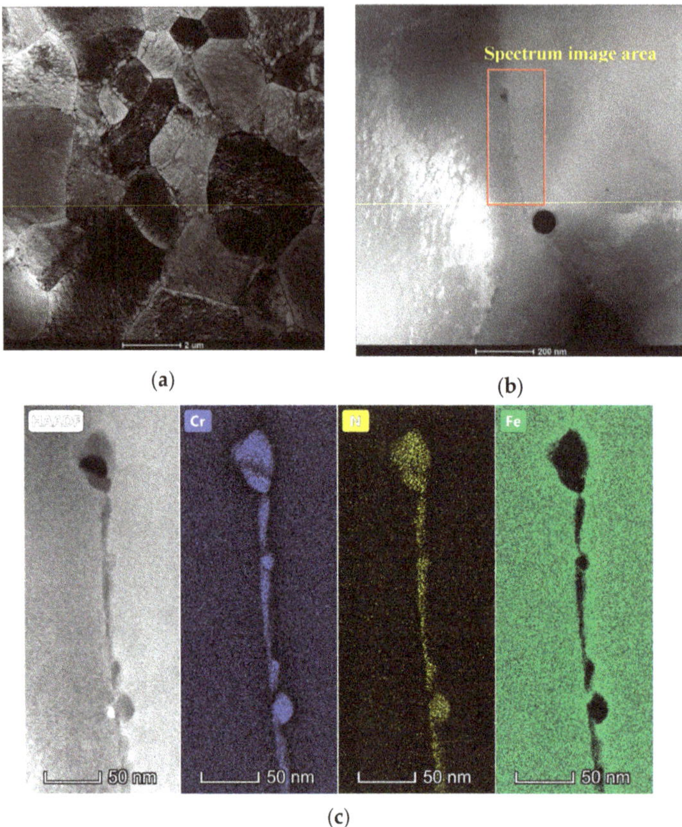

Figure 2. TEM morphology of the S1 sample (**a**) TEM bright field image; (**b,c**) scanning transmission electron microscope-energy dispersive spectrometer (STEM-EDS) mapping image.

The EBSD inverse pole diagram and phase distribution diagram of samples with different solution annealing processes are shown in Figure 3. Table 3 shows the proportion of phases and grain size of samples with different solution annealing processes. As shown in Figure 3, 10% of the σ phase and 88% of the γ_2 phase are precipitated from ferrite by eutectoid transformation [16,17]: $\alpha \rightarrow \sigma + \gamma_2$ at 1050 °C. In the temperature range of 1100–1200 °C, the grain size and the ferrite content increased gradually, while the austenite content decreased gradually with the increment in solution annealing temperature. The γ_2 phase nucleated only along the recrystallized ferrite grain boundary at 1200 °C [18]. The nitrogen content of HDSS powder decreased from 0.36% in powder state to 0.24% in selective laser melted state [15]. Nitrogen is an austenite-forming element, suggesting that its decrease resulted in the decrease of the austenite content after solution annealing, and it is not possible to achieve 50-50% ferrite-austenite ratio due to the nitrogen loss. Figure 4 shows the relationship between nitrogen content and austenite content of UNS S32707 HDSS simulated by JMatPro software. The simulated result shows that the decrease in nitrogen content reduced the austenite content by 10%.

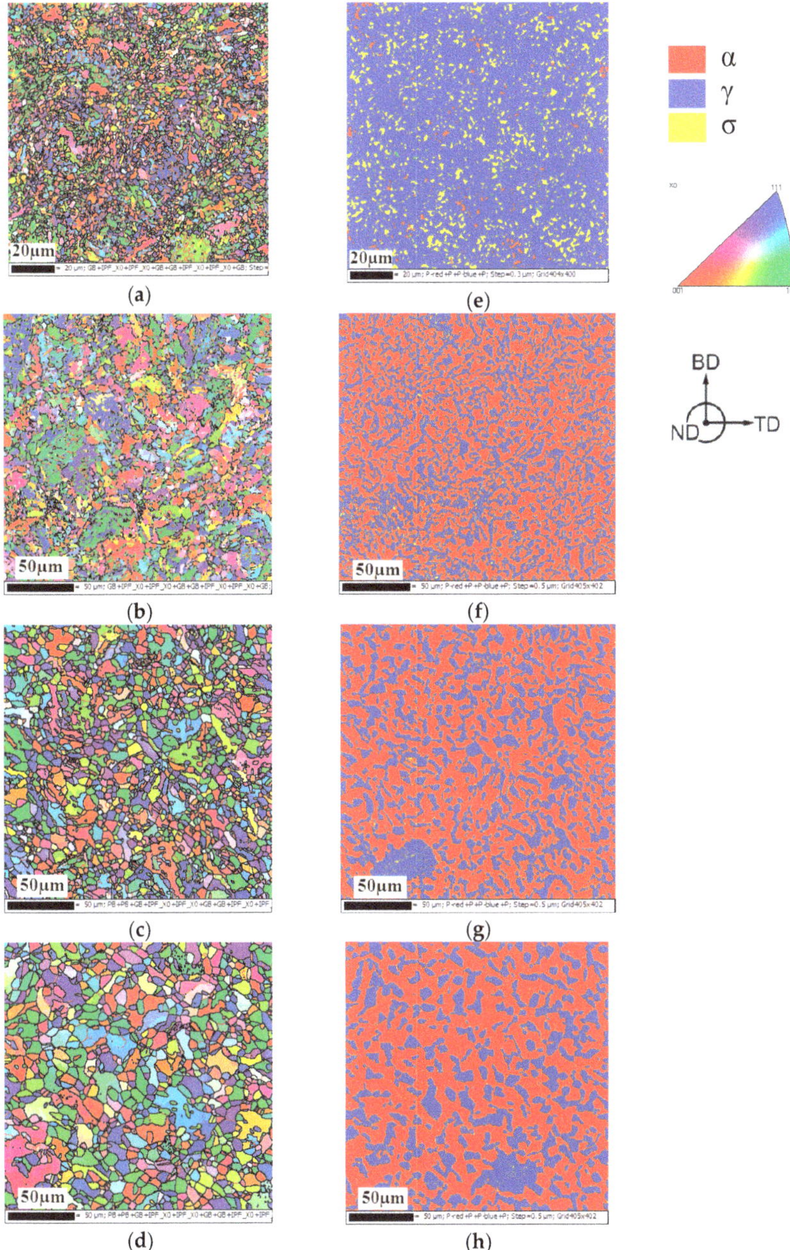

Figure 3. The electron backscatter diffraction (EBSD) diagram of different solution annealing processes (**a–d**) inverse pole figure and (**e–h**) phase distribution figure; (**a,e**) S2; (**b,f**) S3; (**c,g**) S4; (**d,h**) S5.

Table 3. The proportion of phases and grain size of samples with different solution annealing processes.

Sample Number	The Content of Ferrite Phase/vol. %	Average Grain Size of Ferrite/μm	The Content of Austenite Phase/vol. %	Average Grain Size of Austenite/μm	The Content of Sigma Phase/vol. %	Average Grain Size of Sigma Phase/μm
S1	98.5	3.68	0.2	1.66	-	-
S2	1.6	1.24	88.4	2.27	10.0	1.08
S3	59.5	4.21	40.5	2.80	-	-
S4	61.4	5.72	38.6	3.96	-	-
S5	63.4	7.58	36.6	5.25	-	-

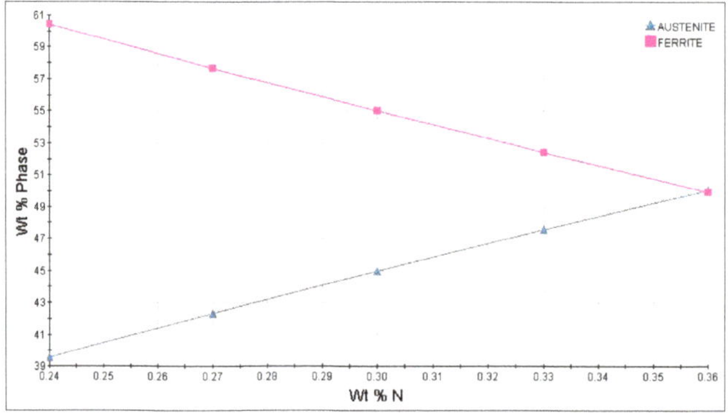

Figure 4. The relationship between nitrogen content and austenite content of UNS S32707 HDSS simulated by JMatPro software.

3.2. Mechanical Properties

Table 4 shows the mechanical properties of specimens processed with different conditions. The S1 specimen has high strength, hardness and relative density, relatively low ductility and toughness [15]. The strength and ductility of the S2 samples decrease rapidly due to the precipitation of the σ phases. In the temperature range of 1100–1200 °C, the strength and hardness decreased gradually while the elongation and section shrinkage increased gradually due to recrystallization and grain growth with the increase of solution annealing temperature. Lastly, the table also shows that the sample solution treated at 1150°C for one hour had better comprehensive mechanical properties compared to the others. Figure 5 is the tensile fracture and impact fracture morphology of the S4 specimen, which is characterized by ductile fracture. There are some pores in the parts prepared by SLM with relative density of 98.2%. The laser energy density, laser scanning strategy, and powder properties will affect relative density. In order to obtain fully compact parts, further densification by hot isostatic pressing can further improve the mechanical properties. Therefore, it is understandable that the mechanical properties prepared by SLM are lower than that of welded and seamless pipe required in ASTM A790 standard.

Table 4. The mechanical properties of specimens processed under different conditions.

Sample Number	Tensile Strength/MPa	Yield Strength/MPa	Elongation/%	Section Shrinkage/%	Impact Absorbing Energy/J	Hardness/HV
S1	1493 ± 6	1391 ± 9	13.2 ± 1	24.1 ± 3	18 ± 3	528.7 ± 4
S2	593 ± 20	-	-	-	-	523.8 ± 8
S3	941 ± 10	665 ± 7	24.6 ± 2	25.8 ± 3	-	321.8 ± 7
S4	901 ± 4	658 ± 10	36.4 ± 2	48.4 ± 2	132 ± 5	291.5 ± 6
S5	893 ± 1	646 ± 3	38.7 ± 2	52.6 ± 3	-	286.7 ± 5
S6	851 ± 7	614 ± 10	29.2 ± 2	59.4 ± 3	128 ± 6	285.4 ± 4
S7	920 (min)	700 (min)	25 (min)	-	-	34 HRC (max)

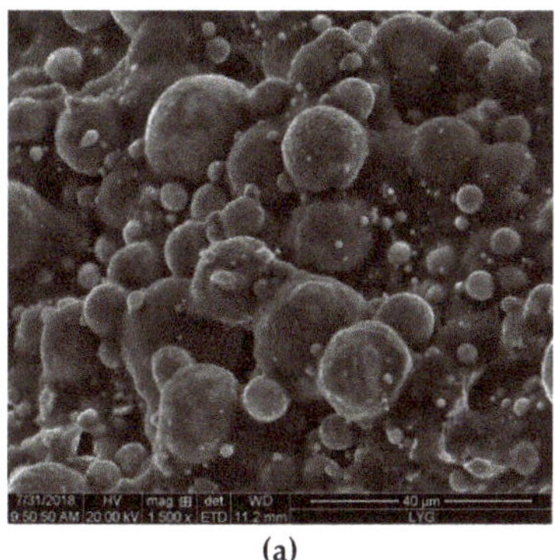
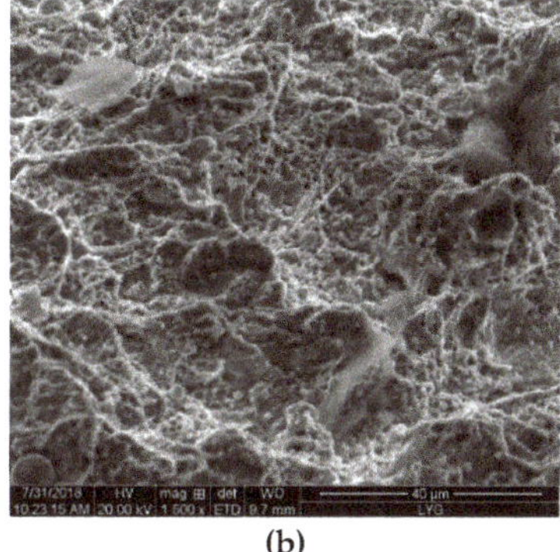

(a)　　　　　(b)

Figure 5. The fracture morphology of the S4 specimen (**a**) tensile fracture; (**b**) impact fracture.

3.3. Corrosion Resistance

Figure 6 shows the potentiodynamic polarization curve of samples processed with different conditions. The results of electrochemical experiments are shown in Table 5. Table 5 shows that the pitting potential of the S3 sample solution treated at 1100 °C for one hour was relatively higher (1196 mV) compared to the others. The ratio of α to γ of the S3 sample was 59.5:40.5. This suggests that good phase arrangement gave the material better pitting resistance. Table 5 also shows that the pitting potential of the S2 sample with σ phase precipitation is the lowest (1055 mV). Zhang et al. [19] studied the transformation mechanisms of the σ phase in UNS S32707aged at nose temperature, the σ phase preferentially formed along the α/α and α/γ phase boundaries and then penetrated into α phase, resulting from the eutectoid reaction $\alpha \rightarrow \gamma_2 + \sigma$. Meanwhile, a few σ phases nucleated at the γ/γ phase boundaries. The precipitation of the σ phase in S2 may have led to poor Cr at α/α, α/γ, γ/γ, α/σ, and γ/σ phase boundaries, which resulted in decreased pitting resistance. Figure 7 shows the EDS spectrum of the α and γ phases in the S3 sample. The element distribution of the α and γ phases is shown in Table 6. In the α phase, Cr, Mo, and other α-forming elements were enriched. In the γ phase, the γ-forming element of Ni was enriched. The results of references [20,21] show that the solubility of nitrogen in ferrite is generally less than 0.05%. The ratio of α to γ of the S3 sample was 59.5:40.5. It can be estimated that the nitrogen content of ferrite and austenite in the S3 sample is 0.05% and 0.52%, respectively. The PREN (PREN = mass% Cr + 3.3 mass% Mo + 16 mass% N) of α and γ were 50.7 and 50.6, respectively. The pitting potential of the S4 and S5 samples is very close and difficult to distinguish. Figure 6 shows that the S4 sample presents a notable re-passivation process after its breakdown with the increment of the potential from 0.6 to 1.0 VSCE, while S5 shows a relatively higher dissolve rate with the increased potential from 0.6 to 1.1 VSCE. This indicates that the passive films of the S4 sample are much more stable than that of S5. Zheng et al. [22] found that more grain boundaries due to grain refinement could improve the chromium diffusion and promoted to the forming of compact passive film for the duplex stainless steel. In this study, the grain size of the S4 sample is smaller than that of S5, which is beneficial to the corrosion resistance of S4 with the refined grain size.

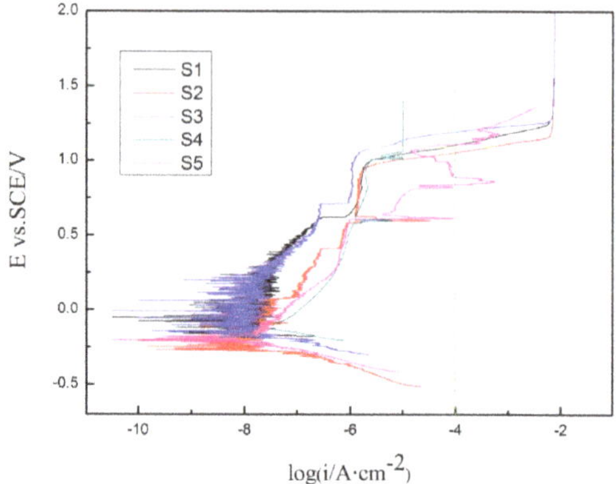

Figure 6. The potentiodynamic polarization curve of samples processed with different conditions.

Table 5. The results of electrochemical experiments.

Sample Number	Pitting Potential/mV
S1	1109
S2	1055
S3	1196
S4	1109
S5	1109

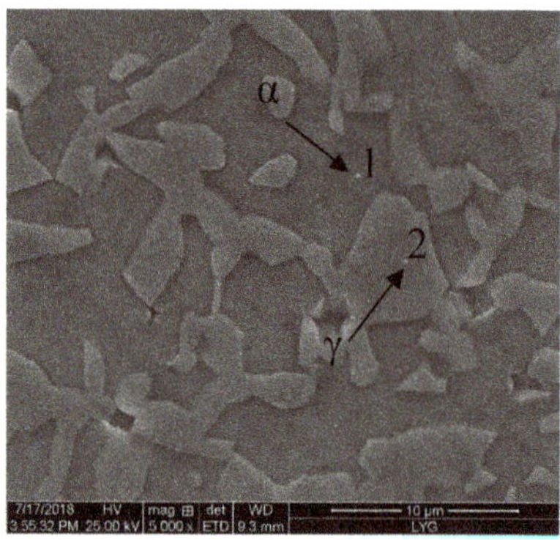

Figure 7. The EDS spectra of the S3 sample.

Table 6. The element distribution of the α and γ phases (mass%).

Phase	Fe	Cr	Ni	Mo	N
α	58.51	29.96	5.49	6.04	0.05
γ	59.41	25.81	9.78	5.00	0.52

Figure 8 shows the full spectrum of XPS on the surface of the S3 sample after solution annealing at different sputtering depths. As shown in Figure 8a, C, O, N, Fe, and Cr were notable, while Mo and Ni were weak before sputtering. Furthermore, Figure 8 shows that the surface of the passivation film is mainly composed of compounds formed by Fe, Cr, N, and O. As shown in Figure 8b,c, the intensity of C1s peak decreases greatly after sputtering, which indicate that the C peak in Figure 8a comes from the contamination of the vacuum chamber. The decrease of N1s peak indicates that N was mainly concentrated on the surface of the passive film. The strong peaks at the positions of Fe2p and Ni2p indicate that Fe and Ni elements mainly existed in the interior of the passive film. Lastly, Figure 8 also shows that the XPS spectra of the samples remain after different sputtering depths (1 nm and 2 nm), indicating that the chemical composition of the passivation film was relatively stable in the thickness of 1 nm to 2 nm.

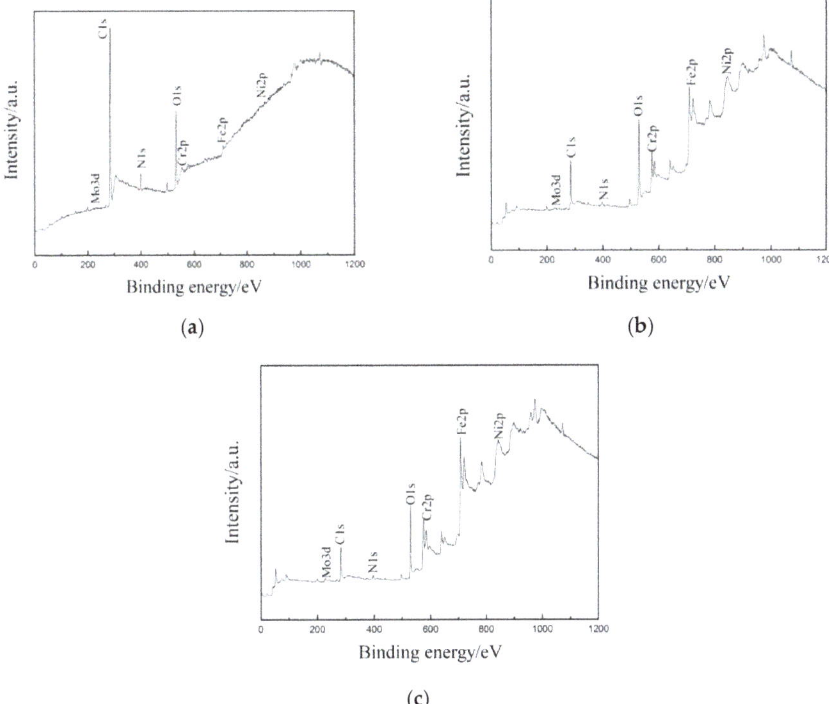

Figure 8. XPS full spectrum of the S3 sample after solution annealing with different sputtering depths (**a**) 0 nm; (**b**) 1 nm; (**c**) 2 nm.

Figure 9 shows the result of the high-resolution XPS spectrum of O1s at different sputtering depths by PeakFit. As shown in Figure 9a, XPS spectra of O1s had three peaks before sputtering, corresponding to binding energies 529.5, 531.3, and 533.1 eV, respectively. According to the binding energy values of O1s reported by Wang et al. [23], the first peak is the characteristic peak of M-O compound, corresponding to O^{2-}, the second peak is the characteristic peak of M-(OH) or M-(OH)$_2$ compound, corresponding to OH^-, and the third peak is the characteristic peak of H_2O. The characteristic peak of 533.1 eV disappeared after sputtering, indicating that the water on the surface of the passivation film came from residual water on the electrode surface. From Figure 9b,c, it can be seen that the O1s high-resolution spectra of sputtered samples only had a characteristic peak of O^{2-} after spectral analysis, which indicates that oxygen in the passive films mainly exists in the form of oxides.

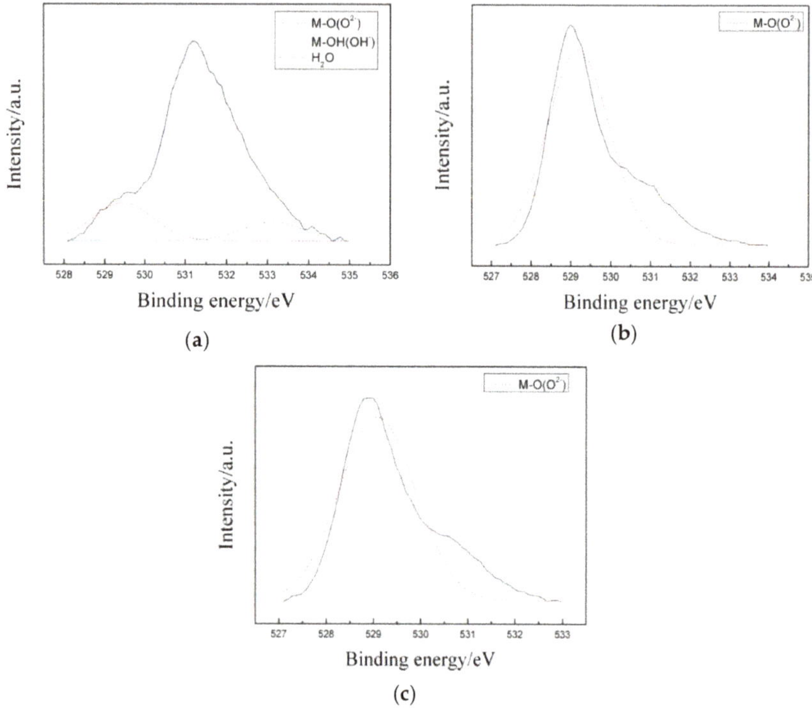

Figure 9. Narrow XPS of O1s with different sputtering depths (a) 0 nm; (b) 1 nm; (c) 2 nm.

Figure 10 shows the high resolution XPS spectra of $Mo3d_{5/2}$ and N1s with different sputtering depths. Mo mainly exists in the form of Mo_4^{2-} with a corresponding binding energy of 231.8 eV before sputtering. When sputtering depth is 1 and 2 nm, Mo mainly exists in MoO_2 form with a corresponding binding energy of 231.2 eV. N mainly exists in the form of NH_4^+ with a corresponding binding energy of 399.8 eV before sputtering. When the sputtering depth is 2 nm, the peak of NH_4^+ disappears and N exists in the form of Cr_2N with a corresponding binding energy of 397.5 eV.

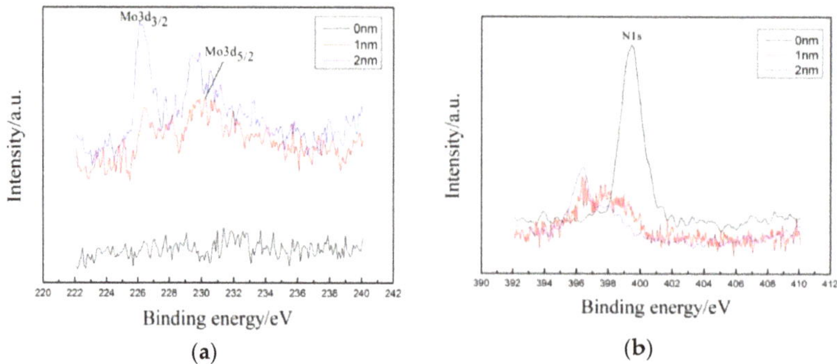

Figure 10. The high resolution XPS spectra of $Mo3d_{5/2}$ (a) and N1s (b) at different sputtering depths.

4. Conclusions

This study has shown that UNS S32707 HDSS prepared by the SLM process can obtain a nearly balanced two-phase structure of ferrite and austenite through the appropriate solution annealing process, which can significantly improve the ductility and toughness. The nitrogen content reduced by 33% in the SLM process, resulting in the imbalance of the two phases and the reduction of the austenite content. The samples with solution treated at 1150 °C for one hour had better comprehensive mechanical properties, the measured tensile strength, the yield strength, the elongation, the section shrinkage, the microhardness, and the impact absorption energy were 901 MPa, 658 MPa, 36.4%, 48.4%, 291.5 HV, and 132 J, respectively.

The samples with solution treated at 1100°C for one hour had better pitting resistance. The pitting potential was 1196 mV. The surface layer of passive film is mainly composed of oxides of Fe and Cr and hydroxide compounds. NH_4^+ adsorbed on the surface of passive film and protected the passive film. The inner layer of the passive film is mainly composed of metal elements, Cr_2N and oxides of Fe, Cr, and Mo.

Author Contributions: F.S. is the first author and analyzed the data and wrote the paper. The experiments were performed by X.C., Z.W., Z.J., and F.M., S.R. and X.Q. conceived, designed, and supervised the experiments. In addition, they contributed to the interpretation of data and editing the paper.

Funding: This study was financially supported by the National Science and Technology Support Program of the Ministry of Science and Technology of China (No.2015BAE03B00), National Natural Science Foundation of China (No.51874038), Fundamental Research Funds for the Central Universities (FRF-AT-18-014), the Natural Science Foundation of Huaihai Institute of Technology (Z2017001), Lianyungang 521 Project (ZKK201805) and Lianyungang Haiyan Project (2018-QD-013).

Acknowledgments: We thank Jiasheng Dong and research fellow Langhong Lou from Shenyang Zhongke Sannai New Materials Co., Ltd. for his help in alloy melting. We thank Shujin Liang from Sino-Euro Materials Technologies of Xi'an Co., Ltd. for his help in PREP. We thank Xiaoming Zhang and Xuehao Zheng from ZKKF (Beijing) Science and Technology Co., Ltd. for TEM and EBSD observations.

References

1. Chail, G.C.; Kangas, P. Super and hyper duplex stainless steels: Structures, properties and applications. *Procedia Struct. Integr.* **2016**, *2*, 1755–1762. [CrossRef]

2. Song, Z.G.; Feng, H.; Hu, S.M. Development of Chinese duplex stainless steel in recent years. *J. Iron Steel Res. Int.* **2017**, *24*, 121–130. [CrossRef]

3. Pilhagen, J.; Sieurin, H.; Sandström, R. Fracture toughness of a welded super duplex stainless steel. *Mater. Sci. Eng. A* **2014**, *606*, 40–45. [CrossRef]

4. Kim, H.J.; Jeon, S.H.; Kim, S.T.; Park, Y.S. Influence of the shielding gas composition on the passive film and erosion corrosion of tube-to-tube sheet welds of hyper duplex stainless steel. *Corros. Sci.* **2015**, *91*, 140–150. [CrossRef]

5. Saeidi, K.; Kevetkova, L.; Lofaj, F.; Shen, Z. Novel ferritic stainless steel formed by laser melting from duplex stainless steel powder with advanced mechanical properties and high ductility. *Mater. Sci. Eng. A* **2016**, *665*, 59–65. [CrossRef]

6. Martín, F.; García, C.; Blanco, Y.; Rodriguez-Mendez, M.L. Influence of sinter-cooling rate on the mechanical properties of powder metallurgy austenitic, ferritic, and duplex stainless steels sintered in vacuum. *Mater. Sci. Eng. A* **2015**, *642*, 360–365. [CrossRef]

7. Trelewicz, J.R.; Halada, G.P.; Donaldson, O.K.; Manogharan, G.P. Microstructure and corrosion resistance of laser additively manufactured 316L stainless steel. *JOM* **2016**, *68*, 850–859. [CrossRef]

8. Akita, M.; Uematsu, Y.; Kakiuchi, T.; Nakajima, M.; Kawaguchi, R. Defect-dominated fatigue behavior in type 630 stainless steel fabricated by selective laser melting. *Mater. Sci. Eng. A* **2016**, *666*, 19–26. [CrossRef]

9. Krakhmalev, P.; Yadroitsava, I.; Fredriksson, G.; Yadroitsev, I. In situ heat treatment in selective laser melted martensitic AISI420 stainless steels. *Mater. Design.* **2015**, *87*, 380–385. [CrossRef]

10. Haase, C.; Bültmann, J.; Hof, J.; Ziegler, S.; Bremen, S.; Hinke, C.; Schwedt, A.; Prahl, U.; Bleck, W. Exploiting process-related advantages of selective laser melting for the production of high-manganese steel. *Materials* **2017**, *10*, 56. [CrossRef]

11. Davidson, K.P.; Singamneni, S. Magnetic characterization of selective laser-melted Saf 2507 duplex stainless steel. *JOM* **2017**, *69*, 569–574. [CrossRef]

12. Davidson, K.P.; Singamneni, S. Selective laser melting of duplex stainless steel powders: An investigation. *Manuf. Processes.* **2016**, *31*, 1543–1555. [CrossRef]
13. Davidson, K.P.; Singamneni, S. Metallographic evaluation of duplex stainless steel powders processed by selective laser melting. *Rapid. Prototyping. J.* **2017**, *23*, 1146–1163. [CrossRef]
14. Hengsbach, F.; Koppa, P.; Duschik, K.; Holzweissig, M.J.; Burns, M.; Nellesen, J.; Tillmann, W.; Troster, T.; Hoyer, K.P.; Schaper, M. Duplex stainless steel fabricated by selective laser melting—Microstructural and mechanical properties. *Mater. Design.* **2017**, *133*, 136–142. [CrossRef]
15. Shang, F.; Chen, X.Q.; Zhang, P.; Ji, Z.C.; Ming, F.; Ren, S.B.; Qu, X.H. Novel Ferritic Stainless Steel with Advanced Mechanical Properties and Significant Magnetic Responses Processed by Selective Laser Melting. *Mater. Trans.* **2019**, *60*, 1096–1102. [CrossRef]
16. Topolska, S.; Labanowski, J. Effect of microstructure on impact toughness of duplex and superduplex stainless steels. *J. Ach. Mater. Manuf. Eng.* **2009**, *36*, 142–149.
17. Chan, K.W.; Tjong, S.C. Effect of secondary phase precipitation on the corrosion behavior of duplex stainless steels. *Materials* **2014**, *7*, 5268–5304. [CrossRef] [PubMed]
18. Maki, T.; Furuhara, T.; Tsuzaki, K. Microstructure development by thermomechanical processing in duplex stainless steel. *ISIJ Int.* **2001**, *41*, 571–579. [CrossRef]
19. Zhang, B.B.; Jiang, Z.H.; Li, H.B.; Zhang, S.C.; Feng, H.; Li, H. Precipitation behavior and phase transformation of hyper duplex stainless steel UNS S32707 at nose temperature. *Mater. Charact.* **2017**, *129*, 31–39. [CrossRef]
20. Ha, H.Y.; Jang, M.H.; Lee, T.H.; Moon, J.O. Interpretation of the relation between ferrite fraction and pitting corrosion resistance of commercial 2205 duplex stainless steel. *Corros. Sci.* **2014**, *89*, 154–162. [CrossRef]
21. He, Y.J.; Guo, X.Y.; Wu, Y.M.; Jiang, J.; Li, J. Effect of solution annealing temperature on pitting behavior of duplex stainless steel 2204 in chloride solutions. *J. Iron Steel Res. Int.* **2016**, *23*, 357–363. [CrossRef]
22. Zheng, Z.J.; Gao, Y.; Gui, Y.; Zhu, M. Corrosion behaviour of nanocrystalline 304 stainless steel prepared by equal channel angular pressing. *Corros. Sci.* **2012**, *54*, 60–67. [CrossRef]
23. Wang, Z.C.; Zhang, Y.Z.; Zhou, S.M. Corrosion compositions of carbon steel under ion-selective coatings by XPS. *J. Chin. Soc. Corros. Prot.* **2001**, *21*, 273–279. [CrossRef]

Article

Densification, Microstructure and Properties of 90W-7Ni-3Fe Fabricated by Selective Laser Melting

Junfeng Li [1,*], Zhengying Wei [1], Bokang Zhou [1], Yunxiao Wu [1], Sheng-Gui Chen [2] and Zhenzhong Sun [2]

[1] State Key Laboratory for Manufacturing System Engineering, School of Mechanical Engineering, Xi'an Jiaotong University, Xi'an 710049, China
[2] School of Mechanical Engineering, Dongguan University of Technology, Dongguan 523808, China
* Correspondence: xjljf2014@stu.xjtu.edu.cn; Tel.: +86-029-8266-5678

Received: 23 July 2019; Accepted: 7 August 2019; Published: 13 August 2019

Abstract: The preparation of refractory tungsten and tungsten alloys has always been challenging due to their inherent properties. Selective laser melting (SLM) offers a choice for preparing tungsten and tungsten alloys. In this work, 90W-7Ni-3Fe samples were prepared by selective laser melting and investigated. Different process parameter combinations were designed according to the Taguchi method, and volumetric energy density (VED) was defined. Subsequently, the effects of process parameters on densification, phase composition, microstructure, tensile properties, and microhardness were investigated. Nearly a full densification sample ($\geq$99%) was obtained under optimized process parameters, and the value of VED was no less than 300 J/mm^3. Laser power had a dominant influence on densification behavior compared with other parameters. The main phases of 90W-7Ni-3Fe are W and γ-(Ni-Fe), dissolved with partial W. In addition, 90W-7Ni-3Fe showed a high tensile strength (UTS = 1121 MPa) with poor elongation (<1%). A high average microhardness (>400 HV$_{0.3}$) was obtained, but the microhardness presented a fluctuation along building direction due to the inhomogeneous microstructure.

Keywords: selective laser melting; 90W-7Ni-3Fe; densification; microstructure; properties

1. Introduction

Tungsten heavy alloys (WHA) have been used in many fields, including aerospace, defense, military, nuclear, electronics, and marine industries, due to their high melting point and density, excellent thermal conductivity, low thermal expansion, good corrosion resistance, and superior comprehensive properties at high temperatures [1,2]. They have also been proven to be ideal facing-plasma materials in the nuclear industry [3,4]. The 90W-7Ni-3Fe allow, a typical tungsten heavy alloy, which consists of tungsten (90 wt.%), Nickel (7 wt.%) and Iron (3 wt.%), possesses good mechanical properties due to the addition of Ni and Fe, while retaining high-density. In addition, Ni and Fe play a key role in the suppression of cracks in the fabrication of tungsten because of the nature of the brittleness of tungsten at room temperature [5]. Traditionally, 90W-7Ni-3Fe is processed by solid and liquid sintering. A mixture of tungsten, nickel, and iron powder was prepared and then sintered at a certain temperature, such as 1400 °C, for a fixed time in a vacuum or inert gas atmosphere [6,7]. Usually, sintered samples need to be treated by isostatic pressing or other heat treatments to reduce their porosity and avoid hydrogen embrittlement [8–10]. However, it is difficult, or even impossible, to obtain complex components of a tungsten heavy alloy by conventional processes. Thus, additive manufacturing (AM) may be an alternative manufacturing process for high melting point refractory metals, such as tungsten and its alloys.

Selective laser melting (SLM), as a metal additive manufacturing technology, has been proven to be a promising technology in the high-accuracy and integrated fabrication of metallic components.

SLM employs a high-energy laser beam to melt metallic powder layer by layer, according to 2D slice data. SLM is a multidisciplinary cross-technology involving mechanical engineering, material science, optics, software, and has attracted much attention from many fields. Deprez et al. [11] produced a complex high-density tungsten collimator. The produced collimator was geometrically accurate, and the tested values of sensitivity and resolution were close to the expected results of the CAD design. The relative density of the SLM pure tungsten prepared by Zhang et al. [12] reached 82%, who found the formation of nanocrystalline in the tungsten samples fabricated by SLM. Zhou et al. [13] analyzed the balling phenomena in the process of SLM-tungsten and proposed a competitive mechanism of spreading and solidification to explain the balling phenomena. Enneti et al. [14] investigated the effects of scan speed and hatch spacing on the relative density of SLM pure tungsten, but the highest relative density was only 75%. The relative density of the SLM tungsten prepared by Wen et al. [15] reached 98.7%, and they studied the effects of process parameters on the surface morphology, microstructure, and properties of SLM pure tungsten. Similarly, Tan et al. [16] also obtained SLM tungsten with a relative density of 98.5% and analyzed the effects of different laser powers and scan speeds on the SLM's surface morphology, microstructure, and properties. All the previous studies on SLM tungsten indicated that the crack phenomenon was inevitable due to the inherent brittle nature of tungsten [17–19]. Therefore, several measures of crack-suppression for SLM pure tungsten were adopted and reported by Li et al. [20] and Iveković et al. [21]. Their results showed that the addition of secondary-phase nanoparticles or Tantalum could reduce cracks in the process of SLM pure tungsten, but a crack-free sample was still not available. In addition, the addition of low-melting-point metals might be beneficial to the fabrication of crack-free tungsten alloy components. Li et al. [22] investigated the fabrication of W-10Cu by SLM and obtained an optimized combination of laser power and scan speed. Their results showed that the forming mechanism of SLM W-10Cu was liquid phase sintering. Iveković et al. [23] obtained a 90W-7Ni-3Fe sample with a high relative density (>95%). Their results indicated that the high densification of 90W-7Ni-3Fe samples required a high energy density. In addition, they observed three major binding mechanisms: liquid phase sintering, partial melting, and complete melting. Preheating contributes to complete melting. After heat treatment, the tensile strength decreased slightly, but the elongation was significantly improved. Li et al. [24] studied the effect of process parameters on the densification and microstructure of 90W-7Ni-3Fe in SLM. They found that a lower scan speed, narrower scan interval, and thinner layer thickness can improve the densification process. Similar findings were also reported by Zhang et al. [25], who established an effective 3D model based on finite element analysis theory and temperature distributions under different process parameters. As mentioned above, tungsten and its alloys are of interest to researchers, but SLM 90W-7Ni-3Fe has been rarely discussed.

In this work, 90W-7Ni-3Fe samples were formed by SLM. The influences of the process parameters on the densification, phase constitution, and microhardness of the 90W-7Ni-3Fe samples were discussed. In addition, the microstructure and tensile properties of the 90W-7Ni-3Fe samples fabricated by SLM were investigated and discussed.

2. Materials and Methods

2.1. Experimental Equipment and Preparation

All experiments were carried out with an SLM 280 HL (SLM solutions, Germany), equipped with two 400 W Nd: YAG lasers. Figure 1a depicts the schematic diagram of SLM 280 HL. SLM 280 HL has two kinds of substrates: 250 mm × 250 mm and 100 mm × 100 mm. A small substrate was utilized in this work. During the process of SLM, tungsten heavy alloy powder particles fell from the powder container under gravity, and then the fallen powder particles were evenly spread on the substrate under the action of a powder scraper. The bi-directional movement of the powder scraper was adopted for the sake of enhancing the building efficiency. All samples with a size of 10 mm × 10 mm × 5 mm were prepared on a 304 stainless steel substrate, which was preheated at 150 °C in order to prevent cracking and warping. In the whole SLM process, the chamber was filled with argon in order to avoid oxidation, and the content of oxygen was kept below 400 ppm. Figure 1b illustrates the building process of the SLM tungsten heavy alloy.

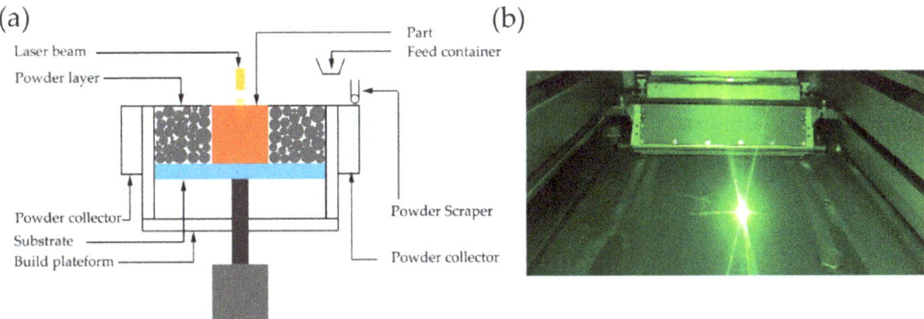

Figure 1. (**a**) Schematic diagram of selective laser melting (SLM) equipment; (**b**) the building process of the SLM.

2.2. Powder Material

In this work, four kinds of powders were prepared in the fabrication of the 90W-7Ni-3Fe powder. Figure 2 illustrates the powder size distribution of the different powders used in this work: pure tungsten powder (D10 = 4.425 μm, D50 = 7.211 μm, and D90 = 11.729 μm), Fe–Ni alloy powder (D10 = 11.420 μm, D50 = 16.478 μm, and D90 = 23.620 μm), nickel powder (D10 = 10.794 μm, D50 = 16.492 μm, and D90 = 24.961 μm), and sub-micro nickel powder (D10 = 0.932 μm, D50 = 1.968 μm, and D90 = 4.002 μm). Considering that the powders had different densities, the average particle size of the tungsten powder was smaller than that of the nickel powder and Fe–Ni alloy, so a relatively uniform powder layer distribution could be obtained. Submicron nickel powder was used to adjust the nickel content in the 90W-7Ni-3Fe alloy. Powders were mixed in a general powder mixer (AM300S-H) under an argon atmosphere was used for 0.5 h to obtain 90W-7Ni-3Fe powder. Figure 3 depicts the morphology of the mixed 90W-7Ni-3Fe powder. It can be observed that the shape of the powder particles was not changed and kept good sphericity. Thus, the 90W-7Ni-3Fe powder had good fluidity, and a uniform and suitable a powder layer could be obtained. Furthermore, fine powder particles adhered to larger particles without falling off (Figure 3b), and the adhesion contributed to the powder's spread and uniformity.

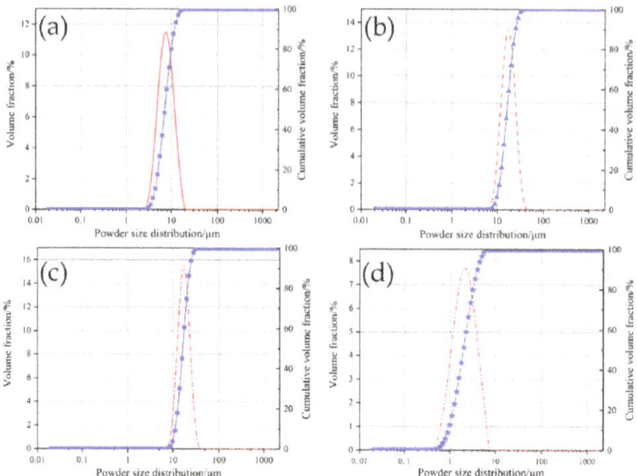

Figure 2. The distribution of Four kinds of powder sizes. (**a**) Pure tungsten powder; (**b**) nickel powder; (**c**) Fe-Ni alloy powder; (**d**) submicron nickel powder.

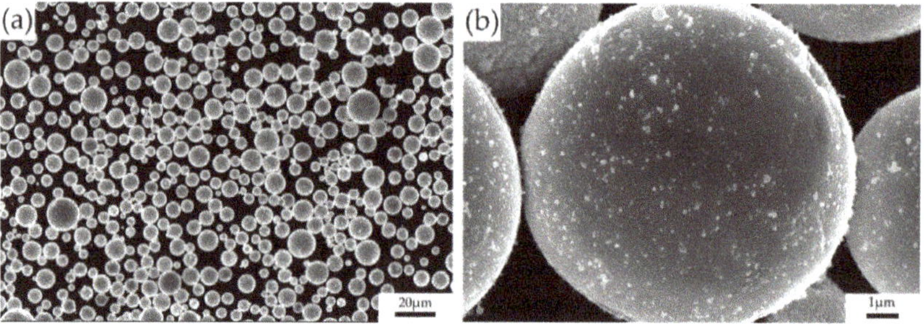

Figure 3. Morphology of the 90W-7Ni-3Fe powder. (**a**) Low magnification; (**b**) high magnification.

2.3. Process Parameters and Conditions

The process parameters used in this work are listed in Table 1. The parameters included four levels of laser power, scan speed, and hatching distance. The powder layer thickness was kept at 30 μm. Various combinations of process parameters were obtained according to the Taguchi experimental design method (Table 2). The scan strategy used in this experiment and building direction is illustrated in Figure 4a. The scan strategy was rotated by 67° between adjacent layers in order to reduce residual stress during the process of SLM. The 90W-7Ni-3Fe alloy samples with dimensions of 10 mm × 10 mm × 5 mm were prepared (Figure 4b). Based on consideration of the combined effect of the process parameters, the volumetric energy density (*VED*) is as follows:

$$VED = \frac{LP}{SS \times HD \times LT},$$ (1)

where *LP* is laser power, *SS* is scan speed, *HD* is hatching distance, and *LT* is layer thickness.

Table 1. Process parameters used in this work.

Process Parameter/Unit	Value
Laser power, LP/W	200, 250, 300, 350
Scan speed, SS/(mm/s)	150, 200, 250, 300
Hatching distance, HD/mm	0.075, 0.09, 0.105, 0.12
Layer thickness, LT/mm	0.03

Table 2. Process parameter combinations and the value of the volumetric energy density (VED).

NO.	LP/W	SS/(mm/s)	HD/mm	VED/(J/mm^3)
1	200	150	0.075	592.59
2	200	200	0.09	370.37
3	200	250	0.105	253.97
4	200	300	0.12	185.19
5	250	150	0.09	617.28
6	250	200	0.105	396.83
7	250	250	0.12	277.78
8	250	300	0.075	370.37
9	300	150	0.105	634.92
10	300	200	0.12	416.47
11	300	250	0.09	533.33
12	300	300	0.075	370.37
13	350	150	0.12	648.15
14	350	200	0.09	777.78
15	350	250	0.075	518.52
16	350	300	0.105	370.37

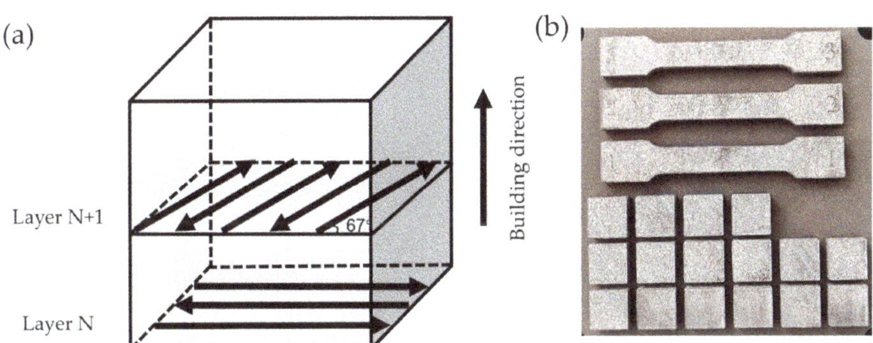

Figure 4. (**a**) Scan strategy used in this experiment; (**b**) 90W-7Ni-3Fe samples fabricated by the designed process parameters.

2.4. Characterization and Test

All the samples were removed from the substrate by wire electrical discharge machining (WEDM). Then, the samples were cleaned with an ultrasonic cleaning machine and dried. After that, the dried samples were ground gradually with sandpaper of different grits, 280#, 400#, 600#, 800#, 1000#, 1200#, 1500# and 2000#. Subsequently, the samples were polished using a diamond suspension of 2.5 and 0.5.

In this work, the theoretical density ρ_T of 90W-7Ni-3Fe can be calculated as [26]

$$\frac{100}{\rho_T} = \sum \frac{W_i}{\rho_i}, \tag{2}$$

where W_i and ρ_i are the mass fraction and the theoretical density of the ith alloy element, respectively. The theoretical density of W, Ni, and Fe are 19.3 g/cm^3, 8.9 g/cm^3, and 7.9 g/cm^3, respectively [27].

According to the Archimedes method, the actual density can be calculated as follows:

$$\rho_a = \frac{M_0}{M_2 - M_1} \times \rho_0,$$ (3)

where M_0 is the mass of sample in the air, M_1 and M_2 are the indications of the balance before and after the sample is placed in the beaker containing water, and ρ_0 is the density of water.

Relative density ρ_{RD} is calculated as follows:

$$\rho_{RD} = \frac{\rho_a}{\rho_T} \times 100\%,$$ (4)

The transverse and vertical morphology of the samples (Figure 5a) were observed under an optical microscope (OM, Nikon MA 200) in order to analyze the defects' characteristics and distribution. The transverse section of the sample was etched with a mixture solution of 10 g KOH: 10 g K_3 [Fe(CN)$_6$]:100 mL H_2O to ensure a clear microstructure. The microstructure was also characterized by an optical microscope (OM, Nikon MA 200). For a better analysis of the microstructure, a scanning electron microscope (SEM, S-4800, Hitachi, Tokyo, Japan) was used, and an energy-dispersive spectrometer (EDS) was adopted for analysis of the element distribution. The phase composition of the sample was tested by an X-ray diffractometer (XRD, Advanced D8, Bruker, Billerica, MA, USA) with a Cu Kα radiation at 40 KV and 40 mA in a 2θ range of 30°–90° by using a step size of 0.02°. Tensile property tests were performed at room temperature using an Instron550R at a constant tensile rate of 1 mm/min. The tensile specimen is illustrated in Figure 5b. The microhardness of the transverse morphology was measured by using a digital microhardness measurement system (MH5) with a load of 0.3 kg and a dwell time of 15 s. Five points were taken at an interval of 0.5 mm along the building direction for all selected samples fabricated under different VEDs.

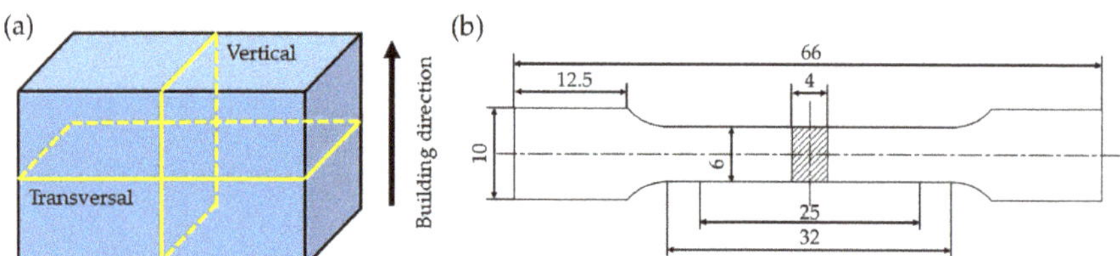

Figure 5. (**a**) Illustration diagram of sample direction; (**b**) dimensions of tensile sample.

3. Results and Discussion

3.1. Densification Analysis

Being porosity-free is of importance to the mechanical properties of the final products. Thus, it is necessary to investigate the effects of the process parameters on the formation of defects in order to obtain high relative density parts. In this section, the relationships between the relative density and process parameters are analyzed. Figure 6 shows the variation of the relative density with different volumetric energy densities. With an increase of VED, the relative density first became higher and tended to be stable. The relative density of the final samples fabricated in this work can were nearly 100% free from cracks. Figure 7 depicts the actual transverse morphology of the samples obtained under different VEDs. With an increase in the value of the VED, the number of defects in the samples gradually decreased. When the input of the VED was insufficient, the low-melting-point metals (Ni/Fe) were melted, and the liquid phase was formed. The un-melted tungsten particle gaps were filled with the liquid phase. However, a low VED indicated a combination of a low laser power (200 W) and a high scan speed (300 mm/s). This result implies that the residence time of the formed liquid phase was too short to fully fill the gaps, so irregular defects were formed, as shown in Figure 8a.

Poor densification was caused by inadequate liquid phase content or a short residence time under the processing conditions. Figure 8b presents the high-magnification transverse morphology of the samples fabricated with a high VED. Nearly full densification could be obtained under the process parameters.

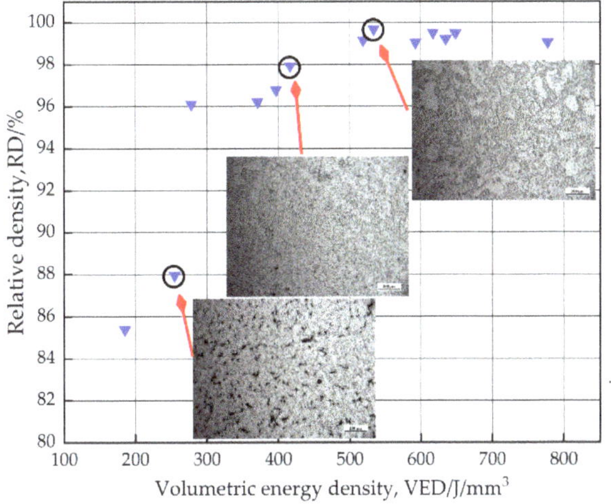

Figure 6. Variation of the relative density with the volumetric energy density. The inserted figure shows the different transverse morphologies corresponding to the VEDs.

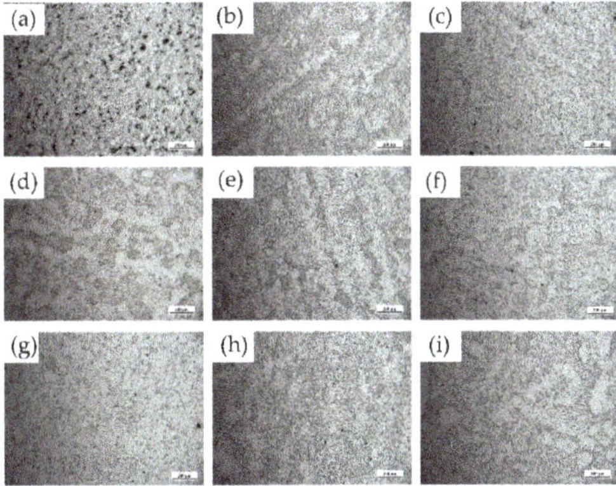

Figure 7. The transverse morphologies of different VEDs. (**a**) 185.19 J/mm^3; (**b**) 277.78 J/mm^3; (**c**) 396.83 J/mm^3; (**d**) 518.52 J/mm^3; (**e**) 592.60 J/mm^3; (**f**) 617.28 J/mm^3; (**g**) 634.92 J/mm^3; (**h**) 648.15 J/mm^3; (**i**) 777.78 J/mm^3.

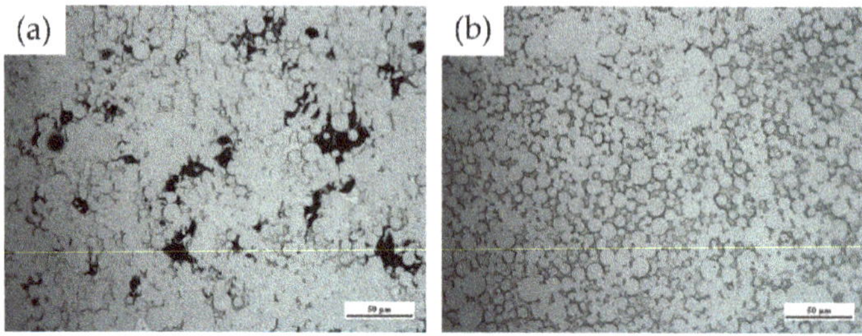

Figure 8. Transverse morphology with a high magnification. (**a**) 253.97 J/mm^3; (**b**) 648.15 J/mm^3.

The pores were nearly fully filled with the formed liquid phase. A combination of high laser power (350 W) and low scan speed (150 mm/s) produced a longer residence time for the liquid phase. Therefore, the best rearrangement characteristics for the tungsten powder particles were observed under the capillary force of liquid, and the densification of samples was improved. When the laser power was 350 W, the viscosity of the formed molten pool decreased, and partial tungsten powder particles could be melted. In this way, an adequate liquid phase and better fluidity for the molten pool were obtained, and the densification process of the samples could be completed.

It can also be seen that under the same VED, there are different relative densities corresponding to distinct transverse morphologies, as shown in Figure 9. This phenomenon could be ascribed to the varying process parameter combinations and similar results of other materials that have been reported [28,29]. Here, the signal-to-noise (S/N) ratio of Taguchi's method was used to compare the effect of every process parameter on the relative density, which means the sensitivity of the relative density to the selected process parameters. In general, the signal-to-noise ratio in Taguchi's method can be classified into three categories: "lower is better", "nominal is best", and "higher is better". In this study, the highest relative density for the sample was expected. Thus, the "higher is better" category was adopted, which can be calculated as follows [30]:

$$\frac{S}{N} = -10\log\left(\frac{1}{n}\sum_{1}^{n}\frac{1}{y_i^2}\right), \tag{5}$$

where n is the number of experiments and y_i is the tested data of relative density for the ith sample.

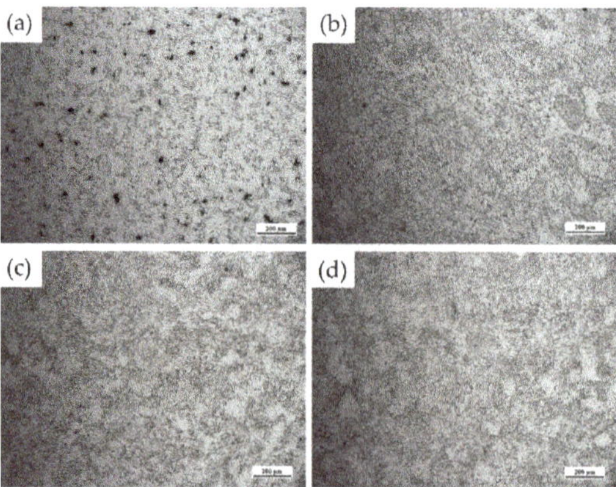

Figure 9. Transverse morphology of the same VED (370.37 J/mm^3).

With the tested data for the relative density under different combinations of process parameters, the calculated results for the S/N ratios are presented in Figure 10 and Table 3. It can be seen that the laser power was identified as the most important factor influencing the final relative density of the samples. The scan speed had a relatively insignificant effect on relative density, and the hatching distance showed the least significant effects among all the selected process parameters. Therefore, the sensitivity of the relative density to the three process parameters decreased according to the following order: laser power > scan speed > hatching distance. This result is consistent with previous results and could be ascribed to the special thermophysical properties of tungsten and tungsten heavy alloys [31].

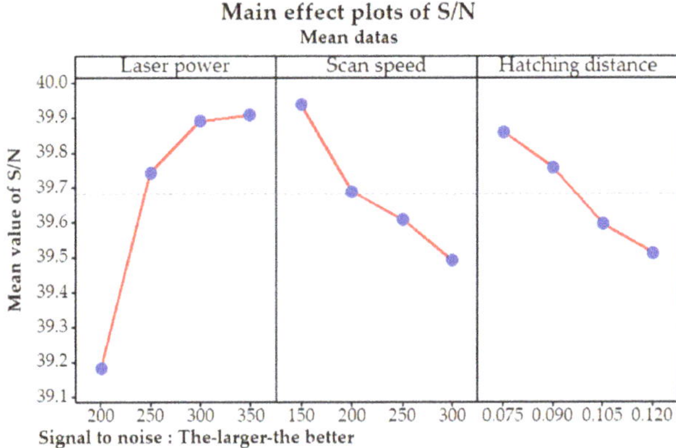

Figure 10. Main effect plots of signal-to-noise (S/N).

Table 3. Response for the signal-to-noise (S/N).

Levels	1	2	3	4	Delta	Rank
Laser power/W	39.18	39.75	39.89	39.91	0.73	1
Scan speed/(mm/s)	39.94	39.69	39.61	39.49	0.45	2
Hatch distance/mm	39.86	39.76	39.60	39.51	0.35	3

In this section, the nearly full densification of the 90W-7Ni-3Fe sample fabricated by SLM was realized. The densification behaviors under different process parameters were analyzed and discussed. Laser power is a dominant factor in the SLM process of the 90W-7Ni-3Fe samples. For the 90W-7Ni-3Fe sample with a high relative density in SLM, a higher laser power ($\geq$250 W) was preferred, and the value of volumetric energy density had to be no less than 300 J/mm^3. The morphology of the 90W-7Ni-3Fe sample presented similar sintering, characteristic in traditional powder metallurgy [32]. The microstructure is mainly composed of a refractory tungsten particle skeleton and a liquid phase formed by low-melting-point metals (Ni/Fe).

3.2. Phase Identification and Microstructure

The XRD patterns of the 90W-7Ni-3Fe samples processed using different volumetric energy densities are shown in Figure 11. The main phases of the 90W-7Ni-3Fe samples fabricated by SLM were W and Ni-Fe solid solution phases, although the volumetric energy density varied from 185.19 J/mm^3 to 777.78 J/mm^3. There seems to have been no significant difference in the phase composition among the samples obtained under different volumetric energy densities. This means that low-melting-point metals (Ni/Fe) still melted, even under low volumetric energy densities.

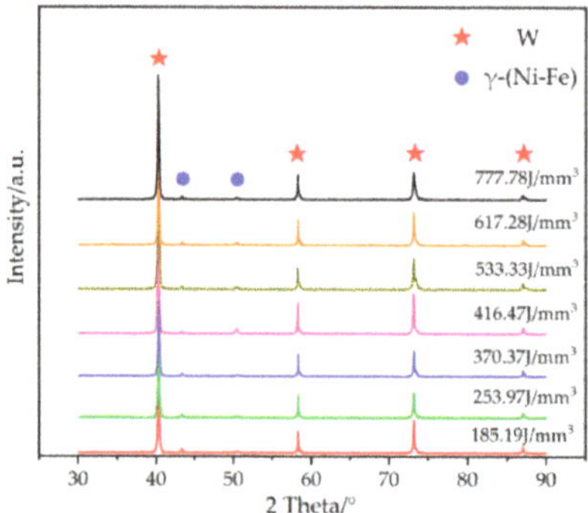

Figure 11. The XRD pattern of different VED.

Figure 12a shows the typical SEM morphology of the 90W-7Ni-3Fe sample fabricated by SLM. Many tungsten particles were distributed in the matrix. The gaps between the tungsten particles were filled with liquid phases formed by low-melting-point metals (Ni/Fe). In addition, partial tungsten particles were contacted under the driving force of the liquid phase during the processes of melting and solidification. Fine tungsten grains can also be found in Figure 12a. The microstructural region can be divided into three main regions: the W particle phase, the fine W dendrite region, and the Ni–Fe matrix region with dissolved W. Scanning electron microscope (SEM) and energy-dispersive spectrometer (EDS) analyses were performed in order to further observe the microstructural constitution and confirm whether the tungsten element had been dissolved into the matrix. EDS analysis results (Figures 12b and 13a) confirmed that the particles were nearly composed of 100% W with little Ni and Fe. EDS analysis results of the Ni–Fe matrix indicated that partial tungsten had been dissolved into the matrix (Figure 13b). This phenomenon could be ascribed to the different solubilities of W, Ni, and Fe in W-Ni-Fe systems. Tungsten has a high solubility in the Ni/Fe matrix, but the solubility of Ni and Fe in tungsten is practically negligible. Moreover, fine W dendrites formed in the matrix when W particles were partially melted and W particles acted as heterogenous nucleation sites. Similar tungsten dendrites were also observed in the WHA samples fabricated by SLM [33,34].

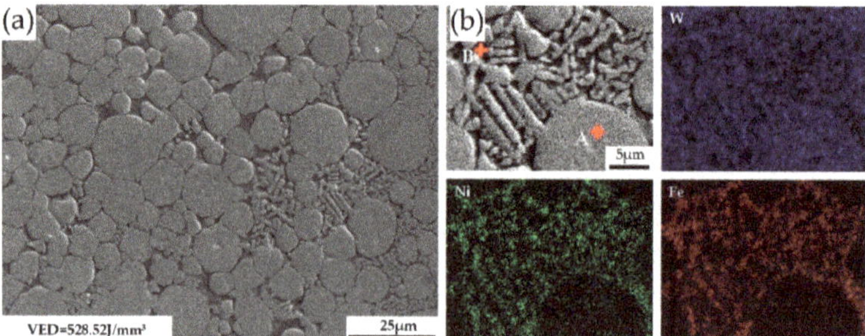

Figure 12. (**a**) The microstructure of 90W-7Ni-3Fe under VED = 518.52 J/mm^3; (**b**) element distribution.

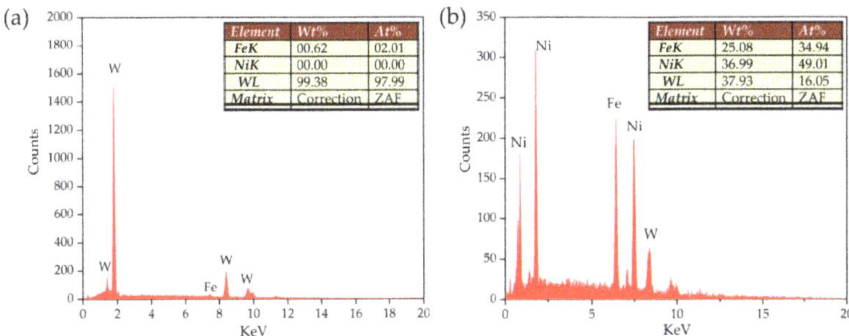

(a)

Element	Wt%	At%
FeK	00.62	02.01
NiK	00.00	00.00
WL	99.38	97.99
Matrix	Correction	ZAF

(b)

Element	Wt%	At%
FeK	25.08	34.94
NiK	36.99	49.01
WL	37.93	16.05
Matrix	Correction	ZAF

Figure 13. Energy-dispersive spectrometer (EDS) element analysis. (**a**) Point A; (**b**) point B.

3.3. Mechanical Properties

3.3.1. Tensile Properties

The tensile sample was fabricated using the optimized process parameters (RD ≥ 99%) and tensile properties of 90W-7Ni-3Fe were investigated. Figure 14 depicts the tensile stress–strain curve of SLM-built 97W-7Ni-3Fe. The ultimate tensile strength (UTS) was 1121 MPa when there was no significant yield platform. The tensile fracture surface was characterized (Figure 15) and analyzed in order to illustrate the tensile fracture mode. In general, for WHA, there exist four kinds of fracture modes: matrix failure, W cleavage, a W–W inter granular fracture, and W-matrix interfacial separation [35]. The coexistence of these four fracture modes is common in tungsten alloys, and the tensile properties of the WHA are determined by the proportion of four fracture modes. From Figure 15a, the W–W inter granular fracture and W-matrix interfacial separation can be found; few (and shallow) dimples and W cleavage are shown in Figure 15b. Therefore, the 90W-7Ni-3Fe tensile samples fabricated by SLM in this work have the characteristic of brittle fracture. Table 4 lists and compares the tensile properties of the 90W-7Ni-3Fe samples in this study and previous studies. The 90W-7Ni-3Fe fabricated in this work presented a higher tensile strength but a similarly poor elongation. Therefore, in order to obtain the comprehensive mechanical properties, and realize the balance between tensile strength and elongation, subsequent heat treatments might be required. Appropriate heat treatments for these SLM-built WHA samples will be investigated in our future work. In addition, the mechanical behavior of samples could be affected by the building direction in SLM [36]. Thus, the relationship between 90W-7Ni-3Fe mechanical properties and building direction will also be discussed in our future work.

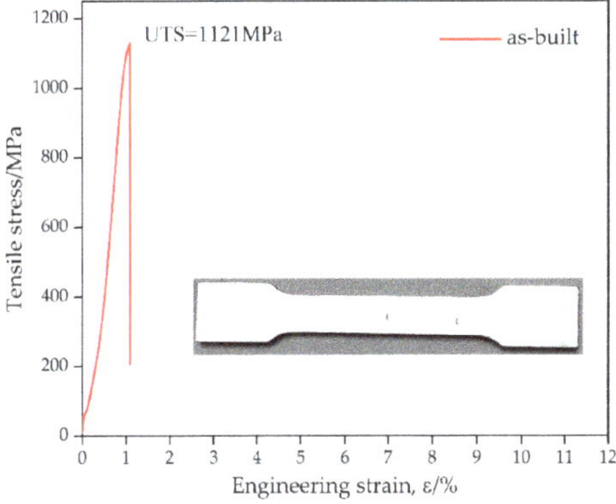

Figure 14. The tensile properties of 90W-7Ni-3Fe fabricated by SLM.

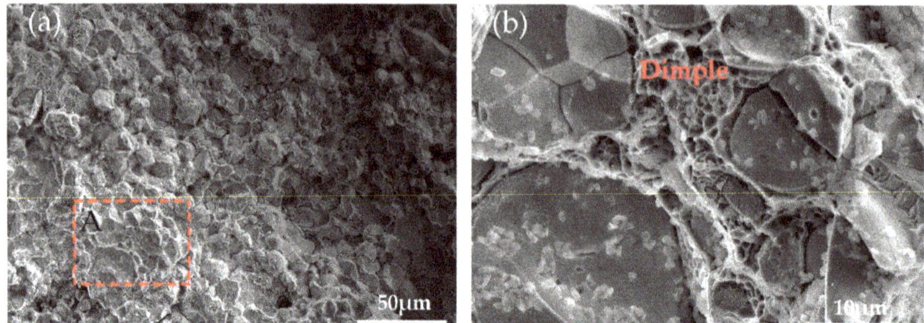

Figure 15. Tensile fracture morphology of 90W-7Ni-3Fe fabricated by SLM. (**a**) Low magnification; (**b**) high magnification of zone A (**a**).

Table 4. Comparison of the tensile properties of 90W-7Ni-3Fe between this work and the published results.

Processing Methods	UTS/MPa	Elongation/%	Source
Selective laser melting, SLM	1121 MPa	<1%	This work
Selective laser melting, SLM	871 MPa	<1%	[23]
Liquid phase sintering, LPS	~1000 MPa	~30%	[10]

3.3.2. Microhardness

Microhardness along the building direction of the sample was measured under the selected process parameter combinations. Figure 16a shows the microhardness distribution of the samples obtained under different volumetric energy densities at five points. The microhardness of the 90W-7Ni-3Fe samples fabricated by SLM presented an apparent fluctuation at every VED. This phenomenon might be ascribed to the inhomogeneous microstructure in the 90W-7Ni-3Fe samples (Figure 12). The average microhardness values of the 90W-7Ni-3Fe samples under different VEDs are shown in Figure 16b. The average microhardness was 406.83HV$_{0.3}$ when the VED was 253.97 J/mm^3, and the average microhardness increased to 467.46HV$_{0.3}$ when the VED was 777.78 J/mm^3. The average microhardness slightly increased with the increase of VED. This result could be explained by the nearly full density of the sample and the fine grains mentioned above. Compared to the result shown in Figure 17b, pores and irregular defects can be found in Figure 17a, which lead to a slight decrease of the average microhardness.

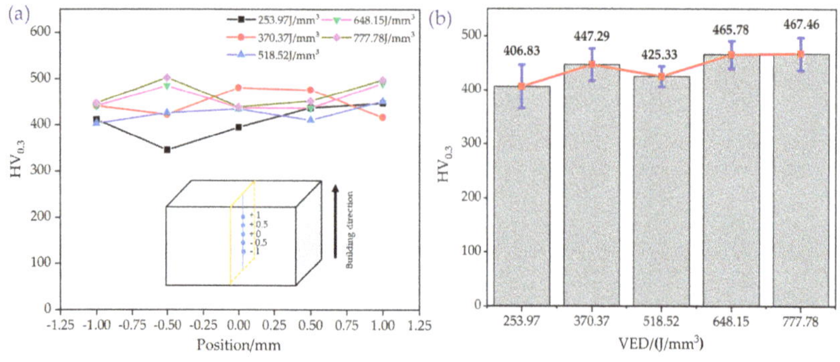

Figure 16. Microhardness under different VEDs. (**a**) Microhardness of each point; (**b**) average microhardness.

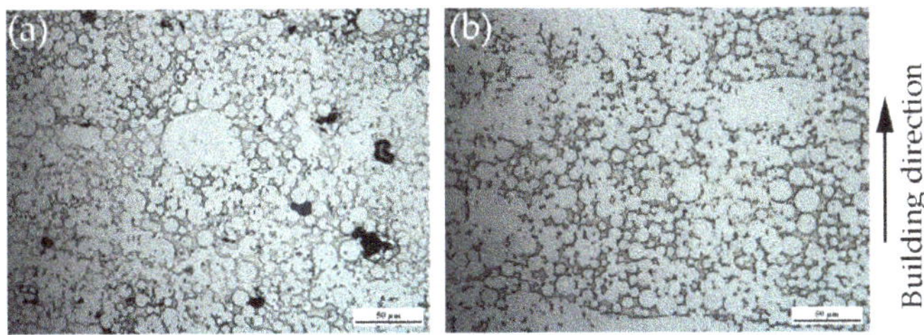

Figure 17. Vertical morphology of the 90W-7Ni-3Fe under different VEDs. (**a**) 253.97 J/mm^3; (**b**) 518.52 J/mm^3.

4. Conclusions

The densification behavior, phase composition, microstructure, tensile properties, and microhardness of 90W-7Ni-3Fe samples fabricated by SLM were investigated in this work. The main conclusions can be drawn as follows:

(1) Nearly a full densification (≥99%) of the tungsten heavy alloy (WHA) sample was obtained by SLM. With the increase of VED, the relative density of WHA increased significantly. The value of VED should be no less than 300 J/mm^3, so that the high relative density of the 90W-7Ni-3Fe components could be obtained. Among the three influencing factors, laser power had the most significant effect on the relative density, and a higher laser power (≥250 W) was preferred. The forming mechanism in this work presented a liquid phase sintering characteristic, which was similar to that in traditional powder metallurgy.

(2) The typical microstructure of WHA was composed of a tungsten particle phase and a γ-(Ni-Fe) matrix phase with dissolved tungsten. Three main regions can be found, including the tungsten particle phase, the fine tungsten dendrite region, and the Ni-Fe matrix region with dissolved W. The tungsten particle gaps were fully filled with the liquid phase, which contributed to the densification behavior. Fine W dendrites formed due to rapid melting and solidification in the process of SLM, and un-melted tungsten particles acted as a heterogenous nucleation site.

(3) The high tensile strength (UTS = 1121 MPa) of WHA was obtained in this work. Compared with traditional liquid phase sintering, SLM realized a significant enhancement of tensile strength but poor elongation. In addition, all samples under different VEDs presented a high microhardness, and the microhardness showed a slight increase with an increase of VED. Microhardness fluctuated along the building direction of the samples. This fluctuation was caused by building defects and an uneven microstructure.

Author Contributions: Conceptualization, J.L.; Data curation, Z.W., S.-G.C. and Z.S.; Investigation, B.Z. and Y.W.; Methodology, J.L., B.Z., and Y.W.; Writing—original draft, J.L.; Writing—review & editing, Z.W. and S.-G.C.

Funding: This research was funded by Science Challenge Project (TZ2018006-0301-01), Guangdong Scientific and Technological Project (2017B090911015) and Dongguan University of Technology High-level Talents (Innovation Team) Research Project (KCYCXPT2016003).

Conflicts of Interest: The authors declare no conflict of interest.

References

1. Jiao, Z.H.; Kang, R.K.; Dong, Z.G.; Guo, J. Microstructure characterization of W-Ni-Fe heavy alloys with optimized metallographic preparation method. *Int. J. Refract. Met. Hard Mat.* **2019**, *80*, 114–122. [CrossRef]
2. Deng, S.H.; Yuan, T.C.; Li, R.D.; Zeng, F.H.; Liu, G.H.; Zhou, X. Spark plasma sintering of pure tungsten powder: Densification kinetics and grain growth. *Powder Technol.* **2017**, *310*, 264–271. [CrossRef]
3. Koedig, M.; Kuehnlein, W.; Linke, J.; Merola, M., Rigal, E.; Schedler, B.; Visca, E. Investigation of tungsten alloys as plasma facing materials for the ITER divertor. *Fusion Eng. Des.* **2002**, *61*, 135–140. [CrossRef]

4. Philipps, V. Tungsten as material for plasma-facing components in fusion devices. *J. Nucl. Mater.* **2011**, *415*, S2–S9. [CrossRef]

5. Gumbsch, P. Brittle fracture and the brittle-to-ductile transition of tungsten. *J. Nucl. Mater.* **2003**, *323*, 304–312. [CrossRef]

6. Akhtar, F. An investigation on the solid state sintering of mechanically alloyed nano-structured 90W–Ni–Fe tungsten heavy alloy. *Int. J. Refract. Met. Hard Mat.* **2008**, *26*, 145–151. [CrossRef]

7. Senthilnathan, N.; Annamalai, A.R.; Venkatachalam, G. Sintering of tungsten and tungsten heavy alloys of W–Ni–Fe and W–Ni–Cu: A review. *Trans. Indian Inst. Met.* **2017**, *70*, 1161–1176. [CrossRef]

8. German, R.M.; Churn, K.S. Sintering atmosphere effects on the ductility of W-Ni-Fe heavy metals. *Metall. Trans. A* **1984**, *15*, 747–754. [CrossRef]

9. Durlu, N.; Çalişkan, N.K.; Bor, Ş. Effect of swaging on microstructure and tensile properties of W–Ni–Fe alloys. *Int. J. Refract. Met. Hard Mat.* **2014**, *42*, 126–131. [CrossRef]

10. Das, J.; Rao, G.A.; Pabi, S.K. Microstructure and mechanical properties of tungsten heavy alloys. *Mater. Sci. Eng. A* **2010**, *527*, 7841–7847. [CrossRef]

11. Deprez, K.; Vandenberghe, S.; Van Audenhaege, K.; Van Vaerenbergh, J.; Van Holen, R. Rapid additive manufacturing of MR compatible multipinhole collimators with selective laser melting of tungsten powder. *Med. Phys.* **2013**, *40*, 012501. [CrossRef] [PubMed]

12. Zhang, D.Q.; Cai, Q.Z.; Liu, J.H. Formation of nanocrystalline tungsten by selective laser melting of tungsten powder. *Mater. Manuf. Process.* **2012**, *27*, 1267–1270. [CrossRef]

13. Zhou, X.; Liu, X.H.; Zhang, D.Q.; Shen, Z.J.; Liu, W. Balling phenomena in selective laser melted tungsten. *J. Mater. Process. Technol.* **2015**, *222*, 33–42. [CrossRef]

14. Enneti, R.K.; Morgan, R.; Atre, S.V. Effect of process parameters on the Selective Laser Melting (SLM) of tungsten. *Int. J. Refract. Met. Hard Mat.* **2018**, *71*, 315–319. [CrossRef]

15. Wen, S.F.; Wang, C.; Zhou, Y.; Duan, L.C.; Wei, Q.S.; Yang, S.F.; Shi, Y.F. High-density tungsten fabricated by selective laser melting: Densification, microstructure, mechanical and thermal performance. *Opt. Laser Technol.* **2019**, *116*, 128–138. [CrossRef]

16. Tan, C.L.; Zhou, K.S.; Ma, W.Y.; Attard, B.; Zhang, P.P.; Kuang, T.C. Selective laser melting of high-performance pure tungsten: Parameter design, densification behavior and mechanical properties. *Sci. Technol. Adv. Mater.* **2018**, *19*, 370–380. [CrossRef] [PubMed]

17. Guo, M.; Gu, D.D.; Xi, L.X.; Du, L.; Zhang, H.M.; Zhang, J.Y. Formation of scanning tracks during Selective Laser Melting (SLM) of pure tungsten powder: Morphology, geometric features and forming mechanisms. *Int. J. Refract. Met. Hard Mat.* **2019**, *79*, 37–46. [CrossRef]

18. Wang, D.Z.; Li, K.L.; Yu, C.F.; Ma, J.; Liu, W.; Shen, Z.J. Cracking behavior in additively manufactured pure tungsten. *Acta Metall. Sin.* **2019**, *32*, 127–135. [CrossRef]

19. Müller, A.V.; Schlick, G.; Neu, R.; Anstätt, C.; Klimkait, T.; Lee, J.; Seidel, C. Additive manufacturing of pure tungsten by means of selective laser beam melting with substrate preheating temperatures up to 1000 °C. *Nucl. Mater. Energy* **2019**, *19*, 184–188. [CrossRef]

20. Li, K.L.; Wang, D.Z.; Xing, L.L.; Wang, Y.F.; Yu, C.F.; Chen, J.H.; Zhang, T.; Ma, J.; Liu, W.; Shen, Z.J. Crack suppression in additively manufactured tungsten by introducing secondary-phase nanoparticles into the matrix. *Int. J. Refract. Met. Hard Mat.* **2019**, *79*, 158–163. [CrossRef]

21. Iveković, A.; Omidvari, N.; Vrancken, B.; Lietaert, K.; Thijs, L.; Vanmeensel, K.; Kruth, J.P. Selective laser melting of tungsten and tungsten alloys. *Int. J. Refract. Met. Hard Mat.* **2018**, *72*, 27–32. [CrossRef]

22. Li, R.D.; Shi, Y.S.; Liu, J.H.; Xie, Z.; Wang, Z.G. Selective laser melting W–10 wt.% Cu composite powders. *Int. J. Adv. Manuf. Technol.* **2010**, *48*, 597–605. [CrossRef]

23. Iveković, A.; Montero-Sistiaga, M.L.; Vanmeensel, K.; Kruth, J.P.; Vleugels, J. Effect of processing parameters on microstructure and properties of tungsten heavy alloys fabricated by SLM. *Int. J. Refract. Met. Hard Mat.* **2019**, *82*, 23–30. [CrossRef]

24. Li, R.D.; Liu, J.H.; Shi, Y.S.; Zhang, L.; Du, M.Z. Effects of processing parameters on rapid manufacturing 90W–7Ni–3Fe parts via selective laser melting. *Powder Metall.* **2010**, *53*, 310–317. [CrossRef]

25. Zhang, D.Q.; Cai, Q.Z.; Liu, J.H.; Zhang, L.; Li, R.D. Select laser melting of W–Ni–Fe powders: Simulation and experimental study. *Int. J. Adv. Manuf. Technol.* **2010**, *51*, 649–658. [CrossRef]

26. Ding, L.; Xiang, D.P.; Li, Y.Y.; Li, C.; Li, J.B. Effects of sintering temperature on fine-grained tungsten heavy alloy produced by high-energy ball milling assisted spark plasma sintering. *Int. J. Refract. Met. Hard Mat.* **2012**, *33*, 65–69. [CrossRef]

27. Bollina, R.; German, R.M. Heating rate effects on microstructural properties of liquid phase sintered tungsten heavy alloys. *Int. J. Refract. Met. Hard Mat.* **2004**, *22*, 117–127. [CrossRef]

28. Joguet, D.; Costil, S.; Liao, H.; Danlos, Y. Porosity content control of CoCrMo and titanium parts by Taguchi method applied to selective laser melting process parameter. *Rapid Prototyp. J.* **2016**, *22*, 20–30. [CrossRef]

29. Sun, J.F.; Yang, Y.Q.; Wang, D. Parametric optimization of selective laser melting for forming Ti6Al4V samples by Taguchi method. *Opt. Laser Technol.* **2013**, *49*, 118–124. [CrossRef]

30. Kong, X.; Yang, L.; Zhang, H.; Chi, G.; Wang, Y. Optimization of surface roughness in laser-assisted machining of metal matrix composites using Taguchi method. *Int. J. Adv. Manuf. Technol.* **2017**, *89*, 529–542. [CrossRef]

31. Leitz, K.H.; Singer, P.; Plankensteiner, A.; Tabernig, B.; Kestler, H.; Sigl, L.S. Multi-physical simulation of selective laser melting. *Metal Powder Rep.* **2017**, *72*, 331–338. [CrossRef]

32. Erol, M.; Erdoğan, M.; Karakaya, İ. Effects of fabrication method on initial powder characteristics and liquid phase sintering behaviour of tungsten. *Int. J. Refract. Met. Hard Mat.* **2018**, *77*, 82–89. [CrossRef]

33. Zhang, D.Q.; Liu, Z.H.; Cai, Q.Z.; Liu, J.H.; Chua, C.K. Influence of Ni content on microstructure of W–Ni alloy produced by selective laser melting. *Int. J. Refract. Met. Hard Mat.* **2014**, *45*, 15–22. [CrossRef]

34. Wang, M.; Li, R.; Yuan, T.; Chen, C.; Zhang, M.; Weng, Q.; Yuan, J. Selective laser melting of W-Ni-Cu composite powder: Densification, microstructure evolution and nano-crystalline formation. *Int. J. Refract. Met. Hard Mat.* **2018**, *70*, 9–18. [CrossRef]

35. Liu, W.; Ma, Y.; Zhang, J. Properties and microstructural evolution of W-Ni-Fe alloy via microwave sintering. *Int. J. Refract. Met. Hard Mat.* **2012**, *35*, 138–142. [CrossRef]

36. Hitzler, L.; Janousch, C.; Schanz, J.; Merkel, M.; Heine, B.; Mack, F.; Öchsner, A. Direction and location dependency of selective laser melted AlSi10Mg specimens. *J. Mater. Process Tech.* **2017**, *243*, 48–61. [CrossRef]

 *metals*

Article

Particle Erosion Induced Phase Transformation of Different Matrix Microstructures of Powder Bed Fusion Ti-6Al-4V Alloy Flakes

Jun-Ren Zhao, Fei-Yi Hung * and Truan-Sheng Lui

Department of Materials Science and Engineering, National Cheng Kung University, Tainan 701, Taiwan
* Correspondence: fyhung@mail.ncku.edu.tw; Tel.: +886-6-275-7575 (ext. 62950)

Received: 9 June 2019; Accepted: 26 June 2019; Published: 28 June 2019

Abstract: In this study, powder bed fusion Ti-6Al-4V alloy flake was subjected to heat treatment at 800 °C for 4 h for inducing the complete transformation of the α' phase into the $\alpha+\beta$ phases. An erosion experiment with 450 μm mean particle diameter of Al_2O_3 particles at a 90° impact on both the as- powder bed fusion (PBF) Ti-6Al-4V and the 4-h 800 °C heat-treated specimens to clarify the particle erosion-induced phase transformation behavior and its effect on mechanical properties. Particle erosion-induced phase transformation to the α phase was observed on both the as-PBF Ti-6Al-4V and the heat-treated specimens. It brought about a sequential formation from the surface to the bottom: (1) a surface softened zone, (2) a hardened zone, and (3) a hardness stabilization zone. The as-PBF Ti-6Al-4V was positively eroded by erosion particles, decreasing strength and ductility. In the case of the heat-treated specimens, we found decreased strength yet an increased ductility.

Keywords: powder bed fusion (PBF); Ti-6Al-4V; erosion; phase transformation; tensile

---

1. Introduction

Titanium alloys are widely applied in environments susceptible to erosion, including blades, turbines, and desalination pipeline [1]. Among them, Ti-6Al-4V is the most representative among the $\alpha + \beta$ titanium alloys [2,3], which possesses excellent specific strength, fracture toughness, corrosion resistance, and bio-compatibility [4]. However, its low thermal conductivity and high reactivity features, which result in its poor machinability characteristics, make it difficult to undertake further exploitation [5]. In addition, specific heat treatment operations are generally required due to the heat hardening phenomena during the cutting process. Additive manufacturing (AM) technology possesses great characteristics as a new manufacturing technique that is able to lower the cost and reduce energy consumption as well, solving current problems in the fabrication of Ti-6Al-4V and producing a near net shape component [6–8]. Powder bed fusion (PBF) process which uses metal powders as a raw material is a type of additive manufacturing (AM) technology. During PBF, metal powders are melted in a specified area with a high-energy laser beam, then it is rapidly solidified at a high cooling rate [9–13], resulting in its possessing the martensitic α' phases, which differs from the classical $\alpha + \beta$ phase structure. In the meantime, the phase transformation takes place when the material is heated.

Erosion wear often results in the failure of mechanical devices and components. Therefore, it is necessary to study the erosion wear phenomenon. However, there are a significant number of research articles on the mechanical properties of PBF Ti-6Al-4V [14,15], but few on the erosion wear. And the little research about the wear of titanium alloys, most of them focusing on water droplet erosion [16–19], not on solid particle erosion. The particle erosion articles had also mainly focused on erosion resistance [20–22], instead of the particle erosion-induced phase transformation. According to our previous researches [23,24], particle erosion wear is able to induce phase transformation and affects the erosion rate. Therefore, we want to clarify the erosion-induced phase transformation mechanism and its effect on the tensile mechanical properties in this study. In this part, erosion particles were used to positively erode the as-PBF Ti-6Al-4V and the heat-treated specimens at a 90° impact angle. The results have significant reference value for relevant industry applications.

2. Experimental Procedure

The PBF process parameters used in this study are shown in Table 1. The specimens were fabricated by using an EOS M290 400 W machine (EOS, Krailling, Germany) in an inert gas (argon) atmosphere. The chemical composition values of as-PBF Ti-6Al-4V measured by the Inductively Coupled Plasma Mass Spectrometry (ICP-MS, THERMO-ELEMENT XR, Alberta, Canada) are shown in Table 2. The specimens were removed from support by electrical discharge machining (EDM) wire cutting and did not go through any post-treatment before the implemented tests. The as-PBF Ti-6Al-4V specimen was labeled AS. The heat-treated specimen was held for 4 h in a tubular furnace in an argon atmosphere at 800 °C and was subjected to air cooling (labeled HT).

Table 1. Process parameters used for the powder bed fusion (PBF) process.

Process parameters	Value
Laser power (W)	170
Scanning velocity (mm/s)	800
Layer thickness (µm)	30
Laser radius (µm)	35
Particle size (µm)	15–45

Table 2. PBF Ti-6Al-4V chemical composition (wt.%).

Al	V	Fe	O	C	N	H	Ti
6.13	3.80	0.30	0.20	0.08	0.05	0.01	Bal.

The erosion test equipment is shown in Figure 1. We used irregular Al_2O_3 particles for erosion test, which had an average particle size of 450 µm. The particle SEM morphology is shown in Figure 2. The specimens were polished with SiC paper (from #80 to #1000) to remove the oxidized layer and were soaked in acetone for ultrasonic cleaning before the erosion test. Two hundred grams of the erosion particles were used under a compressed air flow of 3 kg/cm^2 (0.29 MPa). During the erosion test, the AS and HT specimens were eroded at a 90° impact angle to investigate the particle erosion induced phase transformation phenomenon.

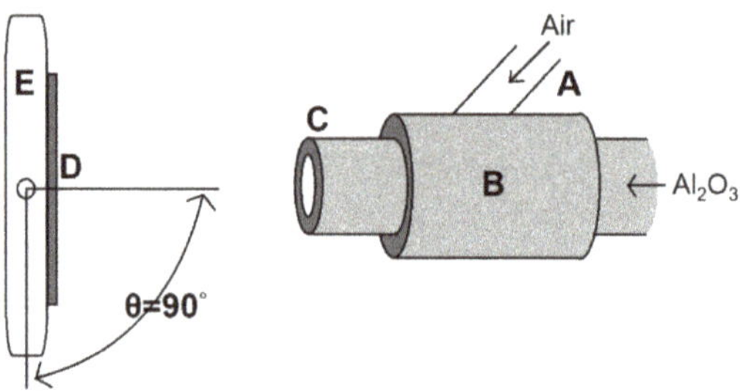

Figure 1. The equipment for particle erosion test. (A: compressed air flow; B: erosion particle (Al_2O_3 particles) supplier; C: erodent nozzle; D: specimen; E: specimen holder; θ: 90° impact angle).

Figure 2. SEM image of Al_2O_3 erosion particles morphology and size.

After the erosion test, the specimens were polished with #80 to #4000 SiC paper, placed into 1 and 0.3 μm Al_2O_3 aqueous solution, and then transferred to a 0.04 μm SiO_2 polishing solution. Finally, they were etched by Keller's reagent (1 mL HF + 1.5 mL HCL + 2.5 mL HNO_3 + 95 mL H_2O). Optical microscopy (OM, OLYMPUS BX41M-LED, Tokyo, Japan) and transmission electron microscopy (TEM; Tecnai F20 G2, EFI, Hillsboro, OR, USA) were used to examine the subsurface microstructure of the erosion specimens.

For analyzing the structural difference between AS and HT specimens before and after erosion, X-ray diffractometry (XRD, Bruker AXS GmbH, Karlsruhe, Germany) was used. To analyze the hardness distribution along the longitudinal direction after the erosion, the hardness measurements were evaluated by the Vickers hardness test (Shimadzu HMV-2000L, SHIMADZU, Kyoto, Japan).

The dimensions of the PBF Ti-6Al-4V tensile specimen are shown in Figure 3. The normal direction (ND) of the specimen was set parallel to the laser direction. The direction vertical to the laser direction was called the side direction (SD). A universal testing machine (HT-8336, Hung Ta, Taichung, Taiwan) was used to perform the tensile test. The crosshead speed was chosen 1 mm/min, corresponding to the initial strain rate of $18.33 \times 10^{-4} s^{-1}$. The parallel section of both the AS and the HT specimens eroded by 200 g of particles under a compressed air flow of 3 kg/cm^2 (0.29 MPa) at the 90° impact angle.

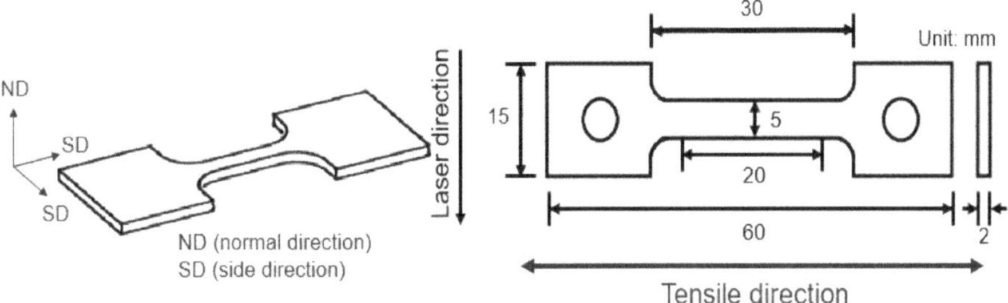

Figure 3. Dimensions of the tensile specimen.

To analyze the influence of the particle erosion induced phase transformation on both the AS and HT specimens, a tensile test was performed at room temperature. There were at least three specimens during each test, for which the tensile test results were averaged.

3. Results and Discussion

3.1. Erosion Induced Phase Transformation Behavior

Figure 4a shows the SD microstructure of the as-PBF Ti-6Al-4V (AS). There is a needle-like structure in AS matrix. The interpretation, according to the literature [11,12], is the extremely high temperature gradient during quenching, which transformed the β phases completely into a needle-like martensitic α′ phase. Figure 4b shows the SD microstructure of the HT specimen (which was subjected to heat treatment at 800 °C for 4 h). The α′ phases were completely transformed into continuous lamellar α + β phases, as shown in our previous research [22].

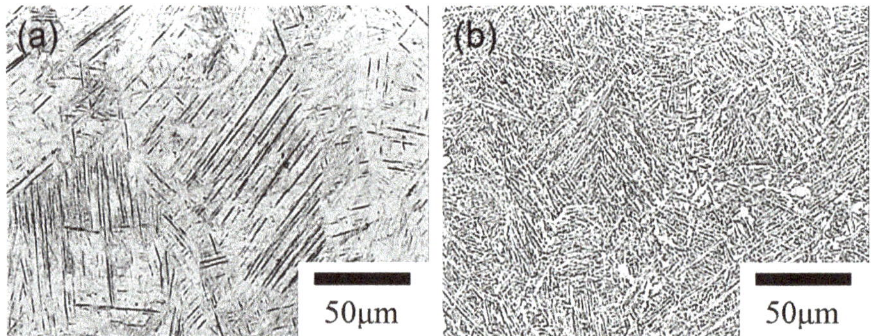

Figure 4. Microstructures of side direction (SD): (**a**) AS (**b**) HT.

The subsurface morphology of AS at 90° impact angle is shown in Figure 5a. We can observe positively eroded pits formed by erosion particles. There was no obvious relationship between the pits and the needle-like structure. Similar pits can be seen on the subsurface morphology of HT at 90° impact angle (Figure 5b), but the interface of the pits is almost located on the white α phase among the lamellar α + β phases. There is no obvious change in optical microscopy (OM) microstructures on both AS and HT specimens after erosion.

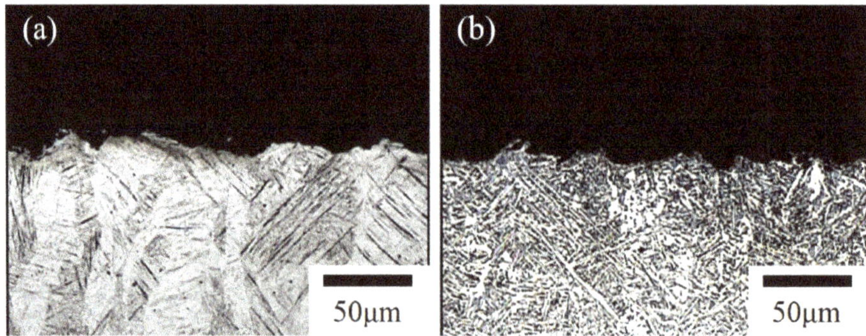

Figure 5. Subsurface morphology at 90° impact angle: (**a**) AS (**b**) HT.

However, the microstructural difference between AS and HT specimens before and after erosion can be obtained through TEM observation. After being positively eroded by endured 90° impact angle, these two specimens are labeled as ASE and HTE, respectively. The α′ phase of AS specimen, as shown in Figure 6a,b, shows the selected area electron diffraction (SAED) pattern of α′ phase. A + β phase comes up in HT specimen (Figure 6c), the thickness of α phase is approximately 1.5 μm. The SAED pattern of β phase is shown in Figure 6d. Figure 7 shows TEM images of AS and HT specimens after erosion. In short, the microstructures after erosion are completely changed. Figure 7b,d illustrate the well-defined ring pattern indicating fine grains within the region selected.

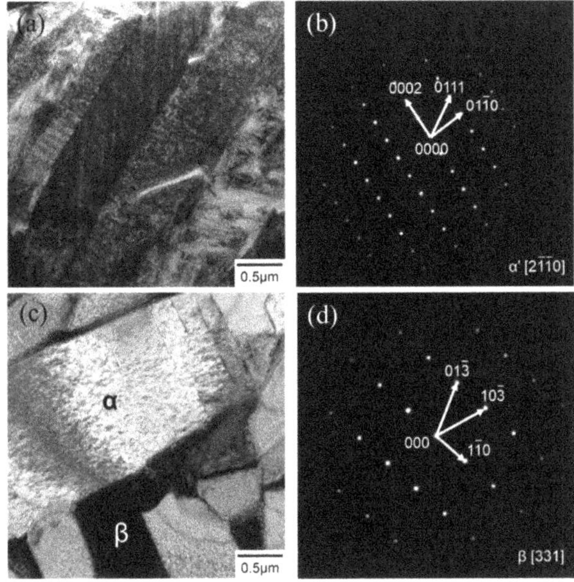

Figure 6. TEM images of specimens before erosion: (**a**) α′ phase microstructure of AS, (**b**) SAED pattern of α′ phase, (**c**) α + β phase microstructure of HT, and (**d**) selected area electron diffraction (SAED) pattern of α + β phase.

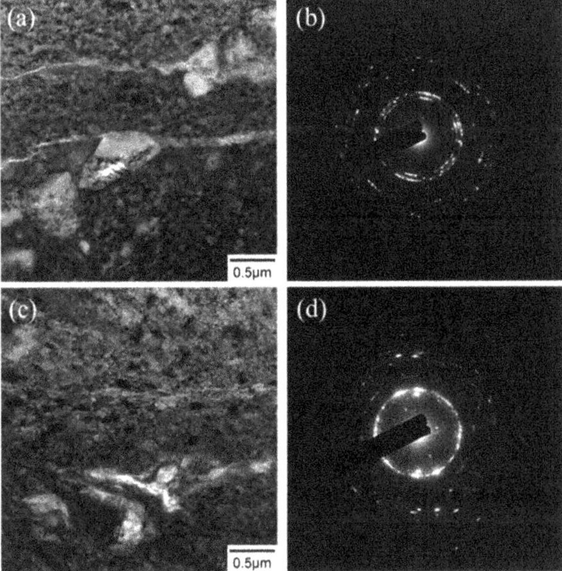

Figure 7. TEM images of specimens after erosion: (**a**) ASE, (**b**) SAED pattern of ASE, (**c**) HTE, and (**d**) SAED pattern of HTE.

Figure 8 displays the XRD diffraction analysis of the AS, and HT after the positively eroded endured 90° impact angle. Note the respective label changes to ASE and HTE. Both Figure 8 and the previous reports [13,22,25,26] suggest the following observation about the presence of the α/α'and the β peaks in HT: when an AS specimen is heated to 800 °C, the α' phases disappear, and then transforms to continuous lamellar α + β phases. Notably, ASE and HTE had two more peaks (210) and (202) than AS and HT. It was confirmed that the martensitic α' phase was generated when the temperature gradient grew extremely high during the cooling process. The α' phase has completely transformed into the α + β phases during heating to and maintaining at 800 °C. Thereby, it is suggested that both of ASE and HTE generated a new α phase after the Al_2O_3 particle erosion-induced phase transformation. The new generation phase ratios of ASE and HTE were 9.2% and 7.4%, respectively, which indicated that more α phase formation was induced in AS than in HT.

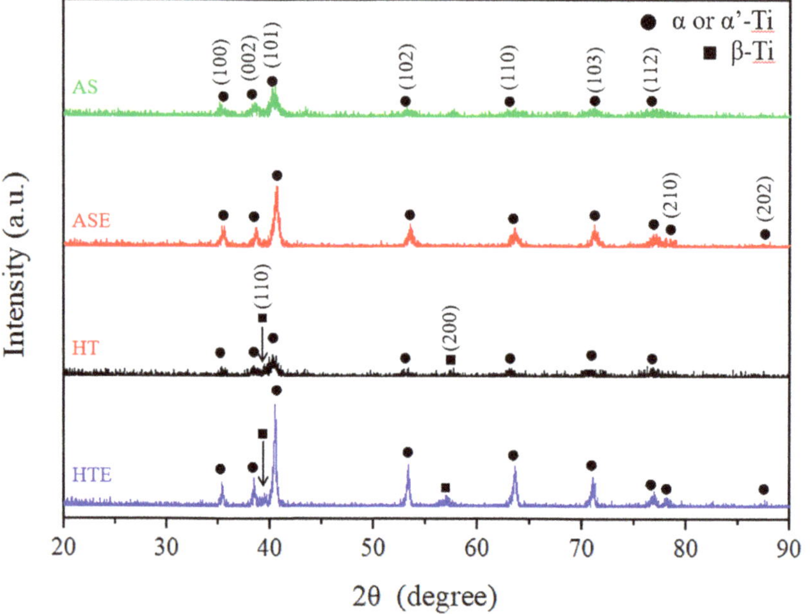

Figure 8. X-ray diffraction pattern of AS, ASE, HT, and HTE.

The hardness distribution along the longitudinal direction of ASE and HTE is shown in Figure 9. The full erosion subsurface can be divided into three zones: (1) the softened zone of the outermost surface, (2) the hardened zone below the softened zone, and (3) the hardness stabilization zone down to the bottom. The softened and hardened zone thicknesses for ASE were 150 μm and 300 μm, respectively. The Vickers hardness of the three zones were 420 to 430 HV, 500 to 540 HV, and 450 to 480 HV, respectively. Similarly, the softened and hardened zone thicknesses for HTE were 250 μm and 200 μm, respectively. The Vickers hardness of the three zones were 360 to 380 HV, 430 to 460 HV, and 410 to 420 HV, respectively. The decline in the hardness of the near-surface zone can be attributed to a compound effect: The AS material was transformed from a hard martensitic α' phase to a soft α phase, and the positive erosion produced a thermosoftened layer. According to previous reports [22,27], the hardness of the double phase Ti-6Al-4V alloy decreases with the rate of the α phase, which can be attributed to a new α phase formation and the thermal compound effect caused by particle impact [23]. The formation of the hardened zone can be attributed to positive erosion caused by the downward pressure of the specimen surface.

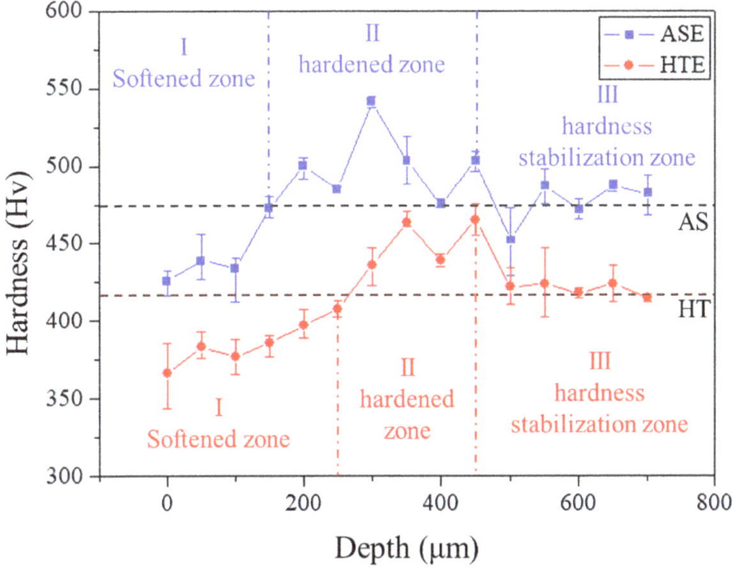

Figure 9. Hardness distribution of ASE and HTE in the longitudinal direction.

In summary, the positive erosion (by Al_2O_3 particles) induced a phase transformation for both AS and HT toward generating a soft α phase. At the same time, the softened zone with a thickness of 150 to 250 μm was formed by the thermal effect of the positively eroded particles, and the hardened zone below the softened zone was formed by a squeeze effect, as shown in Figure 10. Because the AS specimen had a higher rate of the α phase induced by eroded particles, the hardness difference between the softened and the hardened zones grew larger than that for the HT specimen.

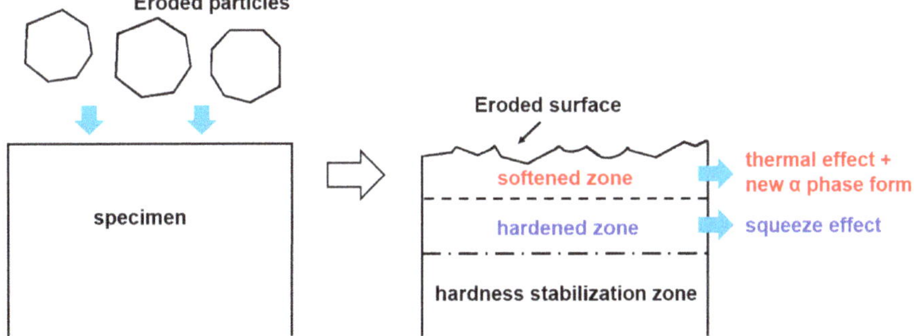

Figure 10. Schematic diagram of particle erosion induced phase transformation.

3.2. Influence of Erosion Induced Phase Transformation on Mechanical Properties

Figure 11 shows the stress–strain curve of AS and HT before and after positive erosion at a 90° impact angle. The tensile strength significantly reduced, and Young's modulus increased after the needle-like α' phase of PBF Ti-6Al-4V completely transformed into the $\alpha + \beta$ phase. The stress–strain curve moved toward the left after the parallel sections of AS and HT eroded by particles. Young's modulus changed from the original 111 ± 5 GPa (AS) and 133 ± 6 GPa (HT) to 130 ± 4 GPa (ASE) and 127 ± 5 GPa (HTE). It means that the stiffness of AS improved after the erosion-induced phase transformation, but the stiffness of HT did not change after the erosion.

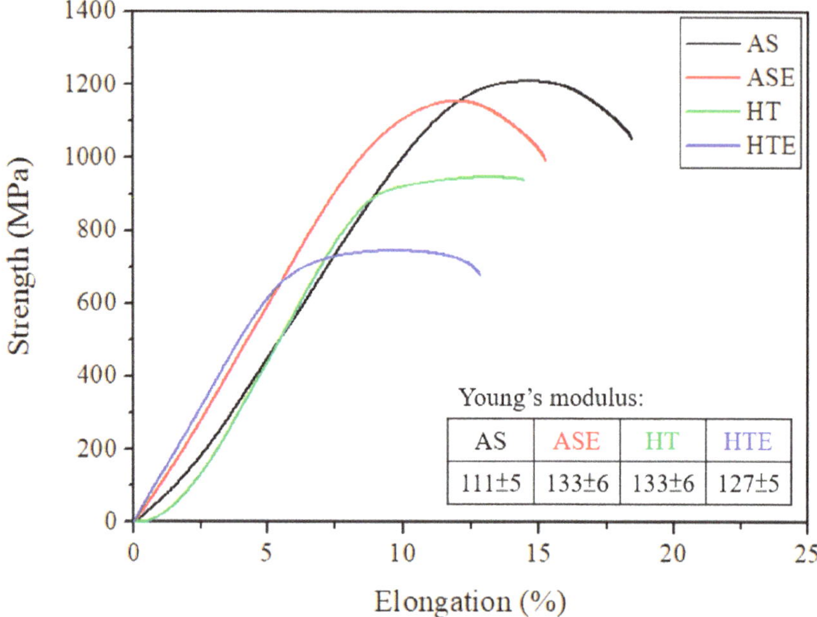

Figure 11. Stress–strain curve of AS, ASE, HT, and HTE.

The tensile mechanical proprieties of AS and HT before and after positive erosion at a 90° impact angle are shown in Figure 12. After heat treatment, the tensile strength of HT became lower than that of AS. This underlines that the phase transformation decreases the tensile strength. After the erosion-induced phase transformation, the tensile strength of ASE and HTE became lower than that of the original AS and HT specimens. This underlines that the generated of the α phase by eroded particles and the change of hardness lead to a decreased of tensile strength; Ductility had a different trend. The ductility of ASE was slightly lower than that of AS. However, the ductility of HTE became higher than that of HT.

To clarify the effect of erosion-induced phase transformation on tensile mechanical properties, the tensile fracture surfaces after erosion were observed. We found no obvious shrinkage phenomenon in ASE and HTE as tensile test specimens. The macroscopic fracture morphology of ASE is flatter, while HTE shows an oblique line of fracture, see Figure 13.

Figure 14a,b show the morphology of the tensile fracture surfaces of AS and HT, respectively. It can be observed that there are Dimpled ductility structures in both specimens, indicating that both the needle-like α′ phase and the lamellar α+β phases contribute ductile fractures. Similarly, the morphology of the softened zone tensile fracture surface of ASE includes an almost dimpled ductility structure (Figure 14c). The difference between ASE and AS and HT specimens is that many particles are located in the dimpled ductility structure. Energy-dispersive X-ray spectroscopy (EDS) indicated a titanium-aluminum phase, with the atomic percentage being 88:12, as listed in Table 3.

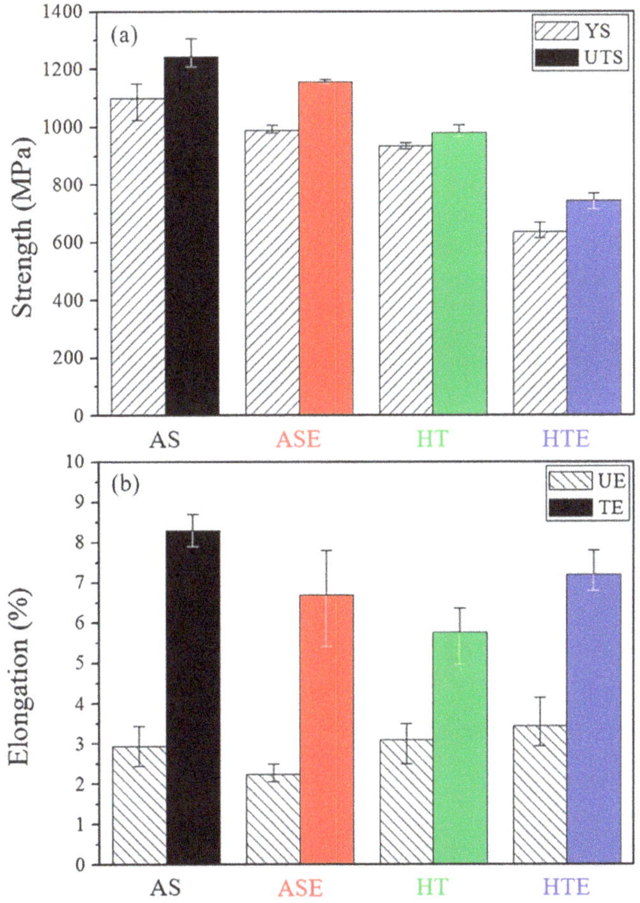

Figure 12. Tensile properties of AS, ASE, HT, and HTE: (**a**) strength (**b**) ductility.

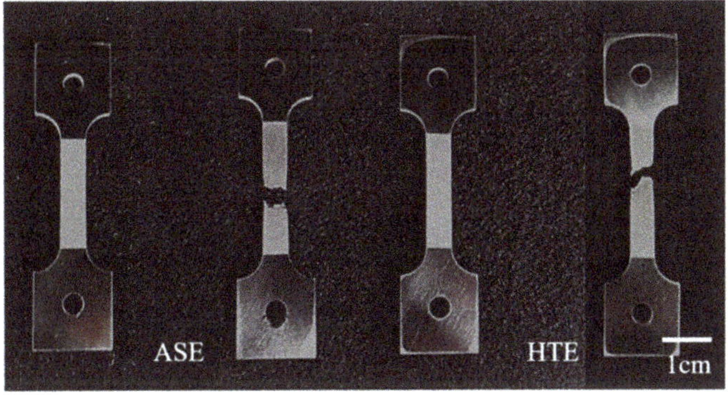

Figure 13. Macroscopic morphology photographs of ASE and HTE before and after tensile test.

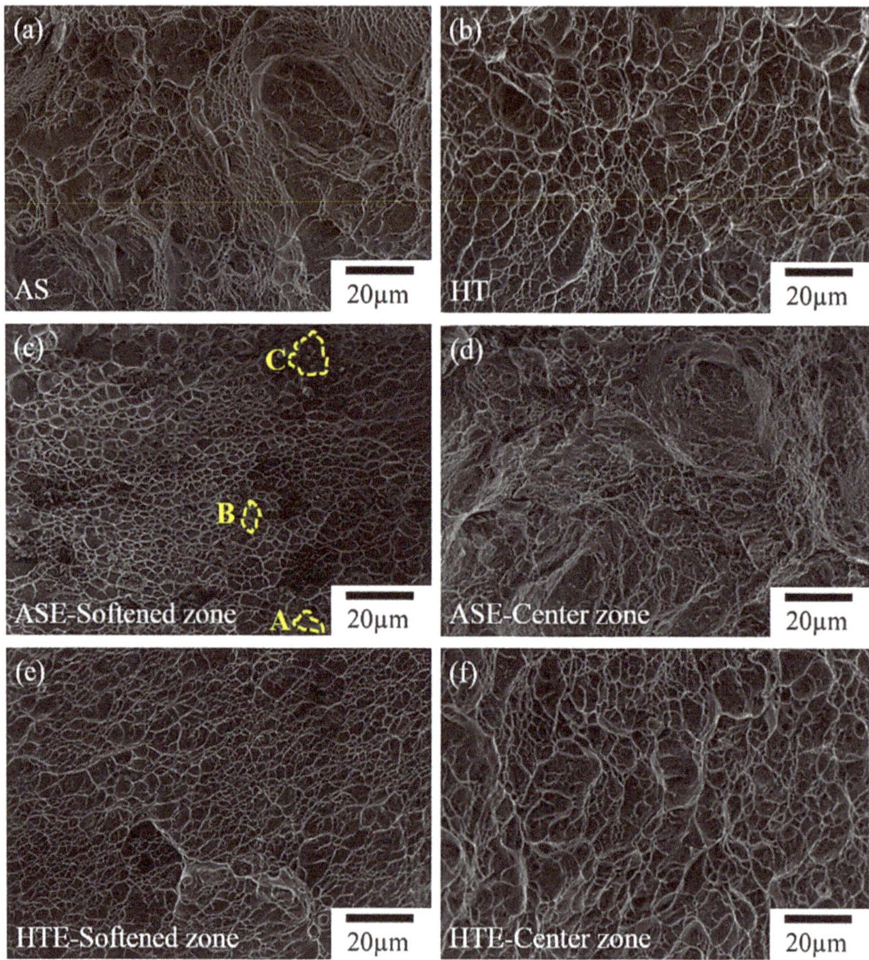

Figure 14. Morphology of tensile fracture surfaces: (**a**) AS (**b**) HT (**c**) ASE softened zone (**d**) ASE center zone (**e**) HTE softened zone and (**f**) HTE center zone.

Table 3. Phase of particles analyzed by EDS.

A		B		C	
Element	At%	Element	At%	Element	At%
Ti	12.38	Ti	11.26	Ti	11.35
Al	86.18	Al	87.76	Al	88.06
V	1.44	V	0.98	V	0.59

Figure 14d displays the morphology of the middle zone tensile fracture surface of ASE. The observed cleavage facets show a different fracture morphology from the softened zone. The tensile fracture surface of HTE is almost the dimpled structure, but there are no titanium–aluminum phase particles in the softened zone of HTE, as shown in Figure 14.

Figure 14f shows the morphology of the middle zone tensile fracture surface of THE. Even if cleavage facets could also be observed, the almost dimpled ductility structures dominated.

Figure 15 is the schematic diagram of the erosion particle impact tensile specimen. When erosion particles eroded, the erosion-induced phase transformation occurred, resulting in decreased strength on both the AS and HT specimens. It is inferred that as the number of erosion particles rises, the tensile strength gradually decreased. In ductility, there are different trends on the AS and HT specimens. After the erosion-induced phase transformation, AS showed titanium–aluminum phase particles in dimpled ductility structures in the softened zone, resulting in a slight decline of strength and ductility. In contrast, there were no titanium–aluminum phase particles in the softened zone of HT, resulting in decreased strength but increased ductility. There are three reasons why the ASE ductility was lower than that of AS, and the HTE ductility was higher than that of HT: 1. Particle erosion-induced phase transformation, 2. the different depth of the softened zone, and 3. the titanium–aluminum phase particles appearing in the softened zone. The α phase was generated by erosion induced phase transformation in the surface softened zone. The titanium–aluminum phase particles present only in the softened zone of ASE formed the α phase.

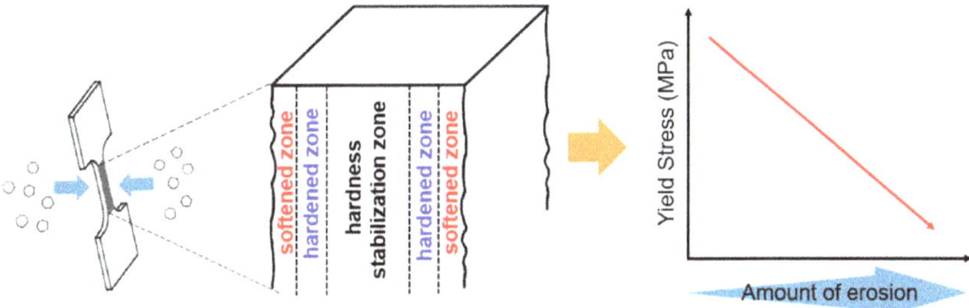

Figure 15. Schematic diagram of the erosion particle impact specimen leading to a decrease in strength.

4. Conclusions

A new α phase was generated by the as-PBF Ti-6Al-4V specimen and the heat-treated specimen after the erosion. The heat affected the composite effect of Al_2O_3 particle erosion, and the erosion surface adopted a serial formation: softened zone, hardened zone, and hardness stabilization zone. The ratio of the new α phase induced by particle erosion in ASE became higher than that in HTE, resulting in a difference of hardness between the softened and the hardened zones.

After the PBF Ti-6Al-4V was subjected to a heat treatment at 800 °C for 4 h, the phase transformation α' α + β resulted in a decrease of the strength. The mechanical properties of both the as-PBF Ti-6Al-4V and the heat-treated specimens were affected by particle erosion-induced phase transformation. The strength and the ductility of the ASE test specimens became lower than those of AS, and titanium–aluminum phase particles were observed in the softened zone. There were no titanium–aluminum phase particles found in HTE. Compared with the HT test specimen, the strength decreased, but the ductility increased.

Author Contributions: Methodology, J.-R.Z.; investigation, J.-R.Z.; data curation, J.-R.Z.; writing—original draft preparation, J.-R.Z.; writing—review and editing, F.-Y.H. and T.-S.L.; supervision, F.-Y.H. and T.-S.L.

Funding: This research received no external funding.

Acknowledgments: The authors are grateful to The Instrument Center of National Cheng Kung University and the Ministry of Science and Technology of Taiwan (Grant No. MOST 107-2221-E-006-012-MY2) for their financial support for this research.

Conflicts of Interest: The authors declare no conflict of interest.

References

1. Shi, Y.; Liu, Y.; Li, X.; Zhang, Y. Effect of Ultrasonic Surface Rolling Process on Solid Particles Erosion Performance of Ti-6Al-4V. *IOP Conf. Ser. Mater. Sci. Eng.* **2018**, *394*. [CrossRef]
2. Ahmadi, M.; Karpat, Y.; Acar, O.; Kalay, Y.E. Microstructure effects on process outputs in micro scale milling of heat treated Ti6Al4V titanium alloys. *J. Mater. Process. Technol.* **2018**, *252*, 333–347. [CrossRef]
3. Oh, S.T.; Woo, K.D.; Kim, J.H.; Kwak, S.M. The Effect of Al and V on Microstructure and Transformation of β Phase during Solution Treatments of Cast Ti-6Al-4V Alloy. *Korean J. Met. Mater.* **2017**, *55*, 150–155.
4. Tao, P.; Li, H.X.; Huang, B.Y.; Hu, Q.D.; Gong, S.L.; Xu, Q.Y. Tensile behavior of Ti-6Al-4V alloy fabricated by selective laser melting: effects of microstructures and as-built surface quality. *Chin. Foundry* **2018**, *15*, 243–252. [CrossRef]
5. Dutta, B.; Froes, F.H.S. The Additive Manufacturing (AM) of titanium alloys. *Met. Powder Rep.* **2017**, *72*, 1–11. [CrossRef]
6. Trevisan, F.; Calignano, F.; Aversa, A.; Marchese, G.; Lombardi, M.; Biamino, S.; Ugues, D.; Manfredi, D. Additive manufacturing of titanium alloys in the biomedical field: Processes, properties and applications. *J. Appl. Biomater. Funct. Mater.* **2017**, in press. [CrossRef]
7. Saboori, A.; Gallo, D.; Biamino, S.; Fino, P.; Lombardi, M. An Overview of Additive Manufacturing of Titanium Components by Directed Energy Deposition: Microstructure and Mechanical Properties. *Appl. Sci.* **2017**, *7*, 883. [CrossRef]
8. Saboori, A.; Tusacciu, S.; Busatto, M.; Lai, M.; Biamino, S.; Fino, P.; Lombardi, M. Production of Single Tracks of Ti-6Al-4V by Directed Energy Deposition to Determine the Layer Thickness for Multilayer Deposition. *J. Vis. Exp.* **2018**, *133*, 1–10. [CrossRef]
9. Yadroitsev, I.; Krakhmalev, P.; Yadroitsava, I. Selective laser melting of Ti6Al4V alloy for biomedical applications: Temperature monitoring and microstructural evolution. *J. Alloys Compd.* **2014**, *583*, 404–409. [CrossRef]
10. Song, B.; Dong, S.; Liao, H.; Coddet, C. Process parameter selection for selective laser melting of Ti6Al4V based on temperature distribution simulation and experimental sintering. *Int. J. Adv. Manuf. Technol.* **2012**, *61*, 967–974. [CrossRef]
11. Kelly, C.N.; Evans, N.T.; Irvin, C.W.; Chapman, S.C.; Gall, K.; Safranski, D.L. The effect of surface topography and porosity on the tensile fatigue of 3D printed Ti-6Al-4V fabricated by selective laser melting. *Mater. Sci. Eng. C* **2019**, *98*, 726–736. [CrossRef] [PubMed]
12. Sun, J.; Yang, Y.; Wang, D. Parametric optimization of selective laser melting for forming Ti6Al4V samples by Taguchi method. *Opt. Laser Technol.* **2013**, *49*, 118–124. [CrossRef]
13. Song, B.; Dong, S.; Zhang, B.; Liao, H.; Coddet, C. Effects of processing parameters on microstructure and mechanical property of selective laser melted Ti6Al4V. *Mater. Des.* **2012**, *35*, 120–125. [CrossRef]
14. Kim, Y.K.; Park, S.H.; Yu, J.H.; AlMangour, B.; Lee, K.A. Improvement in the high-temperature creep properties via heat treatment of Ti-6Al-4V alloy manufactured by selective laser melting. *Mater. Sci. Eng. A* **2018**, *715*, 33–40. [CrossRef]
15. Shunmugavel, M.; Polishetty, A.; Goldberg, M.; Singh, R.; Littlefair, G. A comparative study of mechanical properties and machinability of wrought and additive manufactured (selective laser melting) titanium alloy —Ti-6Al-4V. *Rapid Prototyp. J.* **2017**, *23*, 1051–1056. [CrossRef]
16. Gujba, A.K.; Hackel, L.; Kevorkov, D.; Medraj, M. Water droplet erosion behaviour of Ti-6Al-4V and mechanisms of material damage at the early and advanced stages. *Wear* **2016**, *358–359*, 109–122. [CrossRef]
17. Kamkar, N.; Bridier, F.; Jedrzejowski, P.; Bocher, P. Water droplet impact erosion damage initiation in forged Ti-6Al-4V. *Wear* **2015**, *322–323*, 192–202. [CrossRef]
18. Kamkar, N.; Bridier, F.; Bocher, P.; Jedrzejowski, P. Water droplet erosion mechanisms in rolled Ti-6Al-4V. *Wear* **2013**, *301*, 442–448. [CrossRef]
19. Mahdipoor, M.S.; Kevorkov, D.; Jedrzejowski, P.; Medraj, M. Water droplet erosion behavior of gas nitride Ti6Al4V. *Surf. Coat. Technol.* **2016**, *292*, 78–89. [CrossRef]
20. Pieters, R.; Liu, S. Shortlisted Particle Erosion Resistance of Laser Nitrided Ti-6Al-4V. *Surf. Eng.* **2001**, *17*, 159–162. [CrossRef]
21. Sahoo, R.; Jha, B.B.; Sahoo, T.K.; Mantry, S. Effect of Volume Fraction of Primary Alpha Phase on Solid Particle Erosion Behavior of Ti-6Al-4V Alloy. *Tribol. Trans.* **2015**, *58*, 1105–1118. [CrossRef]

22. Zhao, J.R.; Hung, F.Y.; Lui, T.S.; Wu, Y.L. The Relationship of Fracture Mechanism between High Temperature Tensile Mechanical Properties and Particle Erosion Resistance of Selective Laser Melting Ti-6Al-4V Alloy. *Metals* **2019**, *9*, 501. [CrossRef]
23. Liou, J.W.; Lui, T.S.; Chen, L.H. SiO$_2$ particle erosion of A356.2 aluminum alloy and the related microstructural changes. *Wear* **1997**, *211*, 169–176. [CrossRef]
24. Hung, F.Y.; Chen, L.H.; Lui, T.S. Phase Transformation of an Austempered Ductile Iron during an Erosion Process. *Mater. Trans.* **2004**, *45*, 2981–2986. [CrossRef]
25. Thijs, L.; Verhaeghe, F.; Craeghs, T.; Humbeeck, J.V. A study of the microstructural evolution during selective laser melting of Ti-6Al-4V. *Acta Mater.* **2010**, *58*, 3303–3312. [CrossRef]
26. Do, D.K.; Li, P. The effect of laser energy input on the microstructure, physical and mechanical properties of Ti-6Al-4V alloys by selective laser melting. *Virtual Phys. Prototyp.* **2016**, *11*, 41–47. [CrossRef]
27. Sahoo, R.; Jha, B.B.; Sahoo, T.K.; Sahoo, D. Effect of Microstructural Variation on Dry Sliding Wear Behavior of Ti-6Al-4V Alloy. *J. Mater. Eng. Perform.* **2014**, *23*, 2092–2102. [CrossRef]

Article

Design and Parameter Identification of Wire and Arc Additively Manufactured (WAAM) Steel Bars for Use in Construction

Johanna Müller [1],*, Marcel Grabowski [2], Christoph Müller [3], Jonas Hensel [1],*, Julian Unglaub [2],*, Klaus Thiele [2], Harald Kloft [3] and Klaus Dilger [1]

[1]   Institut für Füge- und Schweißtechnik, TU Braunschweig, Langer Kamp 8, 38106 Braunschweig, Germany
[2]   Institut für Stahlbau, TU Braunschweig, Beethovenstraße 51, 38106 Braunschweig, Germany
[3]   Institut für Tragwerksentwurf, TU Braunschweig, Pockelstraße 4, 38106 Braunschweig, Germany
*   Correspondence: johanna.mueller@tu-braunschweig.de (J.M.); j.hensel@tu-braunschweig.de (J.H.); j.unglaub@stahlbau.tu-braunschweig.de (J.U.); Tel.: +49-531-391-95517 (J.H.)

Received: 6 June 2019; Accepted: 25 June 2019; Published: 27 June 2019

Abstract: Additive manufacturing (AM) in industrial applications benefits from increasing interest due to its automation potential and its flexibility in manufacturing complex structures. The construction and architecture sector sees the potential of AM especially in the free form design of steel components, such as force flow optimized nodes or bionic-inspired spaceframes. Robot-guided wire and arc additive manufacturing (WAAM) is capable of combining a high degree of automation and geometric freedom with high process efficiency. The build-up strategy (layer by layer) and the corresponding heat input influence the mechanical properties of the WAAM products. This study investigates the WAAM process by welding a bar regarding the build-up geometry, surface topography, and material properties. For tensile testing, an advanced testing procedure is applied to determine the strain fields and mechanical properties of the bars on the component and material scale.

Keywords: additive manufacturing; construction; WAAM; welding; steel; ESPI; design

1. Introduction

Advances in automation technology over the last century are the basis for additive manufacturing [1]. The idea of additive manufacturing was originally associated with rapid prototyping in order to test new design strategies in the automotive, aviation, and aerospace industry with a minimum of manufacturing effort. Today, additive manufacturing itself has become a technology that enables the fabrication of geometrically complex construction elements with efficient material usage. Additive manufacturing as a production process is also becoming increasingly interesting for tailored solutions in the construction industry. The beginnings of additive manufacturing in construction started around 10 years ago, mainly focused on extrusion techniques and concrete materials. Pioneering projects were realized by counter crafting [2], d-shape by Enrico Dini, or 3D concrete printing by Buswell et al. [3]. Since then, developments in this field have taken place at ever shorter intervals and the number of people involved in AM in the construction industry is constantly increasing [4,5].

Additive manufacturing of building components in the building industry is not only limited to processes for cement-bound components. The idea of creating metal components by layers of welded seams was already being considered in the 1930s [6]. However, additive manufacturing from metals in construction has only recently become the focus of attention and is the subject of research in the construction industry [7,8]. Arup [9] and Joosten with MX3D [10] demonstrated how steel components can be manufactured by additive manufacturing. Camacho et al., however, showed that the main field of application of metallic AM components currently lies in aircraft construction, automotive engineering, and health technology [11].

The motivation for additive manufacturing in the construction industry results among other things from the low labor efficiency in comparison to automated machines as well as from aspects, such as occupational safety, accident rate, and low construction quality due to inadequate training of employees [12,13]. In addition to these points, increasing material efficiency, minimizing waste, reducing building mass, and, consequently, reducing energy consumption in the construction of the component or building are reasons for additive manufacturing in construction.

The production of components is usually carried out in subsequent steps, which is why the construction industry and the production process can be considered as being very fragmented. For this reason, additive manufacturing should not be regarded as a replacement for previous methods of construction, but as an alternative manufacturing process for components that can only be manufactured by traditional methods at great expense. The manufactured geometries can increase in their complexity without increasing costs and effort in the same measure. The aim is to reduce the sum of all individual parts of a building in order to simplify the construction of the building. By digitalizing the whole design process, the number of necessary steps can be reduced compared to traditional design processes. Design, optimization, and preparation for the additive manufacturing of components is done digitally. Work steps, such as transformation of the three-dimensional digital model into two dimensions for worksite planning, are not necessary. From the planning to the finished component, all information is available digitally and allows the creation of a digital twin, which can be transferred into a building information modelling system (BIM).

For architecture, additive manufacturing changes the limitations of fabrication. It opens up the possibility of adapting the design down to the smallest detail without being subject to the restrictions of today's common manufacturing methods. Particularly in the field of steel construction, the design based on manufacturing from semi-finished products, such as plates and prefabricated profiles. Therefore, there is a high potential of geometric freedom to be achieved by AM in this area.

Processes for the additive manufacturing of steel structures can be subdivided into powder-based (selective laser melting (SLM), laser metal deposition (LMD), electron beam melting (EBM)) or wire-based (WAAM, electron beam freeform fabrication (EBF^3), laser-engineered net-shaping (LENS)) processes [14,15]. In WAAM processes, the wire-shaped additive materials are either supplied to the energy source (gas tungsten arc welding (GTAW), plasma and laser) and are molten, or the wire itself is the electrode, like in gas metal arc welding (GMAW) processes [16].

The advantages of wire-based technologies in comparison to powder-based processes are the higher deposition rate in combination with a better material utilization and lower material cost, making it suitable for applications in the construction sector. Powder-based technologies are more suitable for complex geometries with high requirements on the geometric precision [17]. Further, the component size in the SLM process is limited regarding the component size, since the dimensions of the component cannot exceed the powder bed size. The WAAM process utilizes CNC or robot kinematics and the AM element is, hence, theoretically not limited in size.

WAAM as a field of research gains more and more attention. Numerous investigations focusing on the application of WAAM in lightweight constructions using titanium or aluminum alloys as filler material [18–20]. The driving forces are the automobile and aerospace industry. Other works focusing on steel as construction material, where the fields of application are pipe couplings and flanges, special machines, and plant construction, mainly driven by the naval and crane industry [17]. In the construction sector, studies on the WAAM with steel electrodes of nodal joints and beam reinforcements are available. Their focus is mainly on the constructive design potential or on the generation process itself [7,21]. Other works identified the processing capabilities of WAAM in general by investigating several materials, power sources, and welding processes, such as GMAW, GTAW, and plasma welding. Studies provide guidelines for the additive process selection, aiming at improved material properties. Advanced process modifications, such as workpiece and wire oscillation, modulated power supplies, and use of shielding gas or different cooling strategies, were applied [22–24]. Furthermore, there are publications which investigate the thermal cycles and the reheating of prior layers during the WAAM of wall-like structures using high-strength low-alloy (HSLA) steel [25] and duplex stainless steel [26]. According to Rodrigues et al., during reheating, recrystallization may occur, leading to fine-grained microstructure or coarse grains in the case of high peak temperatures. However, the microstructure and thus the mechanical properties are affected by the thermal cycles and the amount of layers being reheated during the welding of subsequent layers.

WAAM makes it possible to increase the geometric complexity of a component without increased manufacturing effort. At present, there is no unified methodology for the design of component geometry [27]. The most common approaches result either from the manufacturing process or from the optimum use of the material. The first approach is a bottom up process and is based on the path planning of the component to be manufactured [28]. The second approach is a top down process and is usually based on topology optimization. Questions about the aesthetic qualities of the component have not been discussed yet [29]. However, it is of essential importance for architecture and the building industry.

Each working step, from the creation of the basic geometry, through topology optimization and adaptation for the manufacturing process, to the verification of the load-bearing capacity and stability, creates its own geometry. Restrictions should already be defined during the design of the initial geometry in order to meet the architectural requirements of the overall structure. Here, traditional design concepts should be abandoned and possibilities of additive production should be decisive. Good design only appears possible if there is a precise knowledge of the effects of each individual design step. One of these design steps must consider and include the process parameters and boundary conditions for the manufacturing of a component by WAAM. Structures to be produced by WAAM can be subdivided into various substructures, such as bars, surface elements, and volume elements.

In this article, the focus is on steel bars not only as a substructure, but also as an independent structural component. The possibility of production and the appearance of surface topography as well as homogenous material properties are of special interest. Problems, such as sufficient space for the welding torch, inclination of bars, number of crossing points, and resulting build-up, and path planning strategies must be taken into account at the design stage.

In addition to using bars as a substructure in a structural component, WAAM manufactured bars can also be used as reinforcement in reinforced concrete. Figure 1 shows reinforcement arrangements for additively manufactured concrete components. The alignment of the reinforcement is still strongly oriented to the geometry of traditional prefabricated reinforcement elements. WAAM allows a more efficient arrangement of the reinforcement and extends the use of additive manufacturing in concrete construction. Due to the higher degree of freedom of WAAM manufactured bars, reinforcement can be arranged according to the tensile stress curve.

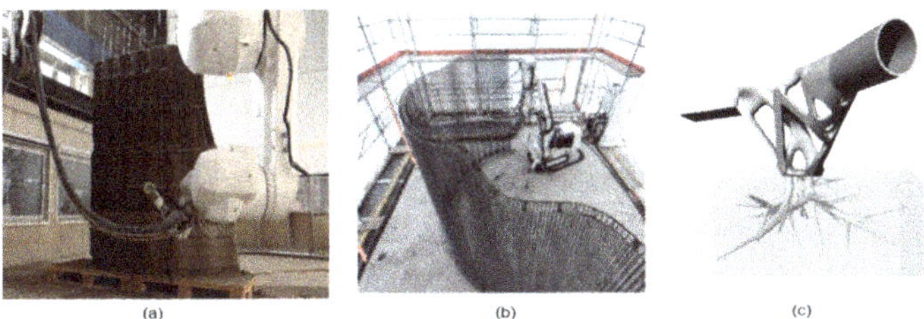

Figure 1. (**a**) Construction of a reinforced concrete wall in the digital building fabrication laboratory (DBFL) at the Institute for Structural Design (ITE). (**b**) Curved mesh of conventional reinforcement bars constructed by an in situ fabricator [30], reproduced with permission from Norman Hack, Ph.D. Thesis: Mesh Mould: A Robotically Fabricated Structural Stay-in-Place Formwork System; published by ETH Zurich, 2018. (**c**) Root-like structure for anchoring steel components in concrete.

An extended application for WAAM-manufactured reinforcement is the anchoring of steel components to reinforced concrete. Figure 1c schematically shows the possible root-like anchoring of a steel construction node to a concrete foundation. The use of bars as reinforcement requires different surface properties than the nodal joints produced by WAAM. The curvature of the surface is decisive for the bond between the concrete and reinforcement. Control of the surface topology would be advantageous, especially with regard to root-like anchoring, as shown in Figure 1c.

In order to use components or production types in the European Union, a European Technical Assessment (ETA) is required. The ETA assesses the performance of a construction product with respect to its essential characteristics. ETA provides a way to CE-certification (European Conformity certification) for construction products that are not or are not fully covered by a harmonized standard. This certification process can only be applied to WAAM components with limitations. The evaluation of process and material irregularities and flaws needs to be addressed in future certification concepts for WAAM in construction. Therefore, existing certification concepts need to be extended. A possible solution is the concept of a digital twin, which holds a set of any virtual information that describes the actual physical manufactured product on different scales [31]. This model is already being used in aircraft production for general components as well as for additive manufactured parts. Currently, this concept will be extended by blockchain technology to ensure the traceability of the process chain steps [32]. It is necessary to adapt this concept to the particularities of the construction industry. The main challenge for a successful AM technology adoption in the construction sector is the certification of AM components for safe use in the buildings of infrastructure. For this purpose, basic knowledge on the relationship between the manufacturing process and material properties of WAAM components first must be determined.

The determination of the material properties, more specifically, the mechanical properties of additively manufactured bars, differs from conventional material testing due to features, such as topography (notches). Further, different manufacturing irregularities, in terms of WAAM weld irregularities and flaws, need to be addressed by the tests and evaluated. For mechanical testing of AM-parts, like walls or cylinders, the well-established way is to section samples in the form of a standard tension test specimen from the AM-structure, which are prepared and tested according to any ISO or ASTM standard [33,34].

Root-like anchoring elements, cast in concrete or reinforcement elements, where an uneven topography or surface irregularities are desired, may show losses in mechanical properties due to notches. Those surface irregularities can be sources for premature failure. So, for applications with cyclic loading, an even surface is more favorable and the bars should be post-processed, i.e., machined. However, those reinforcement elements are not meant to be post-processed, so the influence of the topography on the elastic behavior is essential information when investigating the mechanical properties. So, cutting out standard conform specimen geometries is not an option because the topography information would be lost.

The new approach for the mechanical testing of WAAM-bars, which is introduced here, is to take a standard round tensile test specimen [35] and replace the cylindrical middle part with an additively manufactured bar.

The aim of this work is to investigate the influence of welding process parameters on the surface topography of steel bars and the local strain distribution as a consequence of mechanical loading. Furthermore, the energy inputs and cooling rates of different welding processes are related with the resulting hardness, microstructure, and the mechanical properties.

2. Materials and Methods

The manufacturing process of the steel bars was carried out on a 10 mm thick S355N (1.0545) substrate plate. The welding material itself was a solid wire electrode of unalloyed steel G4Si1 (1.5130) with a diameter of 1.0 mm. The chemical composition of the materials is given in Table 1.

Table 1. Chemical composition of the used materials.

Material	Chemical Composition (wt.-%)								
	C	Si	Mn	P	S	Cr	Mo	Ni	Cu
G4Si1	0.061	0.845	1.66	0.025	0.0139	0.035	0.006	0.044	0.088
S355N [1]	0.2	0.5	0.9–1.65	0.03	0.025	0.3	0.1	0.5	0.55

[1] Nominal composition.

The experimental welding setup was composed of a Fronius Cold Metal Transfer (CMT) Advanced 4000 R power source (Fronius International, Pettenbach, Austria), a handling robot, temperature gauges, and process periphery, as shown in Figure 2. The torch as well as the pyrometer were mounted on the KUKA KR22 robotic handling system (KUKA AG, Augsburg, Germany). The pyrometer was an optris CTlaser 3MH2 with a measuring range from 200 to 1500 °C. The substrate plate was attached to the positioning table.

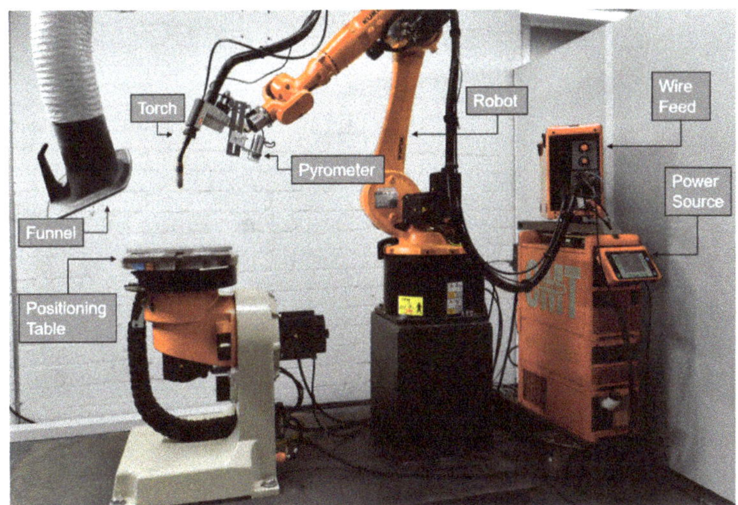

Figure 2. Experimental setup for the WAAM of steel bars [34].

One special feature of this power source is the reversing wire electrode controlling short circuit phases and droplet transfer to the weld pool. That leads to a minimal-spatter ignition due to the almost current-free material transition. The current can be kept very low for a longer time, thus significantly reducing the heat input to the piece [36,37]. This is what makes such modified short arc processes so suitable for WAAM applications.

In this investigation, two different CMT process variants were used. The first was a CMT standard process and the second was the CMT cycle step, which reaches even higher short-circuit percentages than the CMT standard by controlled switch-offs of the arc, leading to a lower heat input [37]. The cycle step also convinces with a precise control, which allows the choice of the exact number of droplet transitions per cycle instead of the welding time, like in CMT standard processes [38]. In addition to the CMT processes, one steel bar was manufactured with an elmatech DV36 L(W) GMAW (denoted in the following as conventional GMAW) power source for comparison. The welding parameters for each process are depicted in Table 2. As shielding gas M21 (82% argon and 18% CO_2) with a constant flow of 12 L/min was used.

Table 2. Welding parameters.

Parameter	Unit	Conventional GMAW	CMT Standard	CMT Cycle Step
Wire Feed	m/min	10.6	5	9.4
Welding time	s	1.5	2.4	1
Material transitions	Droplets/cycle	-	-	100
Current	A	218	158	204
Voltage	V	27.6	11.1	16.4
Energy per layer	kJ	9	4.21	3.35

As shown in Figure 3a, the bars were welded point by point, while the torch was moved up in the build-up direction. According to the deployed process, one point was welded either for a certain time (conventional GMAW, CMT standard) or for a certain number of droplets (CMT cycle step). Subsequently, the torch was automatically moved up for 2, respectively, 1.6 mm (depending on the resulting layer height) and a waiting period followed, until a constant interpass temperature of 200 °C was reached. Then a new cycle began. The $t_{8/5}$-times in the middle layers averaged for both CMT processes was 16 s.

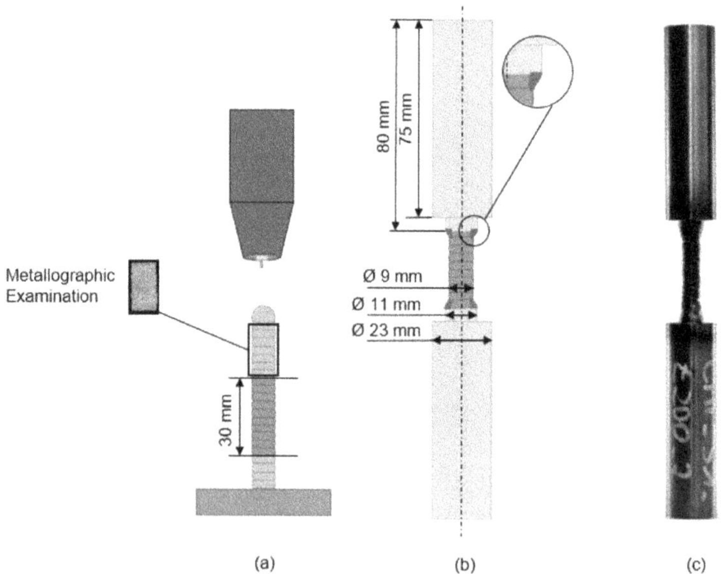

Figure 3. Schematic of the specimen preparation: (**a**) Welding the bar, (**b**) specimen preparation for the tensile tests, and (**c**) ready-to-test specimen.

For characterization of the bars regarding their mechanical properties, the bars had to be in a proper shape for execution of the tensile tests. Especially concerning the clamping cylinder in the tensile test machine, the specimens had to be an even shape to not initiate bending moments into the bar itself or into the machine.

The most obvious specimen shape is a round tensile test specimen in dependence on DIN 50125. Therefore, a piece with a length of 30 mm was cut out of the bar and then welded between two cylindrical ends of standard round tensile test specimens, which were turned to a diameter of 11 mm at one side for better accessibility during the welding of the fillet, as shown in Figure 3b. The joint was welded with a TIG-process using 110 A and no filler material. For all samples, computer tomography scans with a "GE v|tome|×240s Research with micro focus tube" were made. Subsequently, the tensile tests followed. Additionally, for all three samples, the surface was characterized using a laser scanner (optoNCDT 1800, micro-optronic, Langenbrück, Germany), micrographs were prepared, and hardness tests were executed.

The experimental setup for the tensile testing is shown in Figure 4. The specimens were mounted on a tensile testing machine (MTS 810 Servo Hydraulic Testing System, MTS Systems GmbH, Berlin, Germany) and two optical measurement systems were used for the tensile tests. In order to take sufficient account of surface influences and the anisotropic inhomogeneous material behavior, full field strain measurements were taken with the Electronic Speckle Pattern Interferometry (ESPI; Q-300, Dantec Dynamics GmbH, Ulm, Germany). Because the ESPI requires a high positional accuracy relative to the specimen, it is mounted on a 3D-positioning system, which allows the positioning with an accuracy of 1 μm. Additionally, a laser extensometer (P-50, Fiedler Optoelektronik GmbH, Lützen, Germany) was used to verify the measured strain maps during the test. This second measurement was done over the whole length of the welded part of the bars.

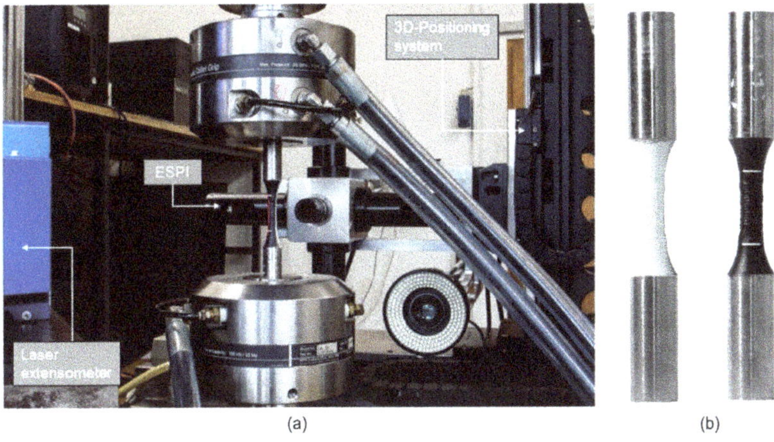

(a)

(b)

Figure 4. (**a**) Experimental setup; (**b**) specimen sprayed with lime powder (left) and black lacquer (right).

ESPI employs a coherent laser that illuminates the specimen from different positions and monitors the change in the intensity of the produced interference due to the displacements with a CCD camera. The setup used here can measure displacements of the surface in all three spatial directions. The measured displacement fields are then numerically differentiated in order to obtain strain maps. One advantage of the ESPI for the measurements on the material scale is its high displacement measurement resolution (up to 10^{-6} m) [39]. Due to this high resolution, ESPI is often used for measurements of displacement and strain due to thermal loading on small parts in the sectors of aerospace technology or electrical engineering. ESPI measurements on aluminide layers on a MAR 247 nickel alloy were also shown to be able to identify localized areas of deformation concentration as a starting point of damage under cyclic loading [40].

The use of standard tensile test specimens for the determination of mechanical parameters was not directly feasible here. Conducting tensile tests until failure with the topography "as-welded" would not allow the calculation of definitive material parameters, except the component's bearing load, as the cross section is not well-defined. Turning the whole specimen leads to a loss in information as the surface would be removed to create a defined measuring section. Therefore, an advanced testing strategy was applied here, which was divided in two stages of tensile tests with an additional metallographic examination parallel to the tensile tests, as shown in Figure 5.

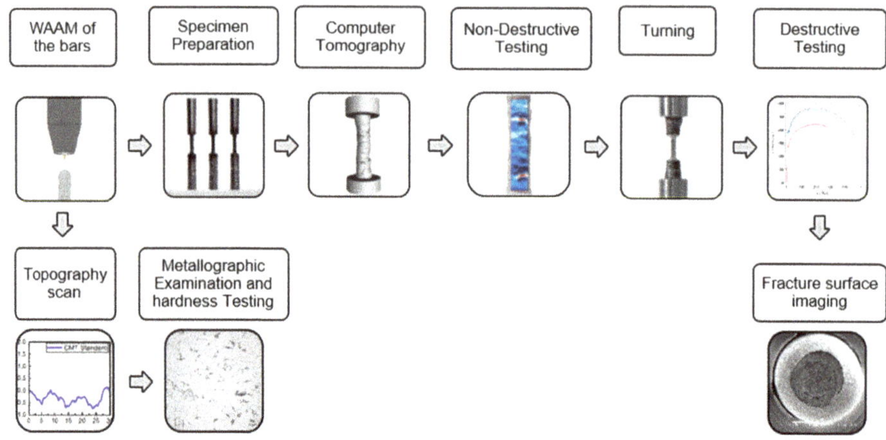

Figure 5. Advanced testing strategy for the identification of strain fields and material parameters.

The non-destructive first stage of testing was applied to the sample with the topography "as-welded" as shown in Figure 5. The aim of the first tests was to analyze the influence of the weld process-dependent surface topography and the inhomogeneous material behavior in the elastic state. The load had to be limited in order to avoid plastic deformations and was identified in preliminary tests to a maximum force of 5 kN. Due to the high sensitivity of the ESPI, the measurements had to be conducted after load step increments of 0.5 kN with holding times of 5 s each. Afterwards, the measured deformations and strains of each load step were summed up. Simultaneously to the ESPI measurement, the laser extensometer measured the strains over the whole length of the welded specimen (30 mm) to verify the strains measured with the ESPI.

Both optical measurements required a different surface preparation. For the ESPI measurement, an optically rough and reflective surface is required. This was achieved by spraying on a layer of lime powder. On the other side of the specimen, a primer with a matt black lacquer was applied to serve as a non-reflective background that could be distinguished from the white measuring marks, which is displayed in Figure 4b.

In order to obtain the material parameters, a second stage of destructive tensile tests was carried out. For these tests, a well-defined transversal section was needed to calculate the resulting stress and to derive the stress–strain diagram with all needed design parameters. For this reason, the middle part of the bars was turned to an even transversal section with a length of 10 mm. To ensure that the failure of the specimen occurred in the desired section and not in the fillet weld, a diameter of 4 mm was chosen for the transversal section. The strains during the test were measured with the laser extensometer. For this, the measurement markers were placed on the outer edge of the transversal section at a distance of ca. 10 mm.

After the destructive testing, the fracture surface was examined using a scanning electron microscope (JEOL JSM 6480, Japan Electron Optics Laboratory, Akishima, Japan) to obtain information about the fracture mechanism.

3. Results

3.1. Comparison of the Build-Up Height and the Resulting Diameters

The resulting bar diameters, including minimum and maximum values, as well as the mean build-up heights of the single layers were examined and are displayed in Figure 6. Furthermore, the energy input per layer was calculated by Equation (1), since there is no welding velocity for calculating the regular energy per unit length:

$$E = U \times I \times t_{weld} \tag{1}$$

For measurement of the welding time per layer (t_{weld}), the current (I), and the voltage (U), a measurement device (HKS weld monitoring system) was connected to the power source. The according energy per layer is also shown in Figure 6.

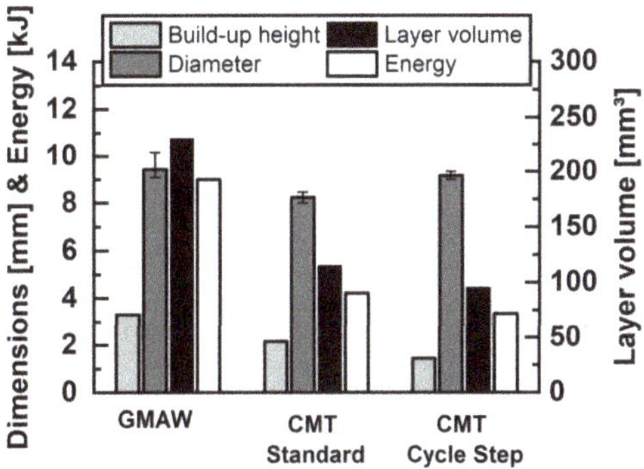

Figure 6. Layer build-up height, diameter, layer volume, and the corresponding energy input of different welding processes.

The conventional GMAW obtained the highest build-up rate with more than a 3.3 mm layer height as well as the thickest bar diameter with almost 10 mm. The CMT cycle step reached 1.4 mm as the lowest layer height and a bar diameter of 9.16 mm. The CMT standard process achieved layer heights of 2.15 mm and diameters of 8.2 mm. Since the build-up height correlates with the energy per layer while the diameter does not, the layer volume was also calculated and is depicted in Figure 6. The layer volume increased with higher energy input per layer.

3.2. Surface Topography

Next to the build-up volumes and diameters of the bars, the topographical properties of the WAAM bars were specified. It can be seen from the results in Figure 7 that the conventional GMAW led to a non-uniform surface with a minimum to maximum range of 1.4 mm, determined along a measuring section of 30 mm. The surface topography of the CMT standard specimen was smoother and the layers were built more regularly. The waviness was in the order of 0.89 mm. In comparison, the CMT cycle step process led to the most uniform and even surface. The range between the minimum and maximum was 0.35 mm.

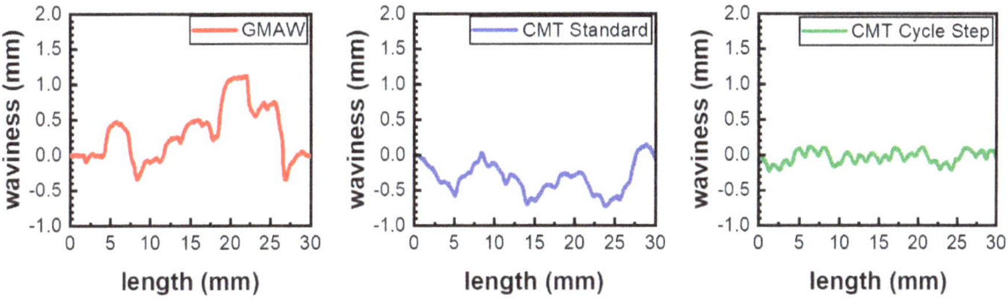

Figure 7. Surface laser scans along the build-up direction.

3.3. Microstructure and Hardness

Subsequently, deposited layers reheat weld metal during the manufacturing process. The reheating causes a change of the microstructure. Especially, the refined geometry of the bars manufactured here lead to heat accumulation and comparably long $t_{8/5}$-times in combination with high peak temperatures. Micrographs showing the weld layer geometry and the resulting hardness are given in Figure 8.

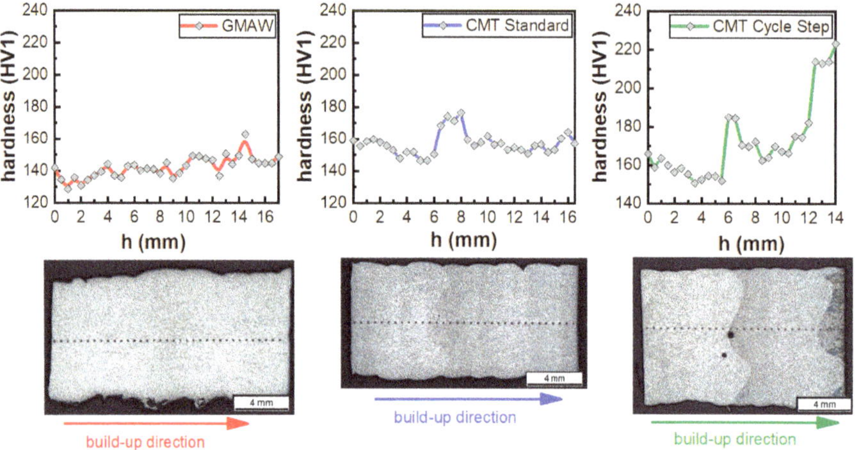

Figure 8. Hardness distribution in the build-up direction.

The averaged hardness increased with decreasing energy input. In the conventional GMAW process (9 kJ per layer), the hardness ranged from 128 to 163 HV1; in the standard CMT process (4.2 kJ per layer), the hardness varied between 146 and 176 HV1; and in the process with the lowest energy input per weld point (3.35 kJ), the hardness increased from 150 HV1 in the lower layers to 223 HV1 in the higher layers. In the pictures on the bottom of Figure 8, one can see that due to the high heat input in the conventional GMAW weld, the layers melted together and an even homogenous secondary microstructure developed. Hence, the hardness did not vary as much as in the CMT cycle step specimen. Looking at the two CMT processes, one characteristic is the hardness increase at the transition zone between two layers. Right when the next layer started, the hardness rose significantly and decreased subsequently.

Figure 9a shows the microstructure of the conventional GMAW specimen consisting of mostly ferrite and small fractions of bainite and perlite. Here, the grains were comparatively large. The grain growth is most likely a result of the peak temperatures well above the austenitization temperature, A_3, and the comparably long holding times.

In general, the microstructure of the CMT processes in the middle and on the top of each layer is a fine-grained bainite-ferrite structure. In the micrographs in Figure 9b,c, the grains appear smaller in the area of high hardness and a higher amount of bainite can be detected. Figure 9c shows at the right side the top layer of the bar, which was welded last. The resulting primary microstructure not affected by subsequent weld layers is composed of ferrite and bainite fractions in an acicular structure. However, the secondary microstructure resulting from both CMT processes did not vary significantly while the high energy input of conventional GMA welding led to a significantly lower, but also more homogeneous, hardness. This is expected to result in a variation of the tensile properties compared with Section 3.6.

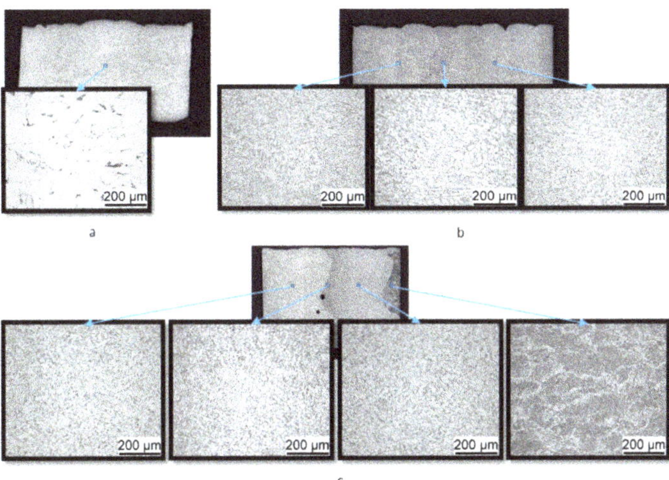

Figure 9. Micrographs of the specimen manufactured with (**a**) conventional GMAW, (**b**) CMT standard, and (**c**) CMT cycle step.

General effects of cooling rates on microstructure and hardness are given in welding time-temperature-transformation (TTT) diagrams. Figure 10 shows the transformation behavior of a G4Si1 at rapid cooling. However, the chemical composition of this welding wire varies from the one used here, but general effects can be explained. Fast cooling rates lead to more bainite and less ferrite, resulting in higher hardness. Low cooling rates result in low hardness and mainly a ferrite/pearlite microstructure. Accordingly, the hardness of deposited G4Si1 can vary depending largely on the welding parameters.

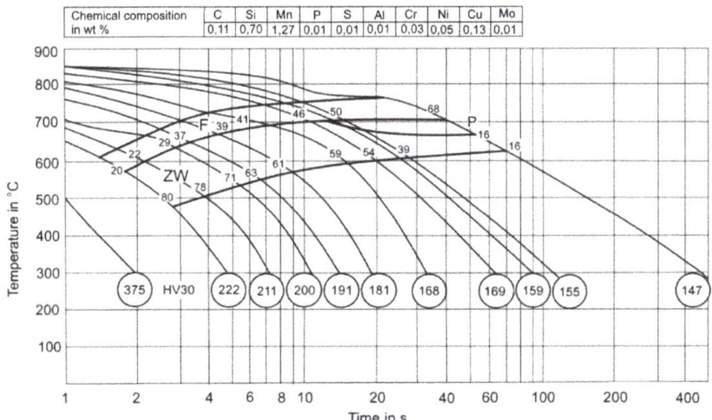

Figure 10. Welding time-temperature-transformation (TTT) diagram for G4Si1 and peak temperatures of 1350 °C, according to [41].

The fine-grained microstructure is an indicator for low heat input while welding subsequent layers and moderate peak temperatures above A_3. The slightly harder regions at the layer interface were reheated less during subsequent thermal cycles. Thus, annealing effects are less prominent. The result is a heterogeneous microstructure in the build-up direction with varying mechanical properties.

3.4. Computer Tomography

The characterization of the three specimens in terms of the porosity and cavities was carried out with computer tomography before the tension testing. The results of the porosity analysis of the CT scans are displayed in Figure 11. On the left side, there is the conventional GMAW bar with a comparatively low amount of pores and porosity. The diameter of the pores ranged from 100 to approximately 1000 μm. The pores in the bar welded by the CMT standard process were both small and rare. The highest detected pore diameter measured between 100 and 200 μm. While the size of the pores in the CMT cycle step bar are comparable to the CMT standard bar, the amount of pores or cavities exceeds that of the CMT standard bar many times over. The differences in terms of porosity are a result of the weld pool size and the varying degassing behavior during welding. Larger weld pools in combination with short welding times (conventional GMAW) cause solidification pores of a distinct size. Smaller weld pools (CMT cycle step) and longer welding times (CMT standard) are favorable for low pore sizes. However, short welding times in combination with small weld pools lead to a high number of small pores.

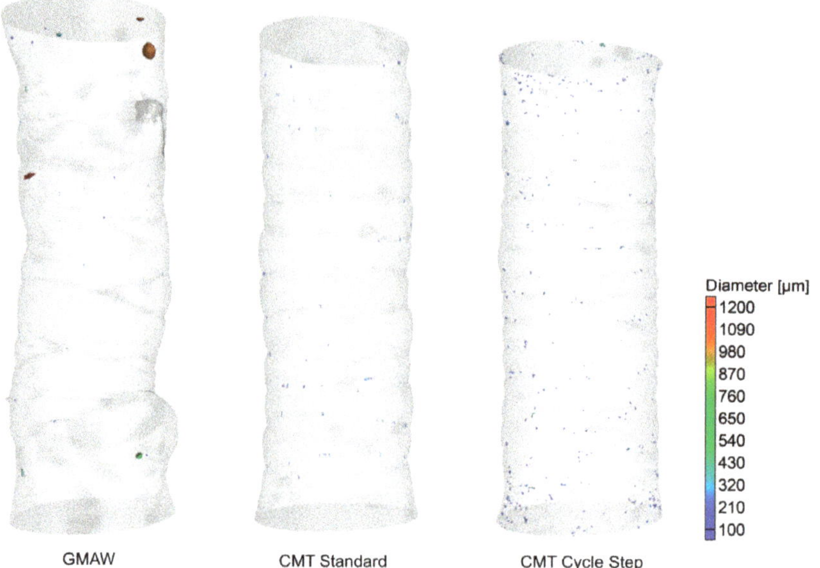

Figure 11. Computer tomography scans with porosity analysis of the conventional GMAW bar, the CMT standard bar, and the CMT cycle step bar.

3.5. Full Field Strain Measurements

The material behavior and the influence of the surface topography were examined by full field strain measurements using ESPI during uniaxial tensile tests in the elastic region. To compare the three specimens, the measured strain maps (longitudinal strains) were related to the global longitudinal strain, ε_g, simultaneously measured with the laser extensometer on each specimen. Figure 12 shows a surface photograph with the ESPI measurement area and the related strain map of each specimen at a load of 5 kN.

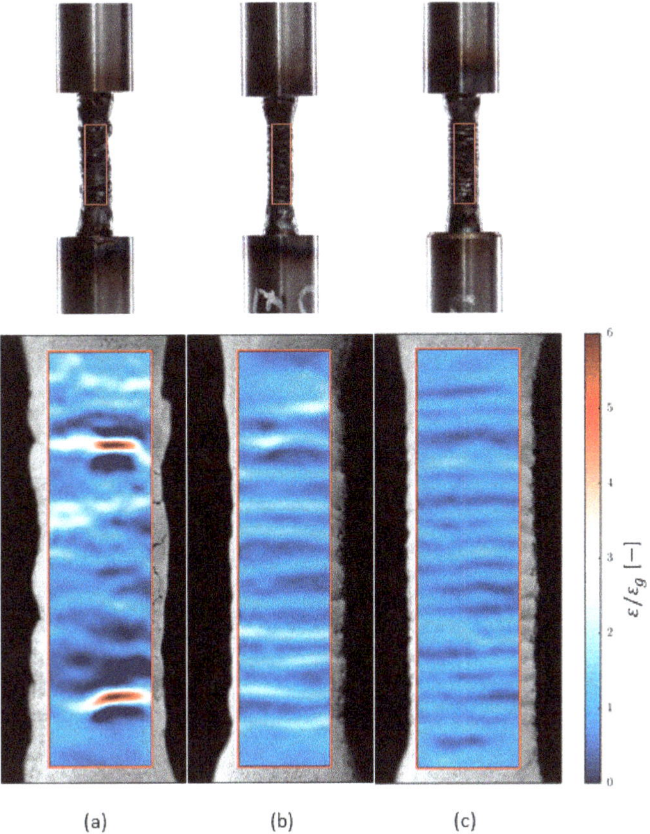

Figure 12. Photograph with the measurement area and local strain maps (longitudinal strain) related to the global strains of (**a**) conventional GMAW, (**b**) CMT standard, and (**c**) CMT cycle step under tensile load.

The ratio from local to global strain ($\varepsilon/\varepsilon_g$) allows detection of areas with different material parameters or strong variations in the surface. The local strains in the conventional GMAW bar showed the largest deviations from the measured global strains with a ratio of up to 6 in two distinct areas of the specimen, as displayed in Figure 12a. The CMT standard bar showed a more even distribution with ratios of up to 2.5. In comparison, the CMT cycle step bar showed the most even strain distribution with a maximum ratio of 1.5. While the layered structure was clearly visible in both CMT bars, the conventional GMAW bar showed two concentrations, which can be attributed to strong deviations in the surface topography. Effects of the microstructure or heterogeneous hardness on local strain distribution were not detectable at this load level.

Due to the high local strain concentrations, especially in the conventional GMAW bar, local plastic deformations might have occurred. In the event of a further load increase, the conventional GMAW bar would most likely fail at one of the localizations. This shows that a uniaxial tension test until failure with this topography would not provide representative results about the mechanical properties, except the ultimate bearing load. Therefore, the advanced testing strategy applied here seems to represent a better overall solution. A rated value similar to the strain ratio presented here could later be used as a notch factor corresponding to the different bars. Therefore, these measurement results can also be used to examine the material behavior of WAAM bars in regards to cyclic loading.

3.6. Mechanical Properties

The second stage of testing consisted of uniaxial tension tests until failure with a further machined specimen geometry, comparable to conventional tensile testing. For each type of bar, one specimen was tested until failure. A laser extensometer was used to measure the integral strains during the test on the length of the transversal section (10 mm). Figure 13 shows the obtained engineering stress–strain curves for all three specimens. The resulting stress was calculated based on the 4 mm diameter in the predefined measurement section.

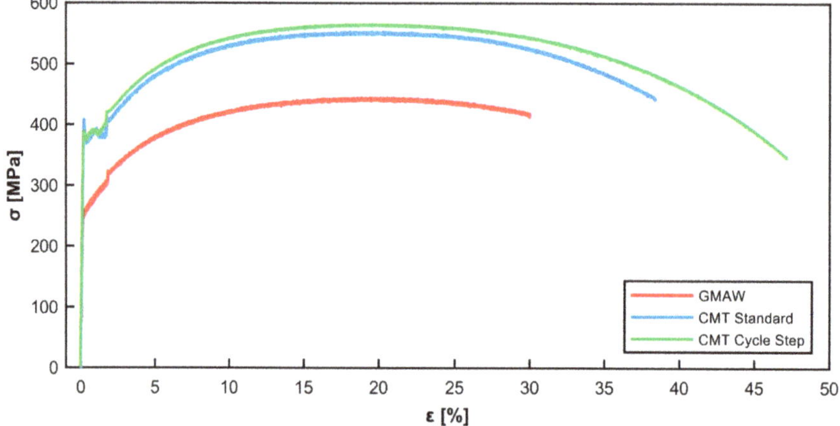

Figure 13. Stress–strain diagram for all three specimens.

The stress–strain curves show a very ductile material behavior with large ultimate elongations, especially for the CMT bars. A clear yield point was also recognizable for all three specimens as is expected from the used unalloyed weld metal. All mechanical properties, as listed in Table 3, were calculated in accordance to DIN EN ISO standards [42]. Table 3 also includes the corresponding properties of the weld metal data sheet from the welding wire used.

Table 3. Mechanical properties of the three bars and the welding wire.

Properties	Unit	Conventional GMAW	CMT Standard	CMT Cycle Step	Weld metal (Data Sheet)
Modulus of Elasticity E	GPa	180	198	191	-
Yield Stress R_{eH}	MPa	265	410	390	≥460
Ultimate Tensile Stress R_m	MPa	445	554	565	530–680
Ultimate Elongation A	%	49.7 [1]	46.3 [1]	4.7 [1]	≥20

[1] Due to the geometry of the specimen, the elongation at break is not comparable to standard tensile test specimens [35].

All the bars showed a linear elastic material behavior with a modulus of elasticity ranging from 180 to 198 GPa. The yield stress, R_{eH}, for all three bars and the ultimate tensile stress, R_m, for the conventional GMAW bar is lower than nominal values from the corresponding weld metal data sheet. For both CMT bars, the R_m lies in the desired range. In comparison, the conventional GMAW bar shows the lowest values in terms of strength properties. The results are in agreement with findings from metallographic investigations and hardness tests.

3.7. Fracture Surface Imaging

After destructive testing of the bars, the fracture surfaces were examined by help of scanning electroscope micrographs. Figure 14 shows the fracture surface of each bar in two different magnifications. The macroscopic images of all the specimens depict significant area reductions resulting from plasticity. This effect is most distinctive at the specimen welded by conventional GMAW. However, all fracture surfaces show dimples at high magnification. On the surface of the conventional GMAW specimen, there are also artefacts, which may be silicate segregations. In comparison to the GMAW specimen, the CMT specimens show fracture surfaces with indications for multiple crack initiation sites. A rougher surface indicates more ductile fracture behavior.

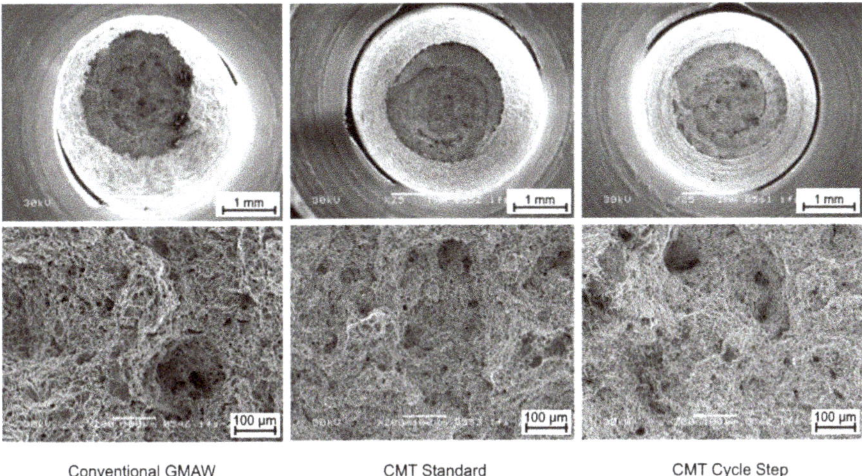

Figure 14. Fracture surface images of the bars after destructive tensile testing.

4. Discussion

Steel bars manufactured by the WAAM process demonstrate the use of AM in construction. For this study, several samples were manufactured with each process to identify the influence of welding parameters. However, only one sample of each process was tested until fracture. The overall objective was the identification of an advanced test routine, which can be applied to further WAAM components as well. The steel bars, as manufactured and tested here, could represent reinforcement elements in concrete applications or standalone, respectively, parts of complex larger steel components structural members. Further, it is demonstrated that the use of manufacturing parameters, such as the welding process and the energy input, may result in a severe influence on the component and material properties. The energy reduced short arc welding processes led to a heterogeneous microstructure with hardness peaks at the layer interfaces, probably resulting from low annealing temperatures between A_1 and A_3. This is usually not favorable as mechanical properties become heterogeneous as well. Higher interpass temperatures could solve this issue also for welding processes with low heat input. However, the resulting lower weld pool viscosity may affect the resulting surface topography as well. The results from the microstructure and hardness testing of WAAM bars as well as the surface fracture images correspond well with theoretical findings from the literature (e.g., TTT-diagrams), weld parameters, and results from tensile tests. The conventional GMAW bar with lower, but homogeneous, hardness showed lower ultimate strength and higher areas of reduction at the fracture compared to the CMT-welded bars. However, the peak temperatures and annealing times should be controlled more intensively in future research as they control the grain size.

The surface topography and the geometric dimensions, here the bar diameter, depend on the process characteristics. A homogeneous diameter is preferable in terms of mechanical material utilization. However, in concrete applications, rougher surfaces may be beneficial due to better load transfer from concrete to reinforcement bars. Cyclically loaded components would benefit from smoother surfaces as observed in the case of CMT-welding. Next, the layer volume increases with increasing energy input, which is beneficial for economical manufacturing times.

CT scans did not show severe inner defects or irregularities in the manufactured steel bars, except a low number of single pores with a diameter of ~1000 μm. The differences in pore size and magnitude are explained by the welding time and weld pool size. However, this must be examined statistically in future research with higher numbers of specimens.

The applied test methodology captured both local elastic component behavior and integral material parameters. The full field strain measurements can identify the inhomogeneous material behavior. Most likely, the heterogeneous strain distribution results from geometric variations, i.e., the change of diameter and local notches at layer interfaces. However, future work will address the material properties of deposited weld metal depending on the applied weld process in more detail. The detected stress concentration is of large importance for plasticity-induced phenomena, such as crack initiation and fatigue.

Various parameter identification methods could be used to quantify the deviations of the material in further research, for example the virtual fields method [43]. The measurements on the material scale presented here are not sufficient to evaluate the behavior of more complex WAAM-components in its entirety and 360° digital image correlation systems could be utilized for the purpose of analyzing the deformation behavior of the entire component during a test, even in regions of large deformations where the ESPI system reaches its limits. When analyzing the component behavior, it is also necessary to evaluate the behavior under compressive loads in regards to buckling as the WAAM components are generally slender components.

5. Conclusions

This study investigated the use of WAAM steel components in construction. The fundamentals of both architectural and structural design were introduced. The experimental investigations made use of steel bars as the geometrically simplest geometry to demonstrate component and material behavior typical for WAAM structures. The WAAM process was linked to the build-up geometry, surface topography, and material properties. The key findings are:

- The layer geometry depends on the weld energy input. High energy leads to heterogeneous surface topography, higher material volumes deposited, and sharp notches at layer interfaces.
- The heat input (weld energy and interpass temperature) correspond with the microstructure. High heat input results in a homogenous microstructure, but overall lower hardness. Low heat input results in hardness peaks at the layer interfaces.
- CT scans revealed internal and external weld irregularities. Porosity depends on weld pool degassing.
- Heterogeneous surface topography and geometry result in heterogeneous strain distributions under tensile loading.
- Integral material properties depend on the weld process. High heat input leads to low hardness, coarse grains, and lower elongation.

Author Contributions: Conceptualization, J.H. and J.U.; methodology, J.H., J.U., and J.M.; investigation, J.M. and M.G.; resources, K.D. and K.T.; writing—original draft preparation, J.M., M.G., and C.M.; writing—review and editing, J.H.; visualization, J.M., M.G., and C.M.; supervision, K.D., K.T., and H.K.; project administration, J.H., K.T., and H.K.

Funding: This research received partial funding from Deutsche Forschungsgemeinschaft, Major Research Instrumentation, Optical Strain Measurement System, No. 221277291.

Acknowledgments: We acknowledge support by the Open Access Publication Funds of the Technische Universität Braunschweig.

Conflicts of Interest: The authors declare no conflict of interest.

References

1. Warszawski, A.; Navon, R. Implementation of Robotics in Building: Current Status and Future Prospects. *J. Constr. Eng. Manag.* **1998**, *124*, 31–41. [CrossRef]
2. Khoshnevis, B.; Dutton, R. Innovative Rapid Prototyping Process Makes Large Sized, Smooth Surfaced Complex Shapes in a Wide Variety of Materials. *Mater. Technol.* **1998**, *13*, 53–56. [CrossRef]
3. Buswell, R.A.; Soar, R.C.; Gibb, A.G.F.; Thorpe, A. Freeform Construction: Mega-scale Rapid Manufacturing for construction. *Autom. Constr.* **2007**, *16*, 224–231. [CrossRef]
4. Buswell, R.A.; Leal de Silva, W.R.; Jones, S.Z.; Dirrenberger, J. 3D printing using concrete extrusion: A roadmap for research. *Cem. Concr. Res.* **2018**, *112*, 37–49. [CrossRef]
5. Tay, Y.W.D.; Panda, B.; Paul, S.C.; Mohamed, N.A.N.; Tan, M.J.; Leong, K.F. 3D printing trends in building and construction industry: A review. *Virtual Phys. Prototyp.* **2017**, *12*, 261–276. [CrossRef]
6. Baker, R. Method of Making Decorative Articles. US1533300A, 14 April 1925.
7. Feldmann, M.; Kühne, R.; Citarelli, S.; Reisgen, U.; Sharma, R.; Oster, L. 3D-Drucken im Stahlbau mit dem automatisierten Wire Arc Additive Manufacturing. *Stahlbau* **2019**, *88*, 203–213. [CrossRef]
8. Reisgen, U.; Willms, K.; Oster, L. Lichtbogenbasierte additive Fertigung – Forschungsfelder und industrielle Anwendungen. In *Additive Serienfertigung: Erfolgsfaktoren und Handlungsfelder für die Anwendung*; Lachmayer, R., Lippert, R.B., Kaierle, S., Eds.; Springer: Berlin, Germany, 2018; pp. 89–106.
9. Galjaard, S.; Hofman, S.; Ren, S. New Opportunities to Optimize Structural Designs in Metal by Using Additive Manufacturing. In *Advances in Architectural Geometry 2014*; Block, P., Knippers, J., Mitra, N.J., Wang, W., Eds.; Springer: Cham, Switzerland, 2015; pp. 79–93.
10. Joosten, S. Printing a Stainless Steel Bridge: An Exploration of Structural Properties of Stainless Steel Additive Manufactures for Civil Engineering Purposes. Master's Thesis, Faculty of Civil Engineering and Geosciences, Delft University of Technology, Delft, The Netherlands, 21 April 2015.
11. Camacho, D.D.; Clayton, P.; O'Brien, W.J.; Seepersad, C.; Juenger, M.; Ferron, R.; Salamone, S. Applications of additive manufacturing in the construction industry—A forward-looking review. *Autom. Constr.* **2018**, *89*, 110–119. [CrossRef]
12. Son, C.-B.; Jang, W.-S.; Lee, D.-E. Effect of changes in the construction economy on worker's operating rates and productivity. *KSCE J. Civ. Eng.* **2014**, *18*, 419–429. [CrossRef]
13. Yossef, M.; Chen, A. Applicability and Limitations of 3D Printing for Civil Structures. In Proceedings of the 2015 Conference on Autonomous and Robotic Construction of Infrastructure, Ames, IA, USA, 2–3 June 2015; pp. 237–246.
14. Gebhardt, A. *Additive Fertigungsverfahren: Additive Manufacturing und 3D-Drucken für Prototyping—Tooling—Produktion*, 5th ed.; Hanser: München, Germany, 2016.
15. Berger, U.; Hartmann, A.; Schmid, D. *3D-Druck—Additive Fertigungsverfahren: Rapid Prototyping, Rapid Tooling, Rapid Manufacturing*, 2nd ed.; Verlag Europa-Lehrmittel—Nourney Vollmer GmbH & Co. KG: Haan-Gruiten, Germany, 2017.
16. Deutscher Verband für Schweißen und Verwandte Verfahren. *Grundlegende wissenschaftliche Konzepterstellung zu bestehenden Herausforderungen und Perspektiven für die Additive Fertigung mit Lichtbogen: Studie im Auftrag der Forschungsvereinigung Schweißen und verwandte Verfahren e.V. des DVS*; DVS Media GmbH: Düsseldorf, Germany, 2018.
17. Henckell, P.; Ali, Y.; Reimann, J.; Bergmann, J.P. Lichtbogenbasierte Additive Fertigung: Modellierung der Bauteileigenschaften. *DVS-Berichte* **2018**, *344*, 104–108.

18. Brandl, E.; Baufeld, B.; Leyens, C.; Gault, R. Additive manufactured Ti-6Al-4V using welding wire: Comparison of laser and arc beam deposition and evaluation with respect to aerospace material specifications. *Phys. Procedia* **2010**, *5*, 595–606. [CrossRef]

19. Gu, J.; Cong, B.; Ding, J.; Williams, S.W.; Zhai, Y. Wire-Arc Additive Manufacturing of Aluminium. In Proceedings of the 25th Annual International Solid Freeform Fabrication Symposium, Austin, TX, USA, 4–6 August 2014; pp. 451–458.

20. Wang, H.; Kovacevic, R. Rapid prototyping based on variable polarity gas tungsten arc welding for a 5356 aluminium alloy. *Proc. Inst. Mech. Eng. Part B J. Eng. Manuf.* **2001**, *215*, 1519–1527. [CrossRef]

21. Lange, J.; Feucht, T. 3-D-Printing with Steel: Additive Manufacturing of Connection Elements and Beam Reinforcements. In Proceedings of the IABSE Symposium 2019 Guimarães, Guimarães, Portugal, 27–29 March 2019.

22. Rodrigues, T.A.; Duarte, V.; Miranda, R.M.; Santos, T.G.; Oliveira, J.P. Current Status and Perspectives on Wire and Arc Additive Manufacturing (WAAM). *Materials* **2019**, *12*, 1121. [CrossRef] [PubMed]

23. Cunningham, C.R.; Flynn, J.M.; Shokrani, A.; Dhokia, V.; Newman, S.T. Invited review article: Strategies and processes for high quality wire arc additive manufacturing. *Addit. Manuf.* **2018**, *22*, 672–686. [CrossRef]

24. Wu, B.; Pan, Z.; Ding, D.; Cuiuri, D.; Li, H.; Xu, J.; Norrish, J. A review of the wire arc additive manufacturing of metals: Properties, defects and quality improvement. *J. Manuf. Processes* **2018**, *35*, 127–139. [CrossRef]

25. Rodrigues, T.A.; Duarte, V.; Avila, J.A.; Santos, T.G.; Miranda, R.M.; Oliveira, J.P. Wire and arc additive manufacturing of HSLA steel: Effect of thermal cycles on microstructure and mechanical properties. *Addit. Manuf.* **2019**, *27*, 440–450. [CrossRef]

26. Hosseini, V.A.; Högström, M.; Hurtig, K.; Bermejo, M.A.V.; Stridh, L.-E.; Karlsson, L. Wire-arc additive manufacturing of a duplex stainless steel: Thermal cycle analysis and microstructure characterization. *Weld. World* **2019**, *63*, 975–987. [CrossRef]

27. Lockett, H.; Ding, J.; Williams, S.; Martina, F. Design for Wire + Arc Additive Manufacture: Design rules and build orientation selection. *J. Eng. Des.* **2017**, *28*, 568–598. [CrossRef]

28. Mehnen, J.; Ding, J.; Lockett, H.; Kazanas, P. Design study for wire and arc additive manufacture. *Int. J. Prod. Dev.* **2014**, *19*, 2–20. [CrossRef]

29. Labonnote, N.; Rønnquist, A.; Manum, B.; Rüther, P. Additive construction: State-of-the-art, challenges and opportunities. *Autom. Constr.* **2016**, *72*, 347–366. [CrossRef]

30. Hack, N.P. Mesh Mould: A Robotically Fabricated Structural Stay-in-Place Formwork System. Ph.D. Thesis, ETH Zurich, Zurich, Switzerland, May 2018.

31. Grieves, M. Digital Twin: Manufacturing Excellence through Virtual Factory Replication. Available online: https://research.fit.edu/media/site-specific/researchfitedu/camid/documents/1411.0_Digital_Twin_White_Paper_Dr_Grieves.pdf (accessed on 24 June 2019).

32. Mandolla, C.; Petruzzelli, A.M.; Percoco, G.; Urbinati, A. Building a digital twin for additive manufacturing through the exploitation of blockchain: A case analysis of the aircraft industry. *Comput. Ind.* **2019**, *109*, 134–152. [CrossRef]

33. Tabernero, I.; Paskual, A.; Álvarez, P.; Suárez, A. Study on Arc Welding Processes for High Deposition Rate Additive Manufacturing. *Procedia CIRP* **2018**, *68*, 358–362.

34. Köhler, M.; Fiebig, S.; Hensel, J.; Dilger, K. Wire and Arc Additive Manufacturing of Aluminum Components. *Metals* **2019**, *9*, 608. [CrossRef]

35. DIN 50125. *Prüfung Metallischer Werkstoffe—Zugproben*; German Institute for Standardization: Berlin, Germany, 2016.

36. Lorenzin, G.; Rutili, G. The innovative use of low heat input in welding: Experiences on 'cladding' and brazing using the CMT process. *Weld. Int.* **2009**, *23*, 622–632. [CrossRef]

37. Grunwald, R.; Mayer, M.; Schörghuber, M. WAAM—Technologie und aktuelle Anwendung. In Proceedings of the DVS Congress 2018, Friedrichshafen, Germany, 17–18 September 2018; pp. 97–103.

38. *CMT Cycle Step von Fronius: Ein neuer Level an Schweißprozess-Kontrolle*; Fronius International GmbH: Wels, Austria, 2018.

39. Creath, K. Phase-shifting speckle interferometry. *Appl. Opt.* **1985**, *24*, 3053–3058. [CrossRef] [PubMed]

40. Kukla, D.; Brynk, T.; Pakieła, Z. Assessment of Fatigue Resistance of Aluminide Layers on MAR 247 Nickel Super Alloy with Full-Field Optical Strain Measurements. *J. Mater. Eng. Perform.* **2017**, *26*, 3621–3632. [CrossRef]

41. Seyffarth, P.; Meyer, B.; Scharff, A. *Großer Atlas Schweiß-ZTU-Schaubilder*, 2nd ed.; DVS Media GmbH: Düsseldorf, Germany, 2018.
42. *DIN EN ISO 6892-1:2017-02, Metallische Werkstoffe—Zugversuch—Teil 1: Prüfverfahren bei Raumtemperatur (ISO 6892-1:2016)*; German Institute for Standardization: Berlin, Germany, 2017.
43. Pierron, F.; Grédiac, M. *The Virtual Fields Method*; Springer: New York, NY, USA, 2012.

 *metals*

Article

The Relationship of Fracture Mechanism between High Temperature Tensile Mechanical Properties and Particle Erosion Resistance of Selective Laser Melting Ti-6Al-4V Alloy

Jun-Ren Zhao, Fei-Yi Hung *, Truan-Sheng Lui and Yu-Lin Wu

Department of Materials Science and Engineering, National Cheng Kung University, Tainan 701, Taiwan;
a2x346yz03@gmail.com (J.-R.Z.); luits@mail.ncku.edu.tw (T.-S.L.); nick22530682@hotmail.com (Y.-L.W.)
* Correspondence: fyhung@mail.ncku.edu.tw; Tel.: +886-6-275-7575 (ext. 62950)

Received: 22 March 2019; Accepted: 26 April 2019; Published: 29 April 2019

Abstract: In this study, selective laser melting (SLM) Ti-6Al-4V is subjected to heat treatment for 4 h at 400 °C, 600 °C, and 800 °C, followed by air cooling. After heat treatment at 400 °C and 600 °C, the ductility was lower (strength increased). This was could be for two reasons: (1) high temperature tensile properties, and (2) particle erosion wear induced phase transformation. Finally, the particle erosion rates of as-SLM Ti-6Al-4V and heat treatment for 4 h at 800 °C (labeled 800-AC) were investigated and compared; the lamellar $\alpha + \beta$ phases in 800-AC are difficult to destroy with erosion particles, resulting in the erosion resistance of 800-AC being higher than that of the martensitic α' needles in the as-SLM Ti-6Al-4V at all impact angles (even the hardness of the 800-AC specimen was lower). The as-SLM Ti-6Al-4V alloy needs heat treatment to have better wear resistance.

Keywords: selective laser melting (SLM); Ti alloy; high temperature tensile; erosion; wear

---

1. Introduction

Selective laser melting (SLM) process is a type of 3D-printer technology used in this study. The process uses metal powder as a raw material, where during the SLM process, metal powders are melted in a specified area with a high-energy laser beam and rapidly solidified at a high cooling rate [1–3].

Ti-6Al-4V is the most representative of the $\alpha + \beta$ titanium alloys [4,5]. The most common Ti-6Al-4V alloys are cast and forged, and their alloy properties have been widely discussed [6,7]. Many SLM Ti-6Al-4V articles have discussed how process parameters and post treatment interact with microstructure and mechanical properties [8,9]. In this research, the process parameter was fixed. The microstructure of the SLM Ti-6Al-4V contains not only the classical $\alpha + \beta$ phases but also the martensitic α' phases (low ductility) because of the high cooling rate [10,11]. For industrial applications, achieving improved ductility through an appropriate heat treatment is necessary. Furthermore, Ti-6Al-4V is commercially used in gas turbine engines up to a test temperature of 350 °C. Industrial applications usually limit Ti-6Al-4V use to 400 °C, so high temperature (250–400 °C) tensile mechanical properties of SLM Ti-6Al-4V are investigated in this study and a relationship between the high temperature failure behavior and erosion wear characteristics is proposed.

In many industrial applications, erosion wear caused by solid particles results in the failure of mechanical devices and components [12,13]. This study also investigates erosion properties (use Al_2O_3 particles) between the martensitic α' phase formed by SLM Ti-6Al-4V and the lamellar double $\alpha + \beta$ phases with heat treatment. Notably, the erosion resistance mechanism was proposed by comparing the relationship between impact angles and erosion rate. According to our previous research studies [14,15], the particle erosion wear is able to induce the phase transformation and affects the erosion rate. The temperature of the eroded surface is more than 400 °C, and the relationship between wear behavior and high temperature tensile properties needs to be clarified [16]. This study is one of the few articles discussing the high temperature strength and particle erosion properties of SLM Ti-6Al-4V alloy. The relevant results have significant reference value for relevant 3D-printer titanium alloys.

2. Experimental Procedure

SLM Ti-6Al-4V (the source of the Ti-6Al-4V powder particles is EOS GmbH (Electro-Optical Systems) was used in this study, for which the chemical composition and process parameters are shown in Tables 1 and 2, respectively. The test specimens were removed by the electrical discharge machining (EDM) by cutting the wire from the support, and no post-treatment was implemented before various tests. As-SLM Ti-6Al-4V is called AS in this study. The original test specimens were held for 4 h in a tubular furnace (Deng Yng, new Taipei City, Taiwan) in argon atmosphere at 400 °C, 600 °C, and 800 °C and subjected to air cooling and induced phase transformation. The test specimens were labeled 400-AC, 600-AC, and 800-AC, respectively.

The normal direction (ND) of the specimen was set as parallel to the laser direction, where the direction vertical to the laser direction was called the side direction (SD). After being polished with SiC paper (from #80 to #5000), Al_2O_3 aqueous solution (1 and 0.3 μm), and a 0.04 μm SiO_2 polishing solution, the specimens were etched with Keller's reagent (1 mL HF + 1.5 mL HCL + 2.5 mL HNO_3 + 95 mL H_2O) to examine the microstructure. The microstructure of the specimens was observed using optical microscopy (OM, OLYMPUS BX41M-LED, Tokyo, Japan), and X-ray diffractometry (XRD, Bruker AXS GmbH, Karlsruhe, Germany) was used for identification of the microstructural phases. Hardness measurements were valuated using the Rockwell hardness test (Mitutoyo, Kawasaki-shi, Japan). The measurement conditions for the HR test followed the C-scale (the indenting load 150 kg), and the mean value for five impressions was taken as the hardness of the corresponding condition.

Table 1. Chemical composition of SLM Ti-6Al-4V (wt.%).

Element	Al	V	O	N	C	H	Fe	Ti
wt.%	6.13	3.80	0.20	0.05	0.08	0.01	0.30	Bal.

Table 2. Process parameters of SLM Ti-6Al-4V.

Particle Size (μm)	Laser Power (w)	Laser Radius (μm)	Scanning Velocity (mm/s)	Layer Thickness (μm)
15–45	170	35	800	30

The dimensions of the SLM Ti-6Al-4V tensile specimen are shown in Figure 1. The tensile test was performed with an universal testing machine (HT-8336, Hung Ta, Taichung, Taiwan), with the crosshead speed at 1 mm/min, which corresponded to the initial strain rate of 18.33×10^{-4} s^{-1}. The AS and the heat-treated different temperatures specimens were subjected to a room temperature tensile test to analyze the mechanical properties of SLM Ti-6Al-4V. There were at least three specimens for each test and the mean value of the test specimens was taken as the tensile results of the corresponding condition.

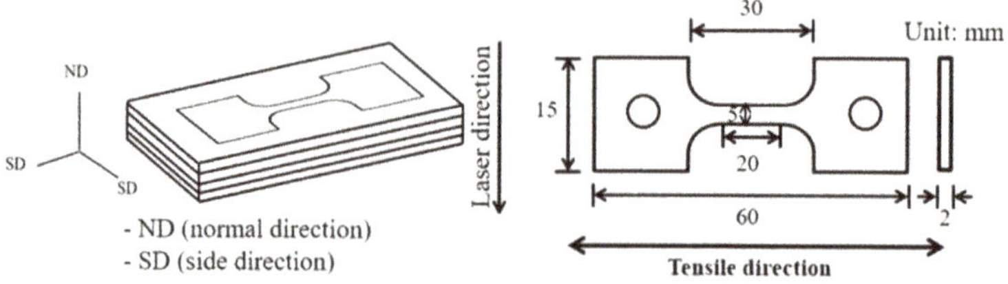

Figure 1. Dimensions of the selective laser melting (SLM) Ti-6Al-4V tensile specimen.

Regarding the high temperature tensile properties, titanium alloys are often applied in a high temperature environment of 250–400 °C [17]; therefore, tensile tests were carried out at 250 °C, 300 °C, 350 °C, and 400 °C to investigate the influence of temperature on the mechanical properties of SLM Ti-6Al-4V. There are at least three specimens for each test and the mean value of the test specimens was taken as the tensile results of the corresponding condition.

The equipment used in the erosion test is shown in Figure 2. Al_2O_3 particles were used, for which the average particle size was approximately 450 μm and scanning electron microscope (SEM) morphology is shown in Figure 3. The specimens were polished with #80 to #1000 SiC papers to remove the oxidized layer and soaked in acetone for ultrasonic cleaning before the erosion test. Then, 200 g of the erosion particles under a compressed air flow of 3 kg/cm^2 (0.29 MPa) were subjected to the erosion test [14,15], for which the impact angles were 15°, 30°, 45°, 60°, 75°, and 90° to compare the erosion behavior of needle-like α′ phase in the AS specimen and the plate-like α + β phase in 800-AC specimen. According to previous reports, the maximum erosion rate of the general ductile material takes place at about 20°–30°, but brittle materials, such as ceramics and glass, have maximum erosion rates at about 90° [14,15]. The erosion rate (ER% = $\delta W/W_{total\ particles}$) of a specimen is defined as its weight loss (δW) divided by the weight of the total erosion particles ($W_{total\ particles}$). Finally, optical microscopy and a scanning electron microscope (SEM, HITACHI SU-5000, HITACHI, Tokyo, Japan) were used to examine the surface and subsurface of the erosion specimens and to determine the erosion mechanism. All heat treatment conditions and measurements are shown in Table 3.

Table 3. Heat treatment condition and measurements.

Sample	Heat Treatment Condition	Room Temperature Tensile Test	High Temperature Tensile Test	Particles Erosion Test
AS	-	✓	✓	✓
400-AC	400 °C/4 h → air cooling	✓	-	-
600-AC	600 °C/4 h → air cooling	✓	-	-
800-AC	800 °C/4 h → air cooling	✓	-	✓

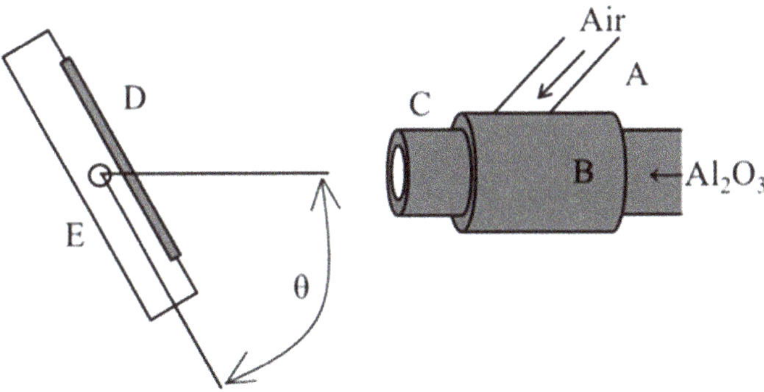

Figure 2. The particle erosion test apparatus. (**A**) Compressed air flow, (**B**) erosion particle supplier, (**C**) erodent nozzle, (**D**) specimen, and (**E**) specimen holder; θ = impact angle.

Figure 3. Scanning electron microscope (SEM) morphology of Al$_2$O$_3$ erosion particles.

3. Results and Discussion

3.1. Microstructures and Phases

Figure 4 shows the microstructures of the as-SLM Ti-6Al-4V subjected to heat treatment at different holding temperatures. Figure 4a shows the normal direction microstructure of AS, where it can be observed that a large number of needle-like phases are surrounded by a network profile with a diameter of approximately 80 μm. Figure 4b shows the same needle-like structure, therefore the AS is a needle-like structure as a whole. The normal direction and side direction microstructures were subjected to heat treatment at 400 °C for 4 h, as shown in Figure 4c,d, respectively. The microstructure of 400-AC is similar to that of AS, which is comprised of needle-like α′ phases as a whole. Similar microstructures can also be seen at 600-AC, as shown in Figure 4e,f. In addition, white plate-like α phases were produced at 600 °C. Figure 4g,h shows that the microstructure of 800-AC is different from that of AS, 400-AC, and 600-AC. At 800 °C, a significant phase transformation occurred, and the α′ phases disappeared and were substituted with continuous lamellar α + β phases comprising white α phases surrounded by black β phases.

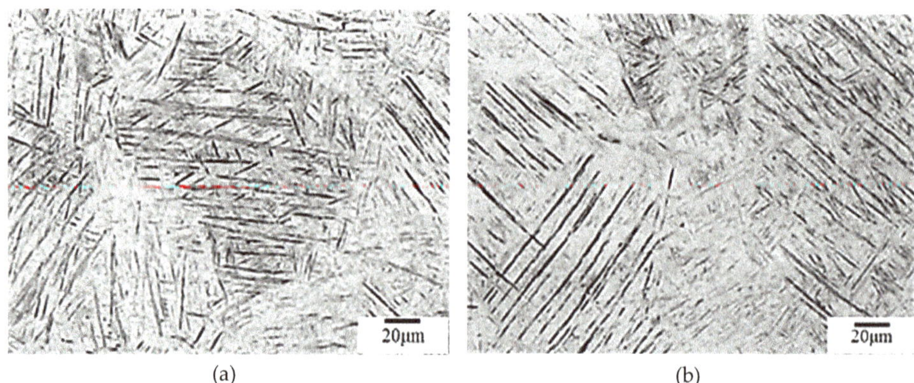

(a)   (b)

Figure 4. *Cont.*

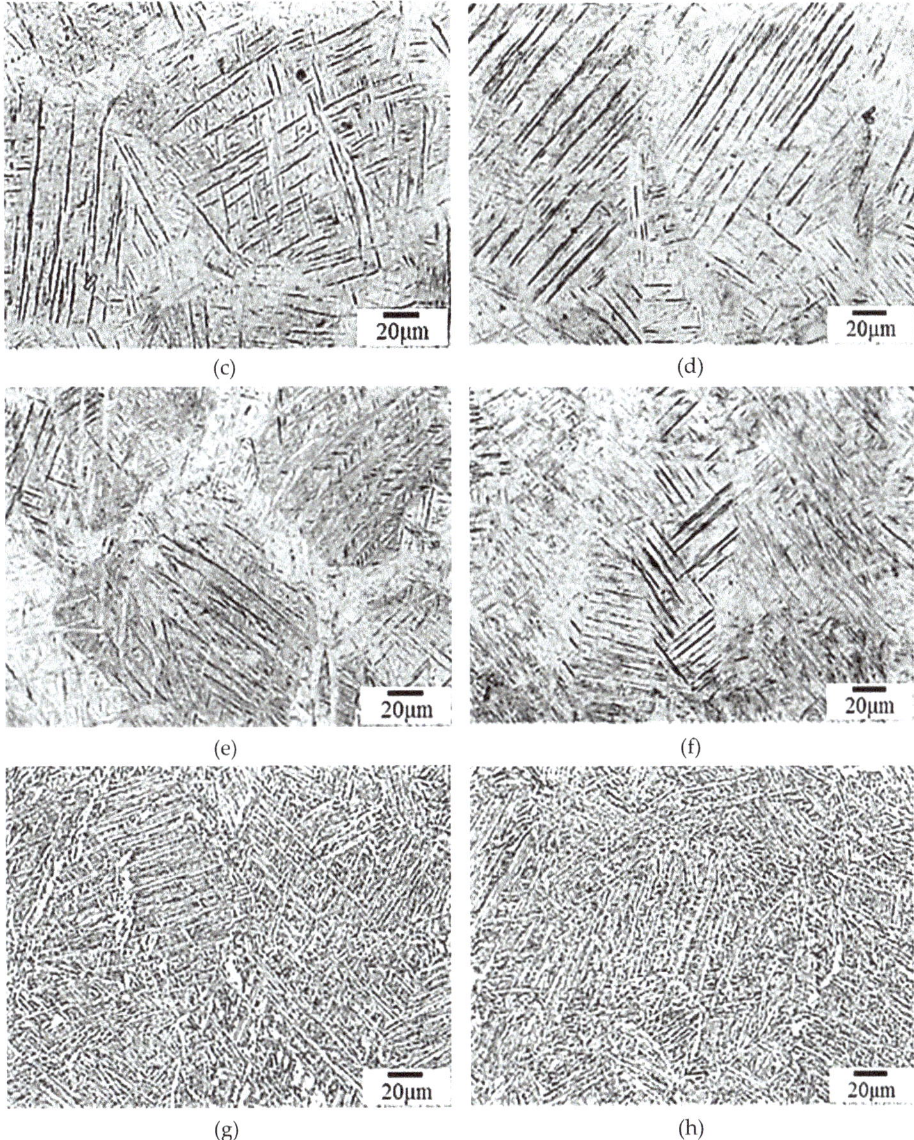

Figure 4. Microstructures in the (**a**) normal direction of as-SLM Ti-6Al-4V (AS), (**b**) side direction of AS, (**c**) normal direction of 400-AC, (**d**) side direction of 400-AC, (**e**) normal direction of 600-AC, (**f**) side direction of 600-AC, (**g**) normal direction of 800-AC, and (**h**) side direction of 800-AC.

The β peak at 600 °C and the presence of the α/α′and β peaks at 800 °C were confirmed with a XRD analysis, as shown in Figure 5. The XRD patterns are referenced from a previous report [18]. This indicates that when the heat treatment temperature increases up to 600 °C, the β-phase will be generated. According to a previous report [19,20], as shown in Figures 4 and 5, there is only an α′ phase in AS. The equiaxed β phases were preferentially formed above the β-transform temperature during the cooling process. Because of the extremely high temperature gradient in quenching [21,22], the β phases were completely transformed into the needle-like martensitic α′ phase. According to the Xu et al. report [19], when the AS specimen was heated to 400 °C, the α phases were precipitated, where the phase transformation mechanism was α′ → α′ + α. When AS specimen is heated to 600 °C, the temperature is higher than the martensitic transformation temperature (575 °C), plate-like α phases precipitate around the α′ phases, and a small amount of β phases precipitate on the α phase boundaries as hold time increases, where the mechanism of phase transformation is α′ → α′ + α + β. When AS specimen is heated to 800 °C, the α′ phases completely disappear, and transformation to the continuous lamellar α + β phase occurs, for which the phase transformation mechanism is α′ → α + β.

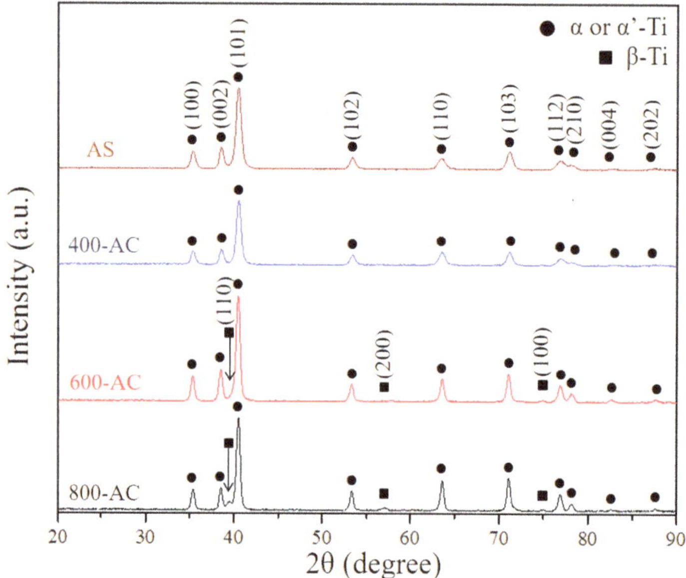

Figure 5. X-ray diffraction pattern of AS, 400-AC, 600-AC, and 800-AC specimens.

3.2. Mechanical Properties

Figure 6 shows the hardness (normal direction) comparison of AS, 400-AC, 600-AC, and 800-AC. The hardness of AS, 400-AC, and 600-AC is similar, but when the heat treatment temperature increases up to 800 °C, the hardness decreases because α′ phases are completely transformed into α + β phases.

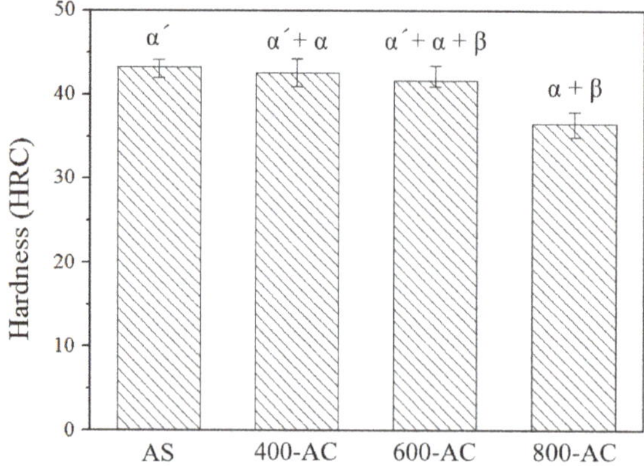

Figure 6. Hardness of AS, 400-AC, 600-AC, and 800-AC (normal direction).

The room temperature tensile properties of AS, 400-AC, 600-AC, and 800-AC are shown in Figure 7, with the mean value of the mechanical properties list in Table 4. It can be seen that the strength of 400-AC is significantly higher than that of AS. According to previous reports [23,24], the tensile strength can be improved by the precipitation of α phases in the grains or on the boundaries, and the residual stress is also reduced by a 400 °C heat treatment, so the strength of 400-AC is higher than that of AS.

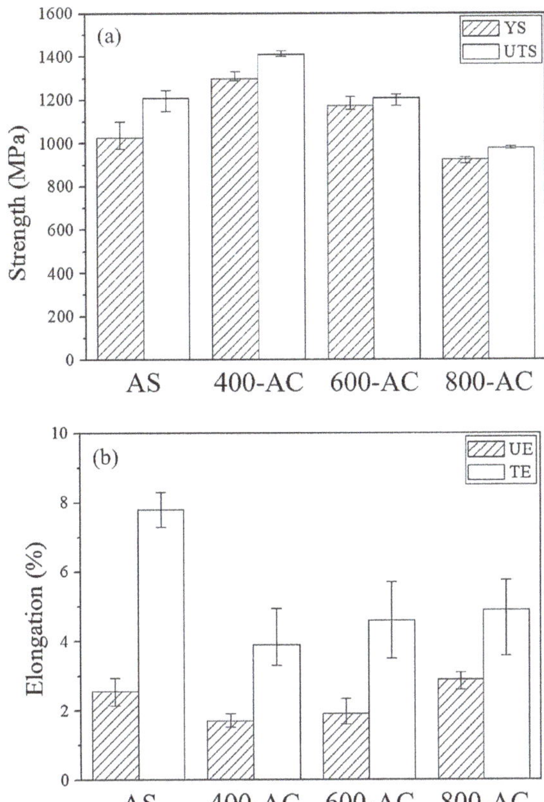

Figure 7. Tensile properties of AS, 400-AC, 600-AC, and 800-AC at room temperature for (**a**) strength and (**b**) ductility.

Table 4. Room temperature tensile properties of AS, 400-AC, 600-AC, and 800-AC. (YS: yield strength; UTS: ultimate tensile strength; UE: uniform elongation; TE: total elongation).

Test Specimen	YS (MPa)	UTS (MPa)	UE (%)	TE (%)
AS	1025	1210	2.6	7.8
400-AC	1300	1410	1.7	3.9
600-AC	1175	1210	1.9	4.6
800-AC	925	980	2.9	4.9

The β phases begin to precipitate, and the strength gradually decreases with increases in the heat treatment temperature, and the strength is similar to AS at 600 °C, but the strength at 800 °C is lower than that of AS because the α' phases are completely transformed into the $\alpha + \beta$ phases. On the other hand, the ductility increases as the heat treatment temperature increases, but is still less than 10% and less than that of AS. The precipitation of the α phases increases the strength but significantly decreases the ductility. Increases in the ductility are attributed to the formation of β phases, and α′ phases are completely transformed into α + β phases upon increases in the heat treatment temperature. In other words, the precipitation of the α phases contributes to the improvement of tensile strength, where the increase in ductility can be attributed to the formation of the β phases.

Figure 8 shows the morphology of the tensile fracture surfaces of AS, 400-AC, 600-AC, and 800-AC at room temperature. Dimpled ductility structures can be observed in all specimens. In addition, some cleavage facets can be observed in 400-AC, which is consistent with its low ductility.

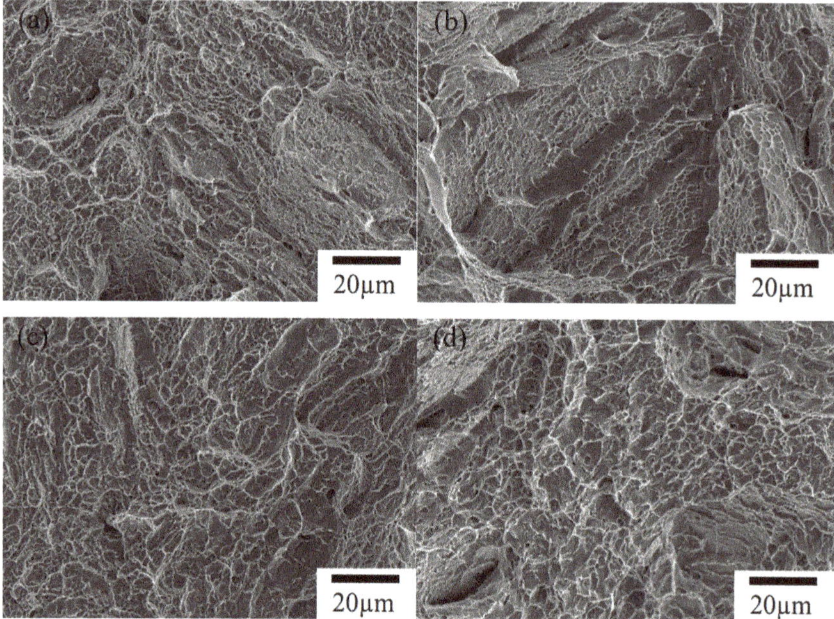

Figure 8. Morphologies of tensile fracture surfaces at room temperature for (**a**) AS, (**b**) 400-AC, (**c**) 600-AC, and (**d**) 800-AC.

Figure 9 shows the high temperature tensile properties of AS, and the mean value of the mechanical properties is listed in Table 5. The strength decreases slowly as the temperature increases. The UTS is close to 1000 MPa and the YS is about 800 MPa at 400 °C, indicating that the phase transformation α′→ α + β is exhibiting decreased strength. Compared to previous reports [25,26], the strength of commercial Ti-6Al-4V has a dramatic decrease whilst the temperature increases. Apparently the as-SLM Ti-6Al-4V could maintain a certain strength under 400 °C. In terms of ductility, there is first an increasing and then a decreasing trend, with the total elongation (TE) close to 10% at 250 °C–300 °C, which means the 3D-printer titanium alloy can be used in medium temperature applications. It is worth noting that the ductility at 400 °C is similar to that of AS, which implies that it is affected by a small α phase precipitation effect. In summary, the tensile properties of 350 °C are most applicable with the highest values in uniform elongation (UE). Figure 10 shows macroscopic morphology photographs of high temperature tensile fracture specimens, though there is not an obvious shrinkage phenomenon in any of the specimens, and there is no hole expansion at the location of the inserted pin. The macroscopic fracture morphology becomes gradually flatter from a zigzag pattern as the tensile temperature increases.

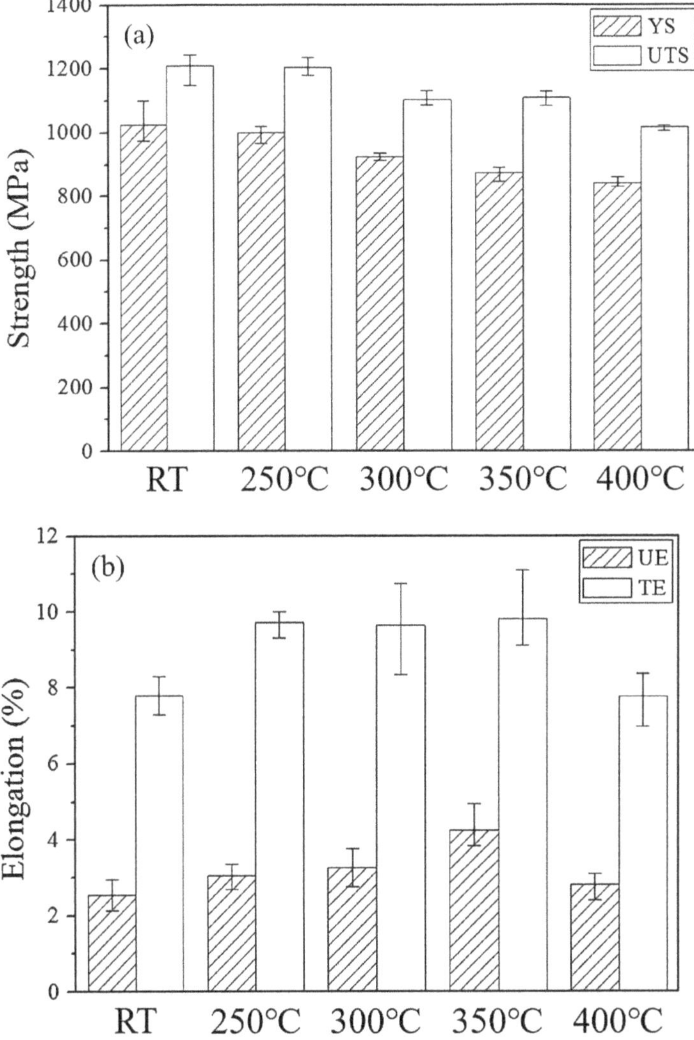

Figure 9. High temperature tensile properties of AS for (**a**) strength and (**b**) ductility.

Table 5. High temperature tensile properties of AS.

Test Specimen	YS (MPa)	UTS (MPa)	UE (%)	TE (%)
AS	1025	1210	2.6	7.8
250 °C	1000	1203	3.1	9.7
300 °C	923	1102	3.3	9.6
350 °C	873	1107	4.3	9.8
400 °C	839	1016	2.8	7.8

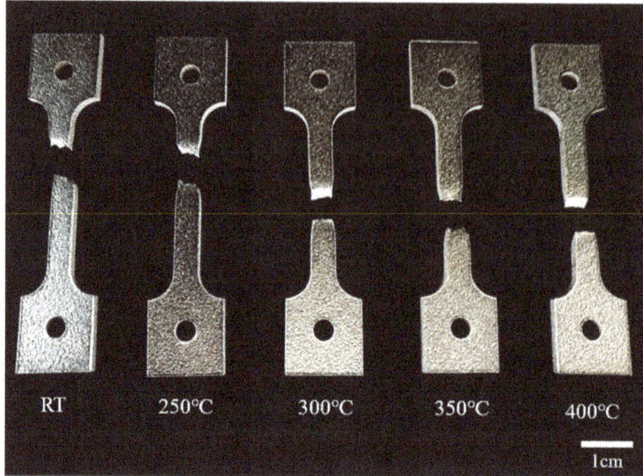

Figure 10. Macroscopic morphology photographs of room temperature and high temperature tensile fracture specimens.

3.3. Particle Erosion Characteristics and Mechanisms

In order to clarify the difference in erosion characteristics of different matrices, this study chose AS and 800-AC specimens exhibiting erosion for comparison. Figure 11 shows the erosion data of AS and 800-AC eroded by Al_2O_3 particles; there are at least three specimens for each test and the mean value of the test specimens was taken as the erosion results of the corresponding condition. The reason for using AS for comparison with 800-AC was due to the fact that the martensitic α' phases of AS were completely transformed into the lamellar $\alpha + \beta$ phases, where the ratio $\alpha:\beta$ was about 68:32. Because AS and 800-AC exhibited completely different microstructures, the effects of the phase difference on the erosion wear properties could be fully investigated. Notably, both AS and 800-AC had maximum erosion rates at 30° impact, where the erosion rates decreased with increases in the impact angle, and the minimum erosion rate at a 90° impact resulted in ductile-cutting dominating the erosion behavior. At all impact angles, the erosion rate of AS was greater than that of 800-AC, indicating the erosion resistance of the continuous lamellar $\alpha + \beta$ phases of 800-AC were better than that of AS. The erosion resistance was positively correlated with the hardness of conventional Ti-6Al-4V alloy [27], but the result was different in this research. The difference in microstructure caused this result, therefore we investigated the surface and subsurface morphologies of AS and 800-AC at different impact angles.

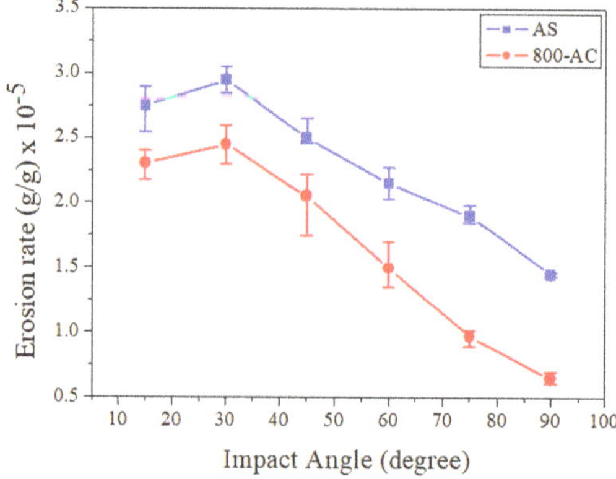

Figure 11. The erosion rate as a function of the impact angles of AS and 800-AC.

Figure 12a shows the surface morphology of AS at 30° impact, where lips and grooving can be seen on the surface. Figure 12b shows the surface morphology of AS at 90° impact, where pits formed by erosion wear can be seen. Figure 12c shows the surface morphology of 800-AC at 30° impact, where compared with the lips on the AS surface, the surface morphology of 800-AC was completely cut by erosion particles and the lips of the erosion subsurface were smaller. Figure 12d shows the surface morphology of 800-AC at 90° impact, where pits formed by erosion wear can also be seen, similar to AS. However, some scratch marks could also be observed on the surface of 800-AC at 90° impact. This phenomenon indicated that the specimen was deformed by squeezing first and then was stripped by the erosion particles.

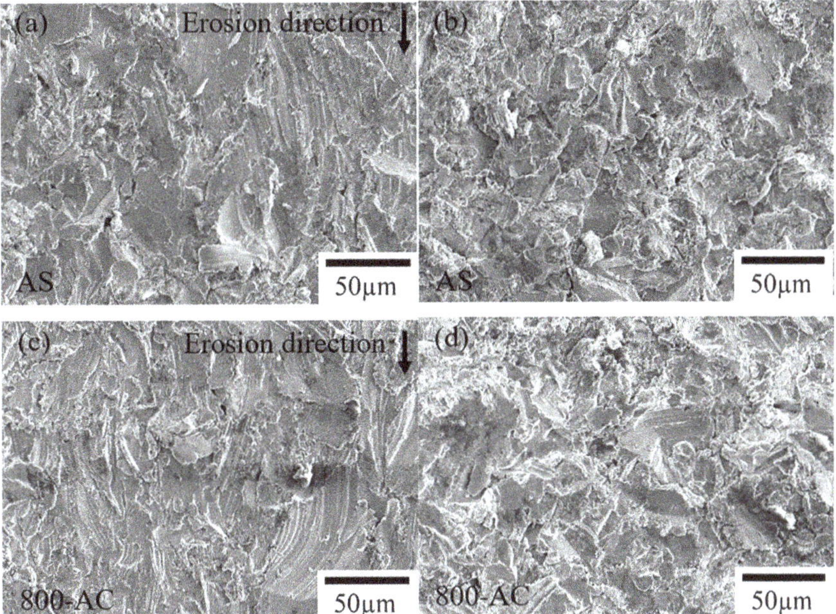

Figure 12. Surface morphologies of AS at (**a**) 30° and (**b**) 90°. Surface morphologies of 800-AC at (**c**) 30° and (**d**) 90°.

Figure 13a shows subsurface morphologies of AS at 30° impact, where lips caused by erosion particles can be found on the subsurface in the circled area. Figure 13b shows subsurface morphologies of AS at 90° impact, where both narrow and deep pits can be observed, and subsurface morphologies of 800-AC at 30° impact, where almost no lips can be observed (Figure 13c). They are replaced by smoother features resulting from erosion particles. Figure 13d shows the subsurface morphologies of 800-AC at 90° impact, where the pits that can be observed in 800-AC are wider and shallower than those in AS. Notably, the surface roughness of erosion subsurface is lower than that of AS-90° impact.

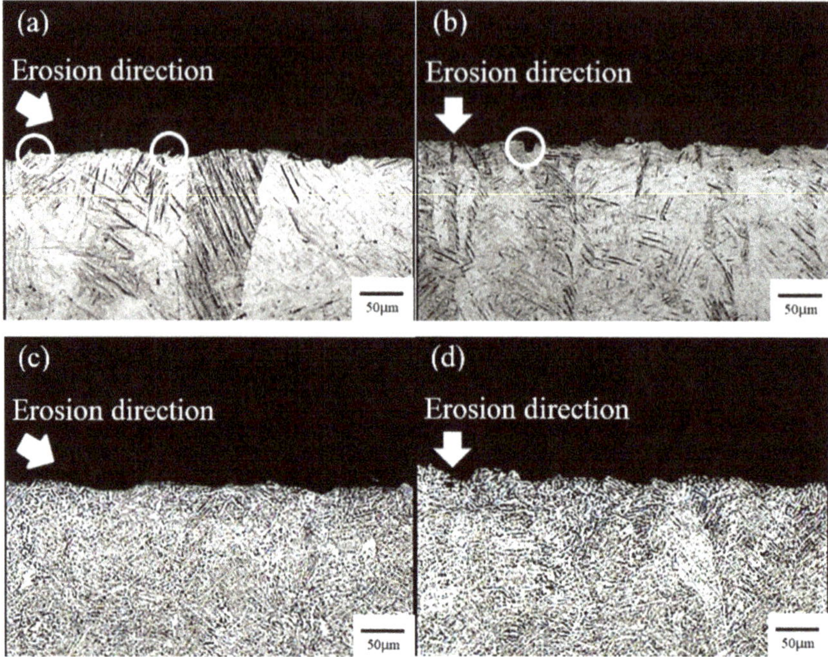

Figure 13. Subsurface morphologies of AS at (**a**) 30° and (**b**) 90°. Surface morphologies of 800-AC at (**c**) 30° and (**d**) 90°.

Figure 14 provides the erosion mechanism schematic diagrams of AS and 800-AC at impacts of 30° and 90°, respectively. Figure 14a is the erosion mechanism of AS at 30° impact, where it can be seen that due to the inability of the needle-like martensitic α' phases to prevent erosion wear due to erosion particles, the lips are formed by the cutting mechanism. When AS is eroded by erosion particles at 90° impact, the surface of the specimen is subjected to the positive impact of erosion particles, which form squeeze features on the erosion wear surface, but the positive impact resistance of AS is weaker than that of 800-AC, thus resulting in the formation of deeper pits, as shown in Figure 14b. When the 800-AC is inclined and eroded by erosion particles at 30° impact, the lamellar $\alpha + \beta$ phases process smaller and fewer lips, as shown in Figure 14c. Figure 14d shows the erosion wear behavior of 800-AC at 90° impact, where some of the continuous lamellar $\alpha + \beta$ phases are scraped off when the erosion particles have a positive impact. However, the damage characteristics are not significant.

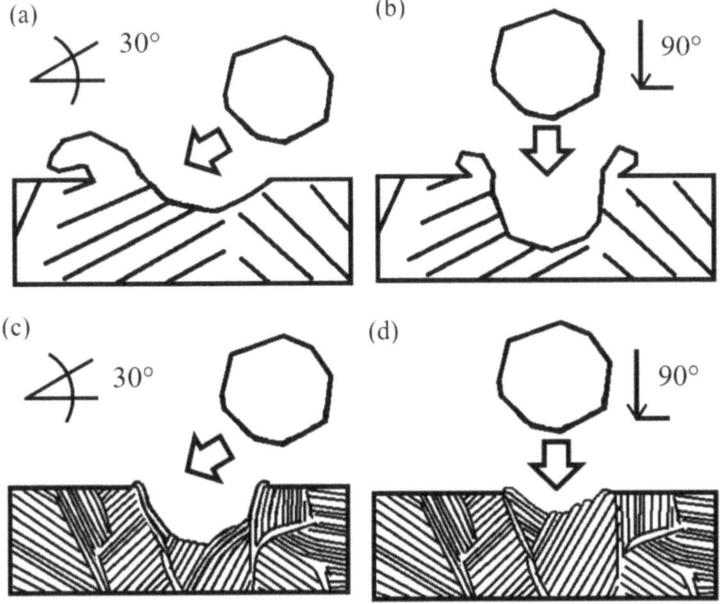

Figure 14. Particle erosion mechanism of AS at (**a**) 30° and (**b**) 90°. Particle erosion mechanism of 800-AC at (**c**) 30° and (**d**) 90°.

According to the discussion above regarding the eroded surface morphology, including the subsurface morphology of the erosion and the erosion mechanism, although the ductility of 800-AC was not as good as that of AS, its continuous lamellar $\alpha + \beta$ phases were tough enough to limit the ability of the erosion particles to scrape off the material at all impact angles. Thus, the erosion resistance of 800-AC was determined to be better than that of AS. The SLM Ti-6Al-4V specimen must be heat treated to have wear resistance.

4. Limitations

The particle erosion induced phase transformation on 3D titanium alloy needs further investigation and analysis.

5. Conclusions

(1) The microstructure of as-SLM Ti-6Al-4V appears as a needle-like martensitic α' phase. Depending on different heat treatment temperatures, there are different phase compositions. The phase composition of 400 °C is $\alpha' + \alpha$, 600 °C is $\alpha' + \alpha + \beta$, and at 800 °C, the α' phase is completely transformed into the lamellar $\alpha + \beta$ phases. The phase composition of 800-AC is $\alpha + \beta$.

(2) For the SLM Ti-6Al-4V specimen, the high temperature strength decreases as the temperature continues to increase after 400 °C, and the ductility increases as the temperature rises. The as-SLM Ti-6Al-4V could maintain a certain strength under 400 °C. Among the specimens, the highest UE is at 350 °C, so the tensile properties at 350 °C are most appropriate for the application of interest.

(3) Both AS and 800-AC had maximum erosion rates at 30° impact, and minimum erosion rates at 90° impact, where ductile failure dominated the erosion behavior. The continuous lamellar $\alpha + \beta$ phases of 800-AC were tough enough to limit the ability of the erosion particles to scrape off the material at all impact angles.

Author Contributions: Methodology, J.-R.Z. and Y.-L.W.; investigation, J.-R.Z. and Y.-L.W.; data curation, J.-R.Z.; writing—original draft preparation, J.-R.Z.; writing—review and editing, F.-Y.H. and T.-S.L.; supervision, F.-Y.H. and T.-S.L.

Funding: This research received no external funding.

Acknowledgments: The authors are grateful to The Instrument Center of National Cheng Kung University and the Ministry of Science and Technology of Taiwan (Grant No. MOST 107-2221-E-006-012-MY2) for their financial support for this research.

Conflicts of Interest: The authors declare no conflict of interest.

References

1. Yadroitsev, I.; Krakhmalev, P.; Yadroitsava, I. Selective laser melting of Ti6Al4V alloy for biomedical applications: Temperature monitoring and microstructural evolution. *J. Alloy Compd.* **2014**, *583*, 404–409. [CrossRef]
2. Song, B.; Dong, S.; Liao, H.; Coddet, C. Process parameter selection for selective laser melting of Ti6Al4V based on temperature distribution simulation and experimental sintering. *Int. J. Adv. Manuf. Technol.* **2012**, *61*, 967–974. [CrossRef]
3. Kelly, C.N.; Evans, N.T.; Irvin, C.W.; Chapman, S.C.; Gall, K.; Safranski, D.L. The effect of surface topography and porosity on the tensile fatigue of 3D printed Ti-6Al-4V fabricated by selective laser melting. *Mater. Sci. Eng. C* **2019**, *98*, 726–736. [CrossRef]
4. Ahmadi, M.; Karpat, Y.; Acar, O.; Kalay, Y.E. Microstructure effects on process outputs in micro scale milling of heat treated Ti6Al4V titanium alloys. *J. Mater. Process. Tech.* **2018**, *252*, 333–347. [CrossRef]
5. Oh, S.T.; Woo, K.D.; Kim, J.H.; Kwak, S.M. The Effect of Al and V on Microstructure and Transformation of β Phase during Solution Treatments of Cast Ti-6Al-4V Alloy. *Korean J. Met. Mater.* **2017**, *55*, 150–155.
6. Zhang, Z.X.; Qu, S.J.; Feng, A.H.; Hu, X.; Shen, J. Microstructural mechanisms during multidirectional isothermal forging of as-cast Ti-6Al-4V alloy with an initial lamellar microstructure. *J. Alloy Compd.* **2019**, *773*, 277–287. [CrossRef]
7. Matsumoto, H.; Velay, V. Mesoscale modeling of dynamic recrystallization behavior, grain size evolution, dislocation density, processing map characteristic, and room temperature strength of Ti-6Al-4V alloy forged in the (α + β) region. *J. Alloy Compd.* **2017**, *708*, 404–413. [CrossRef]
8. Khorasani, A.; Gibson, I.; Awan, U.S.; Ghaderi, A. The effect of SLM process parameters on density, hardness, tensile strength and surface quality of Ti-6Al-4V. *Addit. Manuf.* **2019**, *25*, 176–186. [CrossRef]
9. Wang, Z.; Xiao, Z.; Tse, Y.; Huang, C.; Zhang, W. Optimization of processing parameters and establishment of a relationship between microstructure and mechanical properties of SLM titanium alloy. *Opt. Laser. Technol.* **2019**, *112*, 159–167. [CrossRef]
10. Sun, J.; Yang, Y.; Wang, D. Parametric optimization of selective laser melting for forming Ti6Al4V samples by Taguchi method. *Opt. Laser. Technol.* **2013**, *49*, 118–124. [CrossRef]
11. Song, B.; Dong, S.; Zhang, B.; Liao, H.; Coddet, C. Effects of processing parameters on microstructure and mechanical property of selective laser melted Ti6Al4V. *Mater. Design.* **2012**, *35*, 120–125. [CrossRef]
12. Dai, W.S.; Chen, L.H.; Lui, T.S. A study on SiO_2 particle erosion of flake graphite and spheroidal graphite cast irons. *Wear* **2000**, *239*, 143–152. [CrossRef]
13. Cai, F.; Huang, X.; Yang, Q. Mechanical properties, sliding wear and solid particle erosion behaviors of plasma enhanced magnetron sputtering CrSiCN coating systems. *Wear* **2015**, *324–325*, 27–35. [CrossRef]
14. Liou, J.W.; Lui, T.S.; Chen, L.H. SiO_2 particle erosion of A356.2 aluminum alloy and the related microstructural changes. *Wear* **1997**, *211*, 169–176. [CrossRef]
15. Hung, F.Y.; Chen, L.H.; Lui, T.S. Phase Transformation of an Austempered Ductile Iron during an Erosion Process. *Mater. Trans.* **2004**, *45*, 2981–2986. [CrossRef]
16. Chen, H.; Xu, C.; Chen, J.; Zhao, H.; Zhang, L.; Wang, Z. Microstructure and phase transformation of WC/Ni60B laser cladding coatings during dry sliding wear. *Wear* **2008**, *264*, 487–493. [CrossRef]
17. Baitimerov, R.M.; Lykov, P.A.; Radionova, L.V.; Akhmedianov, A.M.; Samoilov, S.P. An investigation of high temperature tensile properties of selective laser melted Ti-6Al-4V. In Proceedings of the 3rd International Conference on Progress in Additive Manufacturing (Pro-AM 2018), Nanyang Technological University, Singapore, 14–17 May 2018.

18. Da Silva, S.L.R.; Kerber, L.O.; Amaral, L.; dos Santos, C.A. X-ray diffraction measurements of plasma-nitrided Ti–6Al–4V. *Surf Coat Tech.* **1999**, *116–119*, 342–346. [CrossRef]
19. Xu, W.; Brandt, M.; Sun, S.; Elambasseril, J.; Liu, Q.; Latham, K.; Xia, K.; Qian, M. Additive manufacturing of strong and ductile Ti–6Al–4V by selective laser melting via in situ martensite decomposition. *Acta Mater.* **2015**, *85*, 74–84. [CrossRef]
20. Thijs, L.; Verhaeghe, F.; Craeghs, T.; Humbeeck, J.V. A study of the microstructural evolution during selective laser melting of Ti-6Al-4V. *Acta Metall.* **2010**, *58*, 3303–3312. [CrossRef]
21. Lütjering, G. Influence of processing on microstructure and mechanical properties of (α + β) titanium alloys. *Mate. Sci. Eng. A* **1998**, *243*, 32–45. [CrossRef]
22. Do, D.K.; Li, P. The effect of laser energy input on the microstructure, physical and mechanical properties of Ti-6Al-4V alloys by selective laser melting. *Virtual Phys. Prototyping* **2016**, *11*, 41–47. [CrossRef]
23. Tao, P.; Li, H.X.; Huang, B.Y.; Hu, Q.D.; Gong, S.L.; Xu, Q.Y. Tensile behavior of Ti-6Al-4V alloy fabricated by selective laser melting: Effects of microstructures and as-built surface quality. *Chin. Foundry* **2018**, *15*, 243–252. [CrossRef]
24. Vrancken, B.; Thijs, L.; Kruth, J.P.; Humbeeck, J.V. Heat treatment of Ti-6Al-4V produced by Selective Laser Melting: Microstructure and mechanical properties. *J. Alloys Compd.* **2012**, *541*, 177–185. [CrossRef]
25. Bush, R.W.; Brice, C.A. Elevated temperature characterization of electron beam freeform fabricated Ti–6Al–4V and dispersion strengthened Ti–8Al–1Er. *Mater. Sci. Eng. A* **2012**, *554*, 12–21. [CrossRef]
26. Kim, J.; Kim, K.H.; Kwon, D. Evaluation of High-Temperature Tensile Properties of Ti-6Al-4V Using Instrumented Indentation Testing. *Met. Mater. Int.* **2016**, *22*, 209–215. [CrossRef]
27. Sahoo, R.; Mantry, S.; Sahoo, T.K.; Mishra, S.; Jha, B.B. Effect of Microstructural Variation on Erosion Wear Behavior of Ti-6Al-4V Alloy. *Tribol. Trans.* **2013**, *56*, 555–560. [CrossRef]

Article

Effect of Fe Addition on Heat-Resistant Aluminum Alloys Produced by Selective Laser Melting

Shigeto Yamasaki [1,*], **Tomo Okuhira** [1], **Masatoshi Mitsuhara** [1], **Hideharu Nakashima** [1], **Jun Kusui** [2] **and Mitsuru Adachi** [3]

1   Faculty of Engineering Sciences, Kyushu University, 6-1 Kasugakoen, Kasuga, Fukuoka 816-8580, Japan; okuhira.tomo.922@s.kyushu-u.ac.jp (T.O.); mitsuhara@kyudai.jp (M.M.); nakashima.hideharu.792@m.kyushu-u.ac.jp (H.N.)
2   Powder & Paste Headquarters, Toyo Aluminium K. K., Osaka 541-0056, Japan; jun-kusui@toyal.co.jp
3   KOIWAI Co., Ltd., Odawara 256-0804, Japan; m-adachi@tc-koiwai.co.jp
*   Correspondence: yamasaki.shigeto.259@m.kyushu-u.ac.jp; Tel.: +81-92-583-7524

Received: 26 March 2019; Accepted: 20 April 2019; Published: 22 April 2019

Abstract: The effect of Fe addition on the high-temperature mechanical properties of heat-resistant aluminum alloys produced by selective laser melting (SLM) was investigated in relation to the alloy microstructures. Fe is generally detrimental to the properties of cast aluminum alloys; however, we found that Fe-containing alloys produced by SLM had improved high-temperature strength and good ductility. Microstructural observations revealed that the increase in the high-temperature strength of the alloys was due to the dispersion of fine rod-shaped Fe-Si-Ni particles unique to the SLM material instead of the cell-like structure of eutectic Si.

Keywords: aluminum alloys; selective laser melting; high-temperature deformation; microstructure

1. Introduction

Selective laser melting (SLM) is an additive manufacturing method in which three-dimensional structures are formed by repeated melting and solidification of a metal powder by a laser [1]. SLM has been applied to various metal materials [2]. From a metallurgical perspective, the biggest advantage of SLM is that the cooling rate is much faster than that of the conventional casting method [3], and a characteristic microstructure is formed in the metals and alloys produced by SLM due to rapid solidification. In many studies, SLM has been applied to cast Al alloys containing 10–12% Si [4–9]. The microstructure of these alloys features a half-cylindrical melt pool [4,6] and a eutectic Si cellular structure [4–7] formed by quenching solidification due to the repeated laser scanning melting and solidification. This unique microstructure makes the room-temperature strength of SLM alloys higher than that of cast Al alloys [10,11] or Al alloys made by powder metallurgy [12]. The improvement of the room-temperature strength of Al alloys by SLM has been explained by the cellular structure of fine Si particles [12].

An important mechanical property for engine piston materials that are required to be lightweight to reduce inertia is high-temperature strength to withstand high pressure from combustion gases, even for small parts, as well as wear resistance. Heat-resistant cast Al alloys, obtained by adding an element such as Fe, Ni, Cu, or Mg to an Al–Si alloy near to the eutectic composition, are mainly used as a piston material in automobile engines. The demand for materials that retain their mechanical properties at high temperatures (250–350 °C) is increasing owing to the pressure to improve the efficiency and decrease the emissions of automobile engines. The high-temperature mechanical properties of Al alloys have been improved by optimizing the number of additional elements [13] or adding various elements, such as Mn, Ti, Zr, Sc, or Er, to existing heat-resistant Al alloys [14–16].

Fe is usually present in the Al alloy as an impurity. Fe is detrimental to the properties of cast Al alloys, but an Fe content of about 1 wt % or less is intentionally added to Al die-cast alloys to suppress die soldering [17]. If the cooling rate during casting is slow, the addition of Fe causes coarse plate-like β-Al5FeSi to crystallize, substantially decreasing toughness [18,19]. However, because Fe has a low diffusion rate in Al, Fe-containing compounds can be used as a dispersion-strengthening phase at high temperatures. Al alloys produced by pressure sintering of rapidly solidified powders containing large amounts of Fe show excellent high-temperature strength [20,21]. Thus, if an extremely high cooling rate is achieved by SLM, it is expected that Al alloys with a high Fe content could be produced without ductility deterioration. Ma et al. [22] prepared an Al-20Si-5Fe-3Cu-1Mg alloy containing 5 wt % Fe by SLM and characterized its microstructure, although they did not evaluate the mechanical properties. In the present work, we used SLM to produce alloys obtained by adding 3–5 wt % Fe to existing heat-resistant Al alloys and evaluated their high-temperature mechanical properties in relation to the alloy microstructure on the conditions assuming application in engine piston materials.

2. Materials and Methods

Table 1 shows the composition of the alloys prepared in this study. The chemical composition of these alloys was based on a conventional heat-resistant alloy AlSi-12CuNiMg (known as AC8A in Japan) and 3% or 5% Fe added to it (in this paper, all percentages for the chemical composition are wt % unless otherwise stated). The sample, equivalent to the chemical composition of the conventional alloy, is referred to as Base alloy, and the samples containing 3% or 5% Fe are referred to as Base + 3Fe or Base + 5Fe, respectively. The SLM alloy powders, which had a true spherical shape as shown in Figure 1, were prepared by atomization at Toyo Aluminium K.K (Hino, Japan). The compositions of the alloy powders are shown in Table 1.

Table 1. Chemical compositions of the alloys (wt %).

Alloy	Al	Si	Cu	Mg	Ni	Fe
Base	bal.	11.6	0.97	0.96	1.0	0.1
Base + 3Fe	bal.	11.6	0.97	0.96	1.0	3.0
Base + 5Fe	bal.	11.6	0.97	0.96	1.0	5.0

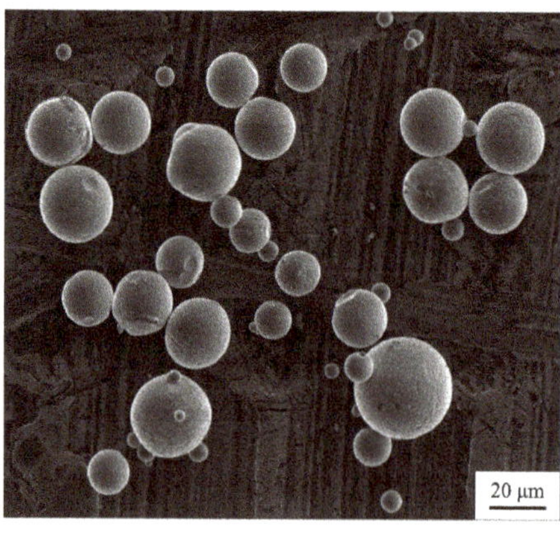

Figure 1. SEM image of the selective laser melting (SLM) powder of Base + 5Fe alloy.

An industrial three-dimensional printer (M280, EOS GmbH, Krailling, Germany) was used for SLM and the details of the manufacturing conditions are summarized in Table 2. Rectangular parallelepipeds of 10 mm × 10 mm × 40 mm or 80 mm (stacking height × width × length) were produced. The energy density, E (J/m^3), was calculated by the following formula [23,24]:

$$E = \frac{P}{hvw} \tag{1}$$

where P (J/s) is the laser power, h (m) is the scan line spacing, v (m/s) is the scan rate, and w (m) is the single-layer thickness of the powder. The energy density is the amount of energy of the irradiation beam per unit volume applied to the powder material and is an important condition for producing a sound SLM body. Also, preheating of the substrate was also necessary to suppress destruction of the SLM body during SLM. The conditions summarized in Table 2 are values found to be optimal from prior examinations for producing the SLM bodies of the present alloy.

Table 2. Parameters of selective laser melting.

Powder Size	45 µm
Layer thickness	30 µm
Stacking speed	approx. 3 mm/h
Stacking height	10 mm
Laser beam size	100 µm
Energy density of laser	7.3×10^{10} J/m^3
Preheat temperature	200 °C

In AlSi-12CuNiMg (AC8A), which was the Base alloy in this work, Cu and Mg are added at 1%, so T6 treatment causes age hardening. When this alloy is practically applied to piston materials, it is used after T6 treatment. Therefore, we prepared a T6-treated sample for comparison with conventional materials. A piston material that is used at a high temperature of about 300 °C for a long time is required to have a stable microstructure at the temperature used. In addition, the rapidly solidified SLM material is a highly supersaturated solid solution (SSSS) in the as-built state. Elements are dissolved in the supersaturation precipitate during high-temperature holding, causing dimensional change of the SLM body. In the case of engine pistons, this causes serious problems. Therefore, annealing was performed at 300 °C in order to make the strength and dimensional changes less likely to occur when used for a long time at high temperatures by causing precipitation from an SSSS and growth of precipitates in advance.

Two series of samples with different heat-treatment sequences for the SLM bodies were prepared. In the first series, the samples were annealed at 300 °C after heat treatment equivalent to T6: 510 °C for 2 h (water quenching) followed by 170 °C for 4 h (air cooling). In the second series, the samples were annealed at 300 °C. The annealing details are described later. The first series of samples is called SLM-T6 and the second is called SLM-annealed. For comparison, two series of mold-cast samples made from Base alloy and Base + 3Fe alloy powders were prepared and subjected to the same heat-treatment sequences. The series subjected to T6 heat treatment is called cast-T6, and the series annealed at 300 °C is referred to as cast-annealed. The cast-T6 Base alloy was a sample equivalent to conventional AlSi-12CuNiMg alloy for practical use. All the samples were annealed at 300 °C for 10 h followed by air cooling before processing to a tensile test piece. The samples were sliced into flat plates 10 mm × 1.5 mm × 40 mm in size with the plate surface perpendicular to the stacking direction. Then, plates were punched into dog-bone-shaped test pieces with a length of 12 mm and a width of 4 mm for the gauge part. The bending caused by punching was smoothed by grinding. The rough surface of the as-produced SLM material was removed in the sample processing. To remove strain, the tensile test pieces were subjected to additional strain-relieving annealing at 300 °C for 2 h. The heat-treatment history of all the samples is summarized in Figure 2. The grain size of the SLM material after heat treatment was about 20 µm, regardless of the heat-treatment conditions.

The tensile test was carried out at 300 °C. The test piece was placed in an electric furnace, and after holding the furnace temperature at 300 °C for about 1 h, the tensile test was started at a strain rate of 6.9×10^{-3} s^{-1}. The microstructures were characterized by observing the backscattered electron (BSE) images obtained by field-emission scanning electron microscopy (SEM; ULTRA 55, Carl Zeiss, Oberkochen, Germany) and the high-angle annular dark field (HAADF) images obtained by scanning transmission electron microscopy (STEM; TITAN Cubed G2, FEI, Waltham, MA, USA). Element mapping by energy-dispersive X-ray spectroscopy (EDS) using STEM was also performed. The samples for the SEM observations were finished by colloidal silica polishing. Thin-film samples for the STEM observations were prepared by focused ion beam microsampling.

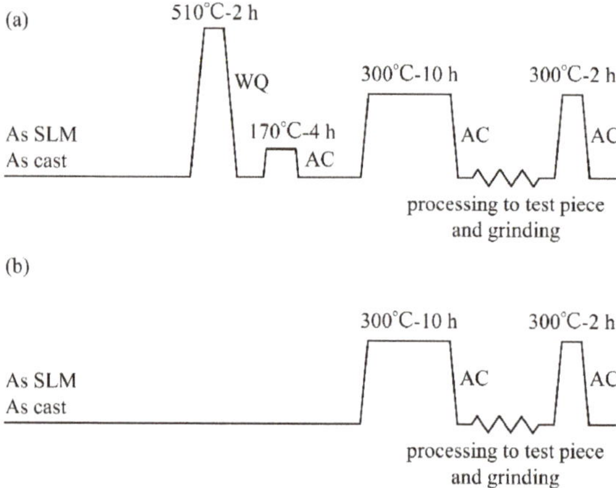

Figure 2. Heat-treatment history of (**a**) SLM-T6 and cast-T6 and (**b**) SLM-annealed and cast-annealed.

3. Results and Discussion

Figure 3 shows the true stress–true strain curves of the SLM-annealed, cast-annealed, SLM-T6, and cast-T6 materials at 300 °C. The ultimate tensile strength (UTS) of the SLM-annealed Base alloy was 77 MPa and the elongation to fracture (EL) was about 31%. On the other hand, the UTS and the EL for the cast-annealed Base alloy were 67 MPa and 11%, respectively. The alloy produced by SLM had good ductility but did not differ much from the cast material in high-temperature strength. However, the SLM-annealed Fe-containing alloys showed higher UTS than the SLM-annealed Base alloy at 300 °C. The UTS of the SLM-annealed Base + 3Fe and Base + 5Fe alloys at 300 °C were 165 and 185 MPa, respectively. The UTS increased with the amount of Fe. The UTS of the cast-T6 Base alloy, which can be considered as the conventional AlSi-12CuNiMg alloy, was 94 MPa. The UTS of the SLM-annealed Base + 5 Fe alloy at 300 °C reached about twice that of the cast-T6 Base alloy. The high-temperature strength of the Base alloy produced by casting was increased by the T6 heat treatment. In contrast, for alloys made with SLM, T6 heat treatment caused a reduction in strength. This deterioration of the UTS was remarkable in the Fe-containing alloys.

The UTS and the EL at 300 °C of the sample prepared in this study and the novel cast Al alloys recently reported [25–28] are summarized in Table 3. The UTS at 300 °C of the Fe-added SLM alloys was comparable to that of novel cast Al alloys, the strength of which is improved by the addition of Sr, Ti, Zr, and V. However, it should be noted that UTS at high temperatures depends on the strain rate. Since the strain rate in this study was larger than the literature value, it was considered that UTS was evaluated somewhat higher. The elongation to fracture of the SLM-annealed Base + 3Fe and Base + 5Fe alloy at 300 °C was 20% and 24%, respectively, values which were larger than those of the cast-T6 Base alloy or other novel cast Al alloys [25,26]. The cast Fe-containing alloys were so brittle that the specimens were destroyed when the they were punched; thus, the tensile test could not be performed for these materials. When a large amount of Fe is added to cast Al alloys, the coarse plate-shaped β-Al$_5$FeSi phase crystallizes out. Pores are formed in the material because the molten metal cannot fill the region surrounding the coarse plate-shaped particles [29], and the toughness is degraded. The Fe-containing alloys produced by SLM showed elongation to fracture of at least 20% at 300 °C. Thus, using SLM allowed Fe to be added without impairing toughness. In general, there is a trade-off between strength and ductility. However, the SLM-annealed Base + 3Fe and Base + 5Fe alloys had high strength and ductility. That is, a material which has a good strength-ductility balance can be produced by using SLM. Similar results have been reported in previous studies and have been explained by the microstructure specific to SLM [11,30].

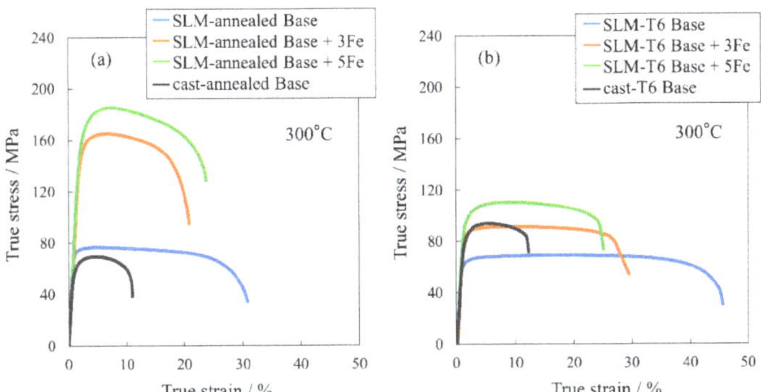

Figure 3. True stress-true strain curves of the SLM material and cast material at 300 °C for (**a**) SLM-annealed and cast-annealed and (**b**) SLM-T6 and cast-T6.

Table 3. Summary of the tensile properties at 300 °C.

Alloy	Production Method	Heat Treatment	Strain Rate (s^{-1})	UTS (MPa)	EL (%)	Ref.
cast-T6 Base (Al-12Si-1Cu-1Mg-1Ni)	cast	T6 + 300 °C-12 h	6.9×10^{-3}	94	12	This work
SLM-annealed Base	SLM	300 °C-12 h	6.9×10^{-3}	77	31	This work
SLM-annealed Base + 3Fe	SLM	300 °C-12 h	6.9×10^{-3}	165	20	This work
SLM-annealed Base + 5Fe	SLM	300 °C-12 h	6.9×10^{-3}	185	24	This work
Al-7Si-1Cu-0.5Mg-0.1Fe-Sr-Ti-Zr-V	cast	T6	1.3×10^{-3}	132	12	[25]
Al-13Si-5Cu-0.6Fe-0.6Mn-Mg-Ni-Ti-Zr	cast	T6	$\sim1.7 \times 10^{-3}$	144	7	[26]
Al-9Si-2Cu-0.5Mg-0.2Ni-Sr-Ti-Zr	cast	as cast	4.0×10^{-4}	172	-	[27]
Al-7Si-1Cu-0.5Mg-0.1Fe-Sr-Ti-Zr-V	cast	as cast	1.0×10^{-3}	189	-	[28]

Using SLM with the Base alloy did not dramatically increase the high-temperature strength; however, the Base + 3Fe and Base + 5Fe alloys were strengthened by SLM because of the addition of Fe. The microstructures of these alloys were observed to reveal the reason for the remarkable strengthening by Fe addition. Figure 4 shows the SEM-BSE images of alloys prepared by casting and SLM. In Figure 4, it should be noted that the cast and SLM materials have different scales. In the cast-annealed Base alloy, coarse particles of Chinese script shape or rod shape were observed. These particles were further coarsened by T6 treatment. In the SLM-annealed materials, second-phase particle cells were observed. In the Base alloy, there was a small amount of bright equiaxed particles, whereas in the Fe-containing alloys, there was a large amount of bright rod-shaped particles. In the SLM-T6 materials, both bright equiaxed particles and bright rod-shaped particles were coarsened and no cellular structure was observed.

Figure 5 shows the STEM-EDS mapping of the constituent elements of the second-phase particles in the SLM-annealed Base and Base + 3Fe alloys. In the SLM-annealed Base alloy, Si, Mg, and Cu-Ni particles formed a cellular structure. In the SLM-annealed Base + 3Fe alloy, in addition to these particles, there were many rod-shaped particles consisting mainly of Fe, Si, and Ni. In the EDS element map, particles only enriched with Si were eutectic Si. Eutectic Si is present in cells formed in the Al-Si binary alloy produced by additive manufacturing [4–7]. In the SEM-BSE images (Figure 4), the particles that were slightly brighter than the Al matrix corresponded to eutectic Si. The Si particles in the SLM-annealed materials were about 200 nm in size, whereas the particles were coarsened to several micrometers in the SLM-T6 materials. The bright, equiaxed particles in the SEM-BSE images of the Base alloy were Cu-Ni particles. Based on the EDS element map, the Fe-Si-Ni particles in the SLM-annealed Base + 3Fe alloy were characteristic rod-shaped particles with a width of 100 nm or less and a length of several hundred nanometers. Based on the shape of the particles, the bright rod-shaped particles in the SEM-BSE images of the Fe-containing alloys corresponded to these Fe-Si-Ni particles.

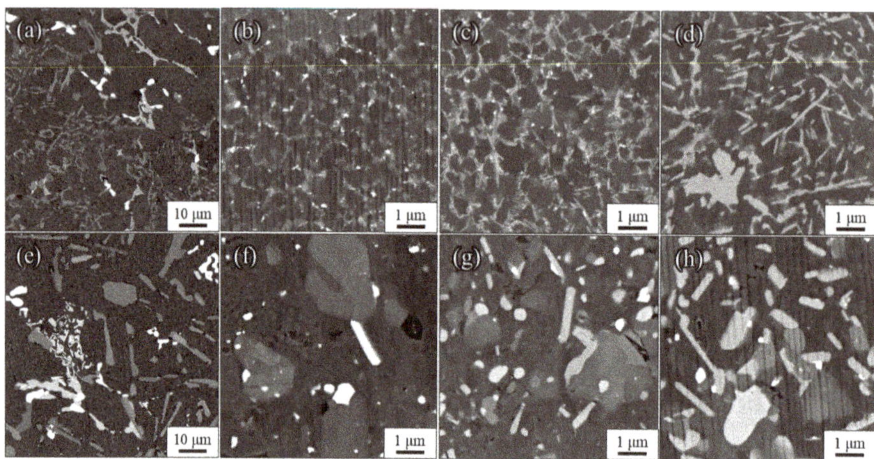

Figure 4. SEM-backscattered electron (BSE) images of cast and SLM alloys. (**a**) Cast-annealed Base, (**b**) SLM-annealed Base, (**c**) SLM-annealed Base + 3Fe, and (**d**) SLM-annealed Base + 5Fe, (**e**) Cast-T6 Base, (**f**) SLM-T6 Base, (**g**) SLM-T6 Base + 3Fe, and (**h**) SLM-T6 Base + 5Fe alloys.

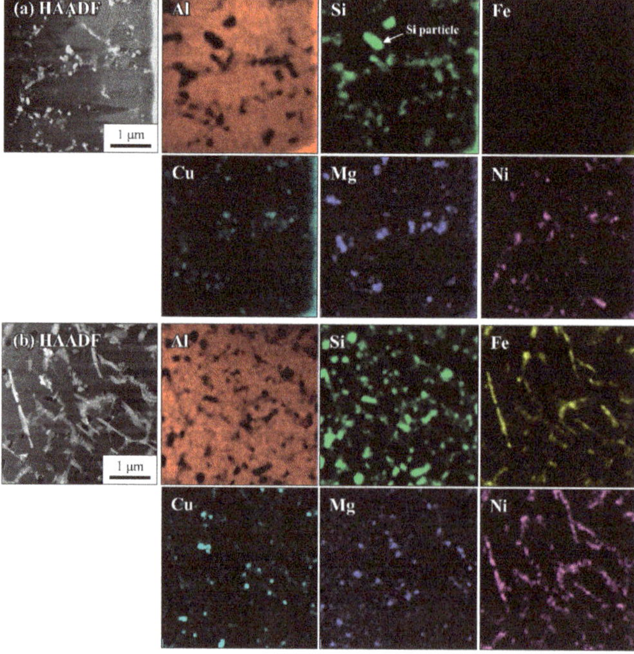

Figure 5. Scanning transmission electron microscopy–high-angle annular dark field (STEM-HAADF) images and corresponding energy-dispersive X-ray spectroscopy (EDS) element maps for (**a**) SLM-annealed Base and (**b**) SLM-annealed Base + 3Fe.

The Fe-Si-Ni particles in the Base + 3Fe alloy were finer than those in the Base + 5Fe alloy. Similar to the Si particles, the rod-shaped Fe-Si-Ni particles were coarsened by T6 treatment. T6 treatment coarsened the Si particles and destroyed the cellular structure of Si characteristic of the SLM materials. However, as shown in Figure 3, the UTS at 300 °C of the SLM-annealed Base alloy and SLM-T6 Base alloy were 77 and 69 MPa, respectively, which were not significantly different. Therefore, in contrast to the room-temperature strength [12], the cell structure of the Si particles did not contribute to an increase in high-temperature strength. This is also clear from that the UTS of the Base alloys produced by SLM and casting were almost the same in annealed condition. The ductility of the SLM-T6 Base alloy was improved compared with the cast Base alloys, though both alloys contained coarse Si particles. The Si particles in the SLM material that exhibited good ductility had an equiaxed shape, whereas those of the cast material had a Chinese script shape or rod shape. Therefore, it is considered that the equiaxial shape of the Si particles contributed to the improvement of the ductility of the SLM material.

In contrast to the Base alloy, the UTSs of the SLM-T6 Base + 3Fe and Base + 5Fe alloys at 300 °C were about 40% lower than the corresponding SLM-annealed alloys. The T6 treatment coarsened the Si and Fe-Si-Ni particles in the Fe-containing alloys. Because the high-temperature strength did not increase only with the fine cellular Si particles, we concluded that the fine dispersion of rod-shaped Fe-Si-Ni particles was the main reason that the SLM-annealed Base + 3Fe and Base + 5Fe alloys had excellent high-temperature strength. These fine particles were present even after annealing at 300 °C for 12 h; thus, they were stable at temperatures up to 300 °C and were useful for dispersion strengthening at high temperatures. In the annealed material, the rod-shaped Fe-Si-Ni particles formed cells together with Si particles. In future research, we intend to examine the effects of cellular particle distribution and cellular morphology on high-temperature strength.

4. Conclusions

- The Base alloy produced by SLM had superior ductility, although the tensile strength at 300 °C was similar to those manufactured by casting.
- The Base alloy manufactured by SLM had a fine cell structure of Si formed by the rapid solidification unique to this method. Although this microstructure did not increase the high-temperature strength of the material, it contributed to improving the ductility.
- Applying T6 treatment, which is commonly used for cast alloys, to the Base alloy produced by SLM destroyed the fine Si cell structure.
- The Fe-containing alloys produced by SLM showed better high-temperature strength while maintaining good ductility. This was caused by Fe being dispersed as fine Fe-Si-Ni particles due to the quenching solidification unique to SLM.

Author Contributions: S.Y. conducted original writing and data analysis; T.O. conducted experimental work; M.M. conducted data analysis; H.N. had a role in supervision; J.K. manufactured powder materials; M.A. conducted molding by selective laser melting.

Funding: This research received no external funding.

Conflicts of Interest: The authors declare no conflict of interest.

References

1. Hitzler, L.; Merkel, M.; Hall, W.; Ochsner, A. A review of metal fabricated with laser- and powder-bed based additive manufacturing techniques: process, nomenclature, materials, achievable properties, and its utilization in the medical sector. *Adv. Eng. Mater.* **2018**, *20*, 1700658. [CrossRef]
2. Herzog, D.; Seyda, V.; Wycisk, E.; Emmelmann, C. Additive manufacturing of metals. *Acta Mater.* **2016**, *117*, 371–392. [CrossRef]
3. Tang, M.; Pistorius, P.C.; Narra, S.; Beuth, J.L. Rapid solidification: Selective laser melting of AlSi10Mg. *JOM* **2016**, *68*, 960–966. [CrossRef]

4. Thijs, L.; Kempen, K.; Kruth, J.-P.; Humbeeck, J.V. Fine-structured aluminium products with controllable texture by selective laser melting of pre-alloyed AlSi10Mg powder. *Acta Mater.* **2013**, *61*, 1809–1819. [CrossRef]

5. Wu, J.; Wang, X.Q.; Wang, W.; Attallah, M.M.; Loretto, M.H. Microstructure and strength of selectively laser melted AlSi10Mg. *Acta Mater.* **2016**, *117*, 311–320. [CrossRef]

6. Liu, M.; Takata, N.; Suzuki, A.; Kobashi, M. Microstructural characterization of cellular AlSi10Mg alloy fabricated by selective laser melting. *Mater. Des.* **2018**, *157*, 478–491. [CrossRef]

7. Li, X.P.; Wang, X.J.; Saunders, M.; Suvorova, A.; Zhang, L.C.; Liu, Y.J.; Fang, M.H.; Huang, Z.H.; Sercombe, T.B. A selective laser melting and solution heat treatment refined Al–12Si alloy with a controllable ultrafine eutectic microstructure and 25% tensile ductility. *Acta Mater.* **2015**, *95*, 74–82. [CrossRef]

8. Prashanth, K.G.; Scudino, S.; Eckert, J. Defining the tensile properties of Al-12Si parts produced by selective laser melting. *Acta Mater.* **2017**, *126*, 25–35. [CrossRef]

9. Suryawanshi, J.; Prashanth, K.G.; Scudino, S.; Ecker, J.; Prakash, O.; Ramamurty, U. Simultaneous enhancements of strength and toughness in an Al-12Si alloy synthesized using selective laser melting. *Acta Mater.* **2016**, *115*, 285–294. [CrossRef]

10. Prashanth, K.G.; Scudino, S.; Klauss, H.J.; Surreddi, K.B.; Löber, L.; Wang, Z.; Chaubey, A.K.; Kühn, U.; Eckert, J. Microstructure and mechanical properties of Al–12Si produced by selective laser melting: Effect of heat treatment. *Mater. Sci. Eng. A* **2014**, *590*, 153–160. [CrossRef]

11. Fousova, M.; Dvorsky, D.; Vronka, M.; Vojtech, D.; Lejcek, P. The use of selective laser melting to increase the performance of AlSi9Cu3Fe alloy. *Materials* **2018**, *11*, 1918. [CrossRef]

12. Chen, B.; Moon, S.K.; Yao, X.; Bi, G.; Shen, J.; Umeda, J.; Kondoh, K. Strength and strain hardening of a selective laser melted AlSi10Mg alloy. *Scr. Mater.* **2017**, *141*, 45–49. [CrossRef]

13. Jung, J.-G.; Lee, S.-H.; Cho, Y.-H.; Yoon, W.-H.; Ahn, T.-Y.; Ahn, Y.-S.; Lee, J.-M. Effect of transition elements on the microstructure and tensile properties of Ale12Si alloy cast under ultrasonic melt treatment. *J. Alloys Compd.* **2017**, *712*, 277–287. [CrossRef]

14. Liu, K.; Chen, X.-G. Improvement in elevated-temperature properties of Al–13% Si piston alloys by dispersoid strengthening via Mn addition. *J. Mater. Res.* **2018**, *33*, 3430–3438. [CrossRef]

15. Hernandez-Sandoval, J.; Garza-Elizondo, G.H.; Samuel, A.M.; Valtiierra, S.; Samuel, F.H. The ambient and high temperature deformation behavior of Al–Si–Cu–Mg alloy with minor Ti, Zr, Ni additions. *Mater. Des.* **2014**, *58*, 89–101. [CrossRef]

16. De Luca, A.; Dunand, D.C.; Seidman, D.N. Microstructure and mechanical properties of a precipitation strengthened Al-Zr-Sc-Er-Si alloy with a very small Sc content. *Acta Mater.* **2018**, *144*, 80–91. [CrossRef]

17. Wang, L.; Makhlouf, M.; Apelian, D. Aluminium die casting alloys: alloy composition, microstructure, and properties-performance relationships. *Int. Mater. Rev.* **1995**, *40*, 221–238. [CrossRef]

18. Cao, X.; Campbell, J. Morphology of β-Al5FeSi phase in Al-Si cast alloys. *Mater. Trans.* **2006**, *47*, 1303–1312. [CrossRef]

19. Dinnis, C.M.; Taylor, J.A.; Dahle, A.K. As-cast morphology of iron-intermetallics in Al–Si foundry alloys. *Scr. Mater.* **2005**, *53*, 955–958. [CrossRef]

20. Rajabi, M.; Vahidi, M.; Simchi, A.; Davami, P. Effect of rapid solidification on the microstructure and mechanical properties of hot-pressed Al–20Si–5Fe alloys. *Mater. Charact.* **2009**, *60*, 1370–1381. [CrossRef]

21. Prusa, F.; Vojtech, D. Mechanical properties and thermal stability of Al-23Si-8Fe-1Cr and Al-23Si-8Fe-5Mn alloys prepared by powder metallurgy. *Mater. Sci. Eng. A* **2013**, *565*, 13–20. [CrossRef]

22. Ma, P.; Jia, Y.; Prashanth, K.G.; Scudino, S.; Yu, Z.; Eckert, J. Microstructure and phase formation in Al-20Si-5Fe-3Cu-1Mg synthesized by selective laser melting. *J. Alloys Compd.* **2016**, *657*, 430–435. [CrossRef]

23. Simchi, A. Direct laser sintering of metal powders: Mechanism, kinetics and microstructural features. *Mater. Sci. Eng. A* **2006**, *428*, 148–158. [CrossRef]

24. Attar, H.; Ehtemam-Haghighi, S.; Kent, D.; Dargusch, M. Recent developments and opportunities in additive manufacturing of titanium-based matrix composites: A review. *Int. J. Mach. Tools Manuf.* **2018**, *133*, 85–102. [CrossRef]

25. Kasprzak, W.; Amirkhiz, B.; Niewczas, M. Structure and properties of cast Al–Si based alloy with Zr–V–Ti additions and its evaluation of high temperature performance. *J. Alloys Compd.* **2014**, *595*, 67–79. [CrossRef]

26. Wang, E.R.; Hui, X.D.; Chen, G.L. Eutectic Al–Si–Cu–Fe–Mn alloys with enhanced mechanical properties at room and elevated temperature. *Mater. Des.* **2011**, *32*, 4333–4340. [CrossRef]

27. Mohamed, A.M.A.; Samuel, F.H.; Al kahtani, S. Microstructure, tensile properties and fracture behavior of high temperature Al–Si–Mg–Cu cast alloys. *Mater. Sci. Eng. A* **2013**, *577*, 64–72. [CrossRef]

28. Shaha, S.K.; Czerwinski, F.; Kasprzak, W.; Friedman, J.; Chen, D.L. Improving High-Temperature Tensile and Low-Cycle Fatigue Behavior of Al-Si-Cu-Mg Alloys Through Micro-additions of Ti, V, and Zr. *Metall. Mater. Trans. A* **2015**, *46A*, 3063–3078. [CrossRef]

29. Khalifa, W.; Samuel, A.M.; Samuel, F.H.; Doty, H.W.; Valtierra, S. Metallographic observations of β-AlFeSi phase and its role in porosity formation in Al–7%Si alloys. *Int. J. Cast Met. Res.* **2006**, *19*, 156–166. [CrossRef]

30. Wang, Y.M.; Voisin, T.; McKeown, J.T.; Ye, J.; Calta, N.P.; Li, Z.; Zeng, Z.; Zhang, Y.; Chen, W.; Roehling, T.T.; et al. Additively manufactured hierarchical stainless steels with high strength and ductility. *Nat. Mater.* **2018**, *17*, 63–70. [CrossRef]

 *metals* 

Article

The Size Effect on Forming Quality of Ti–6Al–4V Solid Struts Fabricated via Laser Powder Bed Fusion

Huixin Liang [1,2], **Deqiao Xie** [1], **Yuyi Mao** [3], **Jianping Shi** [4], **Changjiang Wang** [5], **Lida Shen** [1] and **Zongjun Tian** [1,*]

[1] College of Mechanical and Electrical Engineering, Nanjing University of Aeronautics and Astronautics, Nanjing 210016, China; hxliang@nuaa.edu.cn (H.L.); dqxie@nuaa.edu.cn (D.X.); ldshen@nuaa.edu.cn (L.S.)

[2] Suzhou Kangli Orthopedics Instrument Co. Ltd., Suzhou 215600, China

[3] National Center of Supervision and Inspection on Additive Manufacturing Products Quality (JIANGSU), Wuxi 214028, China; dean.mao@163.com

[4] Jiangsu Key Laboratory of 3D Printing Equipment and Manufacturing, Nanjing Normal University, Nanjing 210042, China; njsfsjp@163.com

[5] Department of Engineering and Design, University of Sussex, Sussex House, Brighton BN19RH, UK; C.J.Wang@sussex.ac.uk

*   Correspondence: tianzj@nuaa.edu.cn; Tel.: +86-025-8489-2520

Received: 4 March 2019; Accepted: 3 April 2019; Published: 6 April 2019

Abstract: Laser powder bed fusion (LPBF) is useful for manufacturing complex structures; however, factors affecting the forming quality have not been clearly researched. This study aimed to clarify the influence of geometric characteristic size on the forming quality of solid struts. Ti–6Al–4V struts with a square section on the side length (0.4 to 1.4 mm) were fabricated with different scan speeds. Micro-computed tomography was used to detect the struts' profile error and defect distribution. Scanning electron microscopy and light microscopy were used to characterize the samples' microstructure. Nanoindentation tests were conducted to evaluate the mechanical properties. The experimental results illustrated that geometric characteristic size influenced the struts' physical characteristics by affecting the cooling condition. This size effect became obvious when the geometric characteristic size and the scan speed were both relatively small. The solid struts with smaller geometric characteristic size had more obvious size error. When the geometric characteristic size was smaller than 1 mm, the nanohardness and elastic modulus increased with the increase in scan speed, and decreased with the decline of the geometric characteristic size. Therefore, a relatively high scan speed should be selected for LPBF—the manufacturing of a porous structure, whose struts have small geometric characteristic size.

Keywords: selective laser melting; Ti–6Al–4V alloy; metallurgical quality; mechanical properties

---

1. Introduction

As a unique type of structural and functional material, a porous structure has unique advantages in fields including filtration and separation [1], energy absorption [2], heat exchange [3], electromagnetic shielding [4], and artificial implants [5], and is widely used in aerospace, automotive, chemical, and biological medical industries. Depending on the demand, the porous structure can be manufactured from materials including metal [6,7], ceramics [8], and polymers [9]. With the deepening of their application, porous structures with function-oriented design have been widely developed to obtain precise and complex structures, such as porous implants [10]. However, it is difficult to fabricate such structures using the traditional processing method. Additive manufacturing (AM) techniques are attracting increasing attention due to their impressive capability to produce precise parts with a controlled architecture [11]. Using a layer-wise building approach and a direct link with a computer-aided design (CAD) model, AM has been described as a crucial production technique to achieve a more controlled porous structure [12–15]. A recently developed AM technique, the laser powder bed fusion process (LPBF), usually called selective laser melting (SLM), can directly create a functional and complex metal part, like a Ti-based porous scaffold for bone tissue engineering, bringing a high degree of freedom to design [16,17]. Most researchers have focused on evaluating the function or forming quality of as-designed porous samples fabricated by LPBF using a set of universal process parameters embedded in the commercial equipment [6,18–22]. However, a porous structure is composed of the struts or walls, which usually have small geometric characteristic size, so a set of process parameters with a universal nature may not be optimal for manufacturing porous structures. A small number of researchers have focused on the impact of process parameters on the forming quality of LPBF-built porous structures. For example, Liu et al. [23] investigated the effect of the scan speed on the forming defects, precision, and mechanical properties of biomedical titanium alloy scaffolds fabricated by LPBF. Ahmadi et al. [24] studied the effects of laser power on features including the surface roughness, strut diameter, relative density, hardness, and elastic modulus of the porous structures. In addition, Jamshidinia and Kovacevic found that the thin wall achieved more heat accumulation during the LPBF process, affecting the forming quality [25]. These findings indicate that the small geometric characteristic size may be a factor affecting the forming quality of LPBF-built porous structures.

On the basis of the above statement, the effect of the geometric characteristic size of the objective part on the forming quality of the solid strut is discussed in this work. The energy density was introduced to reveal the combined action of scan speed, laser power, hatching space, and layer thickness [26], which represented the thermal input during the LPBF process. As the most commonly used material in metal implant fields, Ti–6Al–4V alloy powder was selected for the experiment. Taking the geometric characteristic size into consideration, the influence rule of the factor on the forming quality of solid struts fabricated via LPBF using different scan speeds was investigated from the perspective of thermal transmission. The entire experiment was made up of two steps. Firstly, struts with different geometric characteristic size were fabricated using different combined process parameters. Secondly, porous structures with two strut sizes were fabricated with two scan speeds to verify the analysis in the first step.

2. Materials and Methods

2.1. Materials

Commercial Ti–6Al–4V alloy powder supplied by EOS GmbH was used in the experiment, meeting ISO 5832-3 and ASTM F1472. As an optimized medical material, the trace elements of Ti–6Al–4V ELI such as O, N, H, C, and Fe are relatively low in content (Table 1). As shown in Figure 1 (SEM, ΣIGMA, Zeiss, German), the powder has high sphericity and few satellite spheres with a particle size range of 15–53 μm.

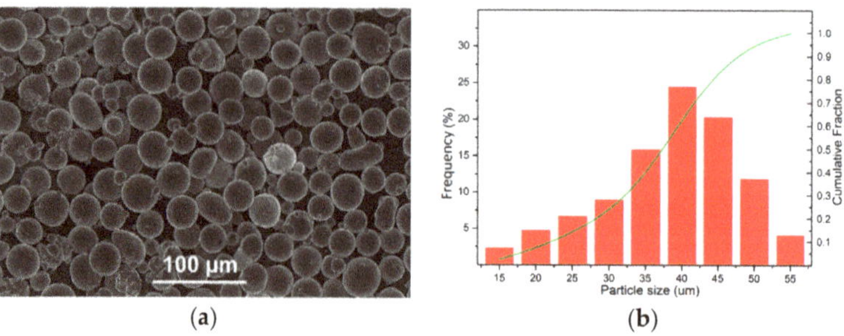

(a)   (b)

Figure 1. (a) SEM micrograph and (b) particle size distribution of Ti–6Al–4V powder.

Table 1. Chemical composition of Ti–6Al–4V powder (wt. %).

Ti	Al	V	O	N	C	H	Fe
Bal.	5.90	3.91	0.12	0.05	0.08	0.012	0.3

2.2. Design and Fabrication

Selective laser melting of a porous structure is characterized by a controllable and precise layer-wise material addition process. This method generates complex structures by selectively melting successive layers of metal powder using a focused and computer-controlled laser beam (Figure 2a). The specimens were fabricated using an LPBF machine (EOSINT M290; EOS GmbH, Munich, Germany), which was equipped with a Yb fiber laser of 400 W with a wavelength range of 1000–1100 nm and a Gaussian spot, and a building chamber filled with argon gas with an oxygen content below 200 ppm.

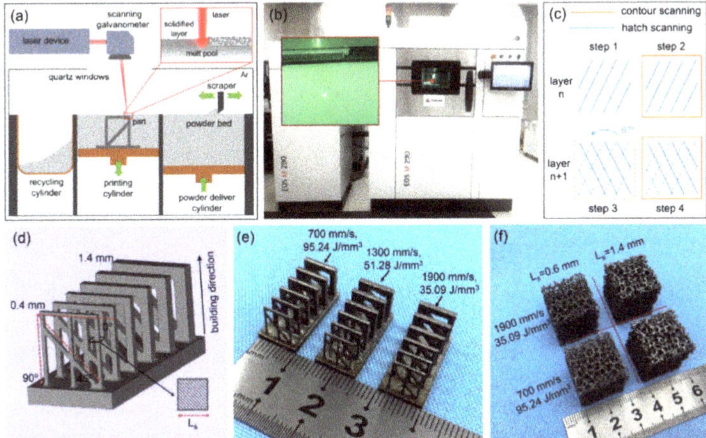

Figure 2. (**a**) Schematic of the laser powder bed fusion process; (**b**) the LPBF machine of EOS M290; (**c**) the selected scanning strategy; (**d**) the as-designed 3D model of solid struts; (**e**) the as-built Ti–6Al–4V solid struts; and (**f**) the as-built Ti–6Al–4V porous structures for the confirmatory experiment.

The volumetric energy density (E_V) was calculated according to Equation (1) [22]:

$$E_V = P/(v \times h \times d),\qquad(1)$$

where P is the laser power, h is the hatching space, v is the scan speed, and d is the layer thickness. This equation takes the most important laser parameters into account and is suitable for calculating the thermal input during the LPBF process. An inside to outside scanning strategy with a meander hatch style was selected. The meander hatch direction rotated by an angle of 67° in the following layer (Figure 2c).

A porous structure is generally made up of struts with relatively small size and various angles. In this work, the sample model consisted of struts with angles of 0°, 45°, and 90°, which had a square section with a side length (L_s) of 0.4 to 1.4 mm (Figure 2d). Three sets of parameters were used in this fabrication process, resulting in different volumetric energy densities of 95.24, 51.28, and 35.09 J/mm³ (Table 2 and Figure 2e). To verify the correctness of the analysis, two kinds of porous structures with as-designed L_s values of 0.6 and 1.4 mm were fabricated via the LPBF process with v values of 700 and 1900 mm/s, respectively (Figure 2f).

Table 2. The selected process parameters for LPBF-fabrication.

Laser Powder/P (W)	Scan Speed/v (mm/s)	Hatch Spacing/h (mm)	Layer Thickness/d (mm)	Laser Spot Diameter/D (mm)
200	700, 1300, 1900	0.1	0.03	0.1

2.3. Measurements and Characterizations

A micro-computed tomography (µCT) scanner (FF35 CT; YXLON International, Hamburg, Germany) with 5 µm resolution was used to scan the samples at 200 kV and 50 µA. The samples were rotated over 360° in steps of 0.18° during the acquisition. Two-dimensional (2D) projection images ($n = 2000$) were then collected. The 3D models of the fabricated samples were reconstructed through slice image data using commercially available software (VG Studio MAX 3.0; Volume Graphics GmbH, Heidelberg, Germany). The same software was also used to detect the size deviation distribution of the as-built samples compared to the 3D model with an error of 5 µm ($n = 3$). The sample section profile and defect distribution (pores) were then extracted from the CT data. After CT scanning, the struts with the same square were cut off together from the base plate via wire electrical discharge machining (WEDM). The relative density of struts with the same square section was measured using the Archimedes method on each set of samples. The Archimedes test results were calculated based on a combination of dry weighing and weighing in pure ethanol and on the theoretical density of 4.43 g/cm^3 for Ti–6Al–4V. Taking open porosity and the highly developed surface of the samples into consideration, the samples were coated with wax after the first dry weighing. The relative density ($\varrho_{\text{relative}}$) of the samples was calculated using Equation (2):

$$\varrho_{\text{relative}} = m_1/[(m_2 - m_3)/\varrho_{\text{ethanol}} - (m_2 - m_1)/\varrho_{\text{wax}}], \tag{2}$$

where m_1 is the mass of the sample without a wax coating in air, m_2 is the mass of the sample with a wax coating in air, and m_3 is the mass of the sample with a wax coating in pure ethanol. Due to the weight of the struts being too small, ten samples of each set were measured together, and the arithmetic mean value of relative density was calculated ($n = 10$). Taking the small size of the struts into consideration, nanoindentation tests on the polished sections of as-built struts with an angle of 90° were performed using a nanoindenter (G200, Agilent, Ltd., Santa Clara, CA, USA) to evaluate the mechanical properties, including the elastic modulus and nanohardness. A loading–unloading test mode was used with a maximum indentation depth of 2000 nm, a loading speed of 10 nm/s, and a hold time of 10 s ($n = 5$). The Oliver-Pharr method [27] was then applied to calculate the nanohardness and elastic modulus. The microstructure of LPBF-produced struts with an angle of 90° was observed using a field-emission scanning electron microscope (S-4800; Hitachi, Ltd., Tokyo, Japan) and a light microscope (GX41; Olympus, Ltd., Tokyo, Japan). To reveal the microstructure, an etchant containing 50 mL distilled water, 25 mL HNO$_3$, and 5 mL HF was used for the polished samples. The difference in phase composition was analyzed by X-ray diffraction (XRD) using an X-ray diffractometer (D/max 2500 PC, Rigaku, Ltd., Tokyo, Japan) with Cu Kα radiation at 40 kV with a beam current of 100 mA. A scan speed of 2°/min was used for the scan range of 30–80° in steps of 0.02°. The porosity (Φ) of the fabricated porous structure was obtained from Equation (3):

$$\Phi = 1 - V_{\text{porous}}/V_{\text{bulk}}, \tag{3}$$

where V_{porous} is the volume of the LPBF-produced structure measured from reconstructed 3D models using commercially available software (Magics 21.0.0; Materialise, Leuven, Belgium) and V_{bulk} is the total volume of the solid cube that has the same outline size as the porous sample ($n = 3$). The main information of all the measuring methods is presented in Table 3.

Table 3. Measuring objectives and methods in this study.

Objective	Method	Apparatus
Size deviation and defect distribution	μCT	YXLON-FF35 CT, 200 Kv-50 μA, VG Studio MAX 3.0
Relative density	Archimedes method	SOPTOP-FA2004, $\varrho_{\text{wax}} = 0.9$ g/cm^3, $\varrho_{\text{ethanol}} = 0.79$ g/cm^3
Porosity of porous structure	$\Phi = 1 - (V_{\text{porous}}/V_{\text{bulk}})$	Materialise-Magics 21.0.0
Mechanical properties	Nanoindentation test	Agilent-G200, $d = 2000$ nm, $v = 10$ nm/s, $t = 10$ s
Phase identification	XRD	Rigaku-D/max 2500 PC, Cu, 40 kV, 100 mA
Microstructure observations	SEM and LM	S-4800-Hitachi, Olympus-GX41

3. Results

3.1. Morphology Features, Relative Density and Defects

The results of the geometric profile error comparison between the as-built samples and the as-designed 3D model are presented in Figure 3 and Table 4. As illustrated in Figure 3, the solid struts with small L_s (0.4, 0.6, 0.8 mm) more easily gained a positive error in size. The higher v made this phenomenon more obvious. The solid struts with larger L_s (>1 mm) had a more stable size. This was in agreement with the results shown in Table 4. Under different v values of 1900, 1300, and 700 mm/s, the struts with an L_s value of 0.4 mm had sizes of 412 ± 20, 465 ± 27, and 489 ± 34 μm, respectively, and the strut with an L_s value of 1.4 mm had sizes of 1432 ± 15, 1421 ± 23, and 1430 ± 51 μm, respectively. For the overhanging struts with an angle of 0°, the actual size was larger than that of the vertical struts. This was because the molten pool would usually generate many tumor forms on the bottom surface, owing to the permeation effect, causing an uneven geometric profile and over-dimension as a condition of low scan speed (Figure 5).

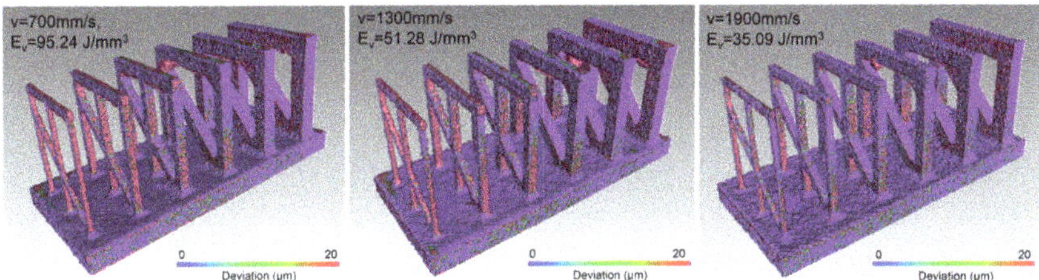

Figure 3. Size deviation distribution of the as-built samples compared to the as-designed 3D model.

Table 4. The average values of measured sizes of the solid struts with L_s values of 0.4 and 1.4 mm (unit: μm, $n = 10$).

No.	L_s (mm)	v (mm/s) 700	1300	1900	Measuring Place
1	0.4	489 ± 34	465 ± 27	412 ± 20	
	1.4	1430 ± 51	1421 ± 23	1432 ± 15	
2	0.4	543 ± 43	507 ± 32	498 ± 26	
	1.4	1427 ± 65	1412 ± 41	1437 ± 38	
3	0.4	705 ± 62	654 ± 54	642 ± 31	
	1.4	1598 ± 76	1553 ± 56	1543 ± 49	

The relative density of the as-built samples is depicted in Figure 4. Under the process conditions of different scan speeds, the struts presented inconsistent variation trends with increasing geometric characteristic size. On one hand, for the struts with small L_s (0.4, 0.6, and 0.8 mm), the relative densities of the samples with a v of 1900 mm/s were highest, separately reaching 96.8%, 96.5%, and 96%. The relative density of samples with a v of 700 mm/s was lowest, only reaching 86%, 90%, and

93%. On the other hand, for the struts with an L_s value of 1.4 mm, the relative density of the samples with a v of 1300 mm/s was highest, reaching 98.1%, and the relative density of samples with a v of 1900 mm/s was lowest at 95%. Thus, it could be seen that the relative density was sensitive to not only scan speed but also to the geometric characteristic size.

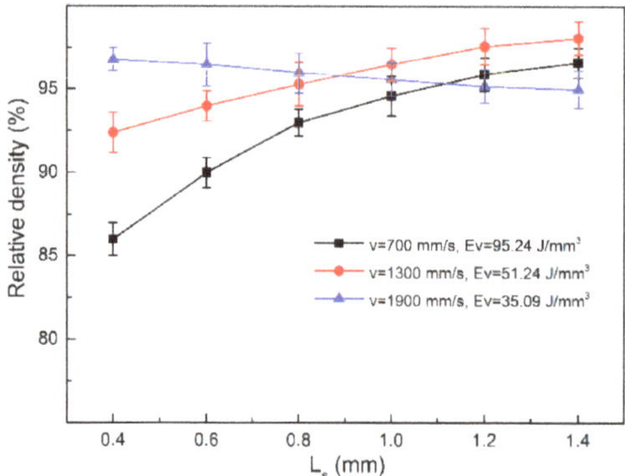

Figure 4. Arithmetic mean value of relative density of struts with different L_s and v values.

In the thresholding CT images shown in Figure 5, the pore distribution of the as-built struts with different dimensions, angles, and scan speeds is roughly demonstrated. This result echoes the relative density, as described in Figure 4. It is apparent that the number of pores decreased with the increase in scan speed for the struts with an L_s value of 0.4 mm, and the struts with an angle of 45° or 0° tended to generate pores. However, the struts with an L_s value of 1.4 mm had the least pores in the condition of a v value of 1300 mm/s.

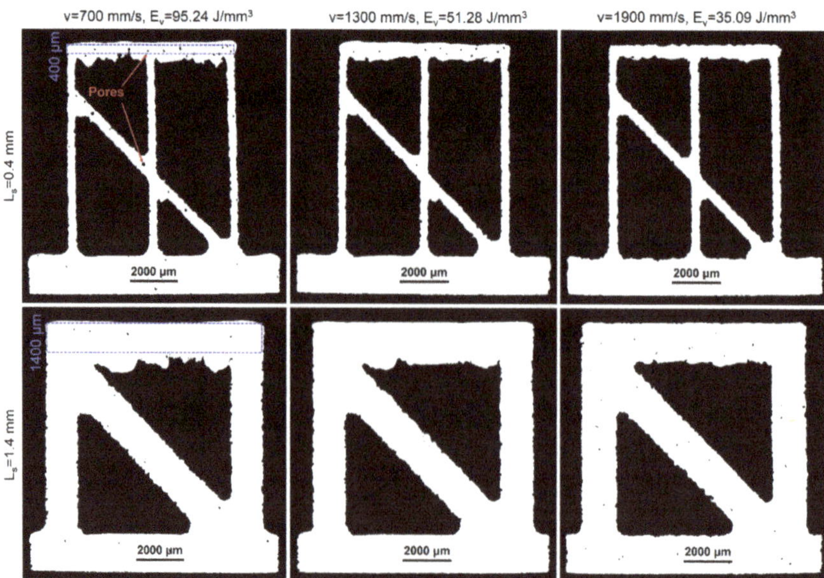

Figure 5. Thresholding single projection CT images of struts with L_s values of 0.4 and 1.4 mm.

3.2. Phase Identification and Microstructure

Figure 6 shows the X-ray diffraction analysis performed to identify phases of the struts with L_s values of 0.4 and 1.4 mm under different process conditions. As shown in Figure 6a, a brief observation of four struts within a wide 2θ range of 30° to 80° revealed that the LPBF-built Ti–6Al–4V samples' phase was mainly composed of α/α' phase. It is known that all of these phases are α' phases, due to the large cooling rate caused by laser melting and solidification [28]. For the struts with an L_s value of 1.4 mm, when the scan speed increased, the diffraction peaks broadened considerably and the intensity decreased, which implied the formation of considerable refined crystal [29]. The XRD characterization with a small 2θ range of 39.5° to 41.5° is depicted in Figure 6b. It shows that the refined 2θ locations of peaks of struts with an L_s value of 1.4 mm generally shifted to the higher 2θ with the increase in scan speed. In addition, under the process condition of a v of 700 mm/s, the peak of the strut with an L_s value of 0.4 mm became thinner than that of the strut with an L_s value of 1.4 mm. This means that the strut with an L_s value of 0.4 mm had a coarser microstructure than did the strut with an L_s value of 1.4 mm. The relatively small geometric characteristic size might have affected the heat dissipation of newly produced solid struts.

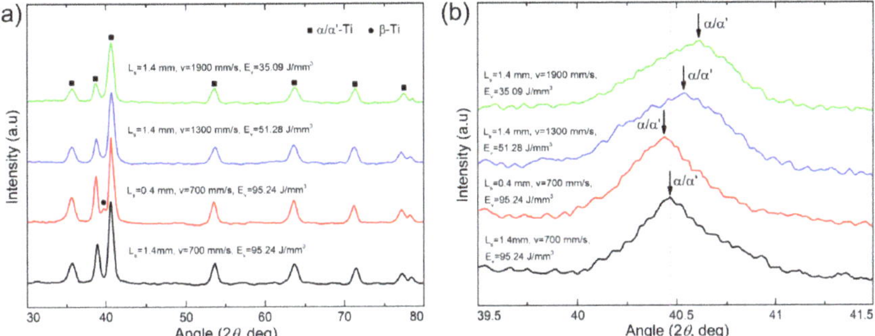

Figure 6. X-ray diffraction analysis results of as-built samples: (**a**) diffraction angle 2θ of 30°–80°; (**b**) diffraction angle 2θ of 39.5° to 41.5°.

To further understand the microstructural differences, observation using an optical metallographic microscope and SEM was conducted on the initial microstructure of the X–Y planes perpendicular to the building direction, and the results are provided in Figures 7 and 8. Figure 7 reveals that the section microstructure was composed of very fine acicular martensites α' and primary columnar β grains with different shapes, which grew along the building direction. The interior of the primary columnar β grains mainly consisted of relatively coarse acicular martensites α' throughout the entire grain. This result is consistent with the findings of the XRD analysis in Figure 6a. Due to the optical metallographic microscope having relatively low magnification and the generation of multilevel martensites α' caused by the repetitive thermal loading of layer-wise laser melting, there was no obvious difference observed in the acicular martensites of struts with different process conditions. However, the primary columnar β grains became thicker and coarser with increasing scan speed. Moreover, as SEM images with 2000× magnification, shown in Figure 8, under the same condition of a v of 700 mm/s, the martensites α' in the strut with an L_s value of 0.4 mm were obviously coarser than that of the struts with an L_s value of 1.4 mm, on the whole.

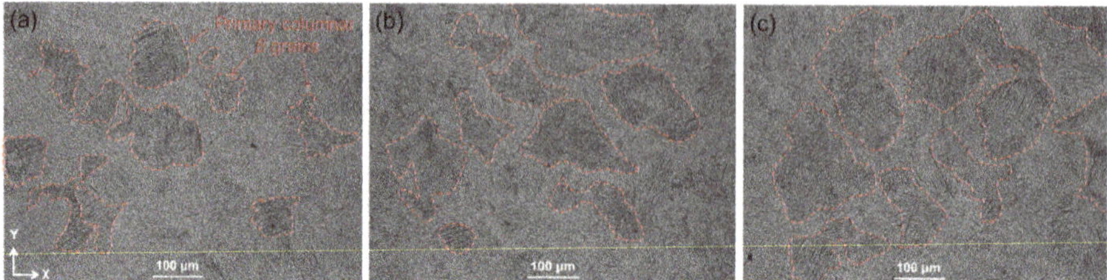

Figure 7. Optical images of microstructures of solid struts on the X–Y plane: (**a**) v = 1900 mm/s, L_s = 1.4 mm; (**b**) v = 1300 mm/s, L_s = 1.4 mm; and (**c**) v = 700 mm/s, L_s = 1.4 mm.

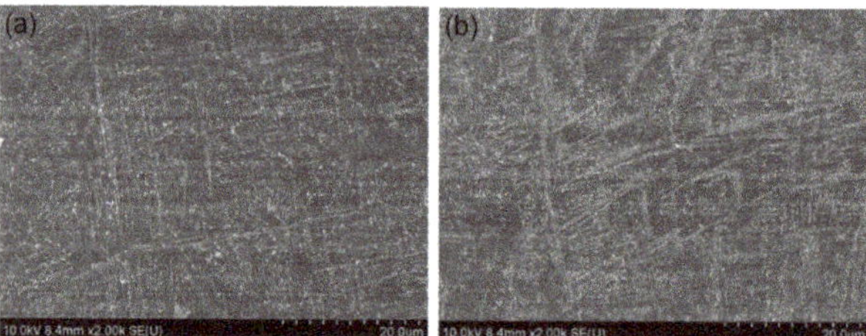

Figure 8. SEM images of the solid struts microstructure of martensites α' on the X–Y plane: (**a**) v = 700 mm/s, L_s = 1.4 mm; (**b**) v = 700 mm/s, L_s = 0.4 mm.

3.3. Mechanical Performance

Figure 9a depicts the nanoindentation load–displacement curves measured on the polished sections of the as-built struts with an angle of 90° and L_s values of 0.4 and 1.4 mm. The indentation depths of the struts with an L_s value of 1.4 mm were larger than those of the struts with an L_s value of 0.4 mm after unloading. As shown in Figure 9b,c, for the struts with a v of 1300 mm/s, the nanohardness and elastic modulus increased with increasing geometric characteristic size. It is noteworthy that the nanohardness and elastic modulus increased very slightly and indistinctively after L_s reached 1 mm. For the struts with an L_s value of 0.4 mm, the nanohardness reached 3.78 to 4.12 GPa and the elastic modulus reached 96.044 to 110.613 GPa. For the struts with an L_s value of 1.4 mm, the nanohardness reached 4.12 to 4.4 GPa and the elastic modulus reached 126.42 to 131.57 GPa. Overall, they all decreased with increasing scan speed.

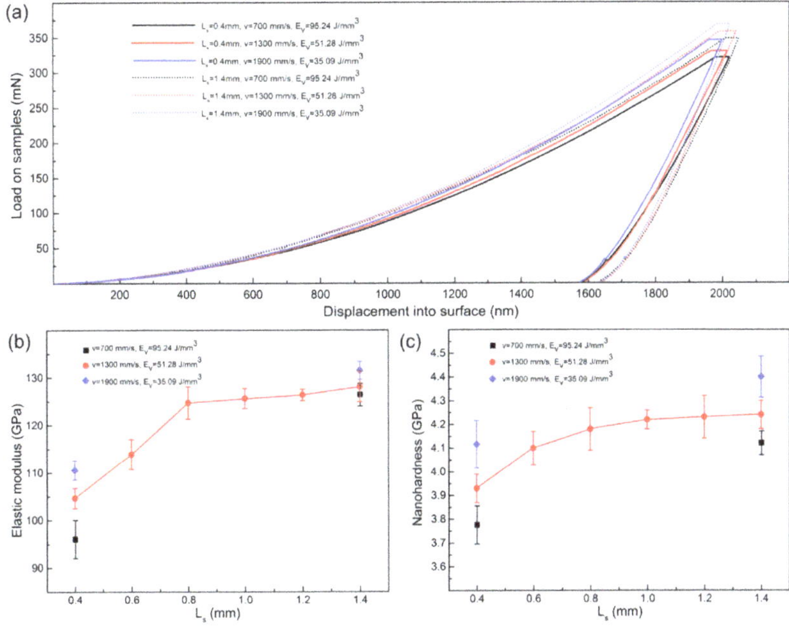

Figure 9. (**a**) Loading–unloading curves; (**b**,**c**) the calculated elastic modulus and nanohardness from the nanoindentation tests.

4. Discussion

The existing experimental results suggested that the geometric characteristic size within a certain range would affect the physical characteristics of an LPBF-fabricated porous structure, including the morphology features, relative density, microstructure, and mechanical properties. This phenomenon can be explained through the simplified thermal transmission model in a laser powder bed fusion process (Figure 10). As a classical additive manufacturing method using a powder bed, three main thermal transferring forms occur during the LPBF process: thermal conduction from molten pool to part, thermal conduction from part to substrate, and thermal radiation from part to atmosphere [30]. It is important to note that thermal conduction between the part and the powder is negligible because the thermal conductivity coefficient of the metal powder is much smaller than that of the metal part [31]. Therefore, the as-built part could be regarded as a thermal container. According to Equation (1), the energy density is inversely proportional to the scan speed. The energy density represents its thermal source strength and the geometric characteristic size in a certain range reflects its own thermal storage volume and thermal-sinking capability. These two aspects will influence the temperature distribution of the as-built part. Higher energy density means that the molten pool will receive more heat from the laser beam, resulting in higher peak temperature, longer keeping time, and decreased undercooling degree [28]. A geometric characteristic size that is too small means that the molten pool and as-built part have a poor cooling condition, and less thermal storage volume and interfacial area with atmosphere and substrate, which results in a lower cooling rate. However, the phenomenon may gradually emerge when the geometric characteristic size is small enough (<1 mm).

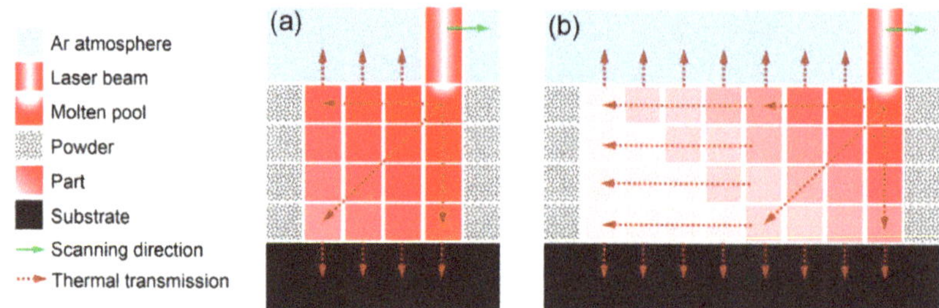

Figure 10. Diagram of the thermal transmission model in the selective laser melting process. (**a**) Small geometric characteristic size; (**b**) large geometric characteristic size.

As shown in Figure 10, a higher energy density means that the molten pool has less surface tension and better wettability, absorbing more powder. Due to the layer-wise manufacturing method, the side surface generally bonds a large quantity of incompletely melted particles, which may need post treatment for further applications. Based on the above analysis, a geometric characteristic size that is too small may more easily cause the as-built part to gain more obvious size errors under the high-scan-speed conditions. This is because the newly forming part will remain at a high temperature for a longer time under the poor cooling condition. Then, the side wall of the part will bond more particles, resulting in an obvious size error. The relative density of the LPBF-built part is associated with pore deficiency. Too high a temperature of the molten pool with too high energy density will cause the evaporation of elements, generating pores with rounded shapes. Too high a cooling rate of the molten pool with too low an energy density will cause the incomplete fusion of powder, also generating pores with sharp shapes and incompletely melted particles. In addition, the poor cooling condition relating to too small a geometric characteristic size will aggravate pore generation under the condition of high energy density.

It is known that for metal materials, hardness is closely related to the microstructure. There is also a loose corresponding relationship between hardness and strength, namely, high hardness corresponding to high strength. High hardness is attributed to the refinement of the α/α' phase and β phase caused by rapid solidification, and much dislocation generation caused by residual stress in additive-manufactured Ti–6Al–4V parts [32]. The elastic modulus is associated with the residual stress level to a certain extent [33]. Based on the proposed thermal model, the higher energy density and small geometric characteristic size cause more thermal input and a poor cooling condition, respectively. This will weaken the grain refinement strengthening effect and decrease the residual stress, causing a decline in hardness and strength. A relatively low scan speed and small geometric characteristic size correspond to a low elastic modulus, which is partially due to the decrease of the residual stress [32].

As is known, the relative density is associated with the pore deficiency, and the porosity is associated with the size of the as-built struts of a porous structure. As shown in Figure 11, different from the porous structure with an L_s value of 1.4 mm, the relative density and porosity of the porous structures with an L_s value of 0.6 mm and a v of 700 mm/s were obviously less than those for the structures with a v of 1900 mm/s. This finding indirectly indicated that the struts of the porous sample with an L_s value of 1.4 mm achieved more pores and larger positive size error under the v of 700 mm/s. In other words, a porous structure with an L_s value of 0.6 mm and a v of 1900 mm/s had a better forming quality than did those with a v of 700 mm/s. These experimental results are in accordance with the results shown in Figures 3 and 4, confirming the validity of the above analysis to a certain extent.

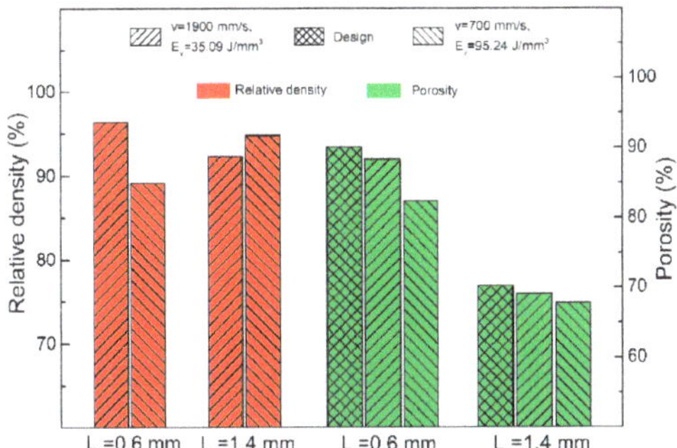

Figure 11. A comparison of the relative density and porosity for samples in the verification experiment.

5. Conclusions

In this study, taking the geometric characteristic size into consideration, samples consisting of solid struts with different section characteristic sizes (0.4 to 1.4 mm) were fabricated by selective laser melting with three scan speeds (700, 1300, and 1900 mm/s). The morphology features, relative density, microstructure, and mechanical properties were measured and a simple thermal model was presented to systematically explain the effect of geometric characteristic size under different scan speed conditions. The following was determined:

(1) The geometric characteristic size influenced the physical characteristics of the LPBF-produced struts by affecting the cooling condition. A small geometric characteristic size resulted in a poor cooling condition. The effect may only become obvious when the geometric characteristic size is small (<1 mm).

(2) The relative density of the solid strut with an L_s value of 0.4 mm reached the highest value of 96.8% when the v was 1900 mm/s, but the relative density of the solid strut with an L_s value of 1.4 mm reached the highest value of 98.1% when the v was 1300 mm/s. The solid strut with smaller geometric characteristic size had a more obvious size error. To a certain degree, the nanohardness and elastic modulus increased with increasing scan speed, and decreased with the decline of the geometric characteristic size when the geometric characteristic size was smaller than 1 mm.

(3) In contrast to the solid bulk forming process, for a superior forming quality, a higher scan speed should be set for the LPBF-manufacturing of porous structures, especially for structures with small geometric characteristic size in solid struts.

Author Contributions: Experiment design and writing, H.L. and D.X.; data analysis of μCT, Y.M.; data analysis of microstructure and nanoindentation, J.S. and C.W.; writing—review and editing, L.S. and Z.T.

Funding: This research was funded by the Advanced Research Project of Army Equipment Development (301020803), the Key Research and Development Program of Jiangsu (BE 2015161), the Young Scientists Fund of the National Natural Science Foundation of China (51605473), the Jiangsu Provincial Research Foundation for Basic Research, China (BK 20161476), the Science and Technology Planning Project of Jiangsu Province of China (BE 2015029) and the Science and Technology Support Program of Jiangsu (BE 2014009-1, BE 2014009-2, BE 2016010-3).

Conflicts of Interest: The authors declare no conflicts of interest.

References

1. Jena, A.; Gupta, K. Characterization of pore structure of filtration media. *Fluid Part. Sep. J.* **2002**, *14*, 227–241.
2. Chen, D.; Kitipornchai, S.; Yang, J. Dynamic response and energy absorption of functionally graded porous structures. *Mater. Des.* **2018**, *140*, 473–487. [CrossRef]
3. Hutter, C.; Büchi, D.; Zuber, V.; Rudolf von Rohr, P. Heat transfer in metal foams and designed porous media. *Chem. Eng. Sci.* **2011**, *66*, 3806–3814. [CrossRef]
4. Xu, Z.; Hao, H. Electromagnetic interference shielding effectiveness of aluminum foams with different porosity. *J. Alloys Compd.* **2014**, *617*, 207–213. [CrossRef]

5. Krishna, B.V.; Bose, S.; Bandyopadhyay, A. Low stiffness porous Ti structures for load-bearing implants. *Acta Biomater.* **2007**, *3*, 997–1006. [CrossRef] [PubMed]

6. Liang, H.; Yang, Y.; Xie, D.; Li, L.; Mao, N.; Wang, C.; Tian, Z.; Jiang, Q.; Shen, L. Trabecular-like Ti-6Al-4V scaffolds for orthopedic: Fabrication by selective laser melting and in vitro biocompatibility. *J. Mater. Sci. Technol.* **2019**. [CrossRef]

7. Gao, C.; Wu, P.; Yang, Y.; Feng, P.; Shuai, C.; He, C.; Bin, S.; Guo, W. Biodegradation mechanisms of selective laser-melted Mg–x Al–Zn alloy: Grain size and intermetallic phase. *Virtual Phys. Prototyp.* **2017**, *13*, 59–69.

8. Ohji, T.; Fukushima, M. Macro-porous ceramics: Processing and properties. *Int. Mater. Rev.* **2012**, *57*, 115–131. [CrossRef]

9. Dhandayuthapani, B.; Yoshida, Y.; Maekawa, T.; Kumar, D.S. Polymeric Scaffolds in Tissue Engineering Application: A Review. *Int. J. Polym. Sci.* **2011**, *2011*, 1–19. [CrossRef]

10. Wang, X.; Xu, S.; Zhou, S.; Xu, W.; Leary, M.; Choong, P.; Qian, M.; Brandt, M.; Xie, Y.M. Topological design and additive manufacturing of porous metals for bone scaffolds and orthopaedic implants: A review. *Biomaterials* **2016**, *83*, 127–141. [CrossRef] [PubMed]

11. Leong, K.F.; Cheah, C.M.; Chua, C.K. Solid freeform fabrication of three-dimensional scaffolds for engineering replacement tissues and organs. *Biomaterials* **2003**, *24*, 2363–2378. [CrossRef]

12. Parthasarathy, J.; Starly, B.; Raman, S. A design for the additive manufacture of functionally graded porous structures with tailored mechanical properties for biomedical applications. *J. Manuf. Process.* **2011**, *13*, 160–170. [CrossRef]

13. Dinda, G.P.; Song, L.; Mazumder, J. Fabrication of Ti-6Al-4V scaffolds by direct metal deposition. *Metall. Mater. Trans. A Phys. Metall. Mater. Sci.* **2008**, *12*, 2914–2922. [CrossRef]

14. Ponader, S.; Von Wilmowsky, C.; Widenmayer, M.; Lutz, R.; Heinl, P.; Körner, C.; Singer, R.F.; Nkenke, E.; Neukam, F.W.; Schlegel, K.A. In vivo performance of selective electron beam-melted Ti-6Al-4V structures. *J. Biomed. Mater. Res. Part A* **2010**, *92*, 56–62. [CrossRef]

15. Salmi, M.; Tuomi, J.; Paloheimo, K.; Paloheimo, M.; Björkstrand, R.; Mäkitie, A.A.; Mesimäki, K.; Kontio, R. Digital design and rapid manufacturing in orbital wall reconstruction. In *Innovative Developments in Design and Manufacturing*, 1st ed.; Reddy, J.N., Ed.; CRC Press: London, UK, 2009; pp. 357–360.

16. Kruth, J.P.; Levy, G.; Klocke, F.; Childs, T.H.C. Consolidation phenomena in laser and powder-bed based layered manufacturing. *CIRP Ann. Manuf. Technol.* **2007**, *56*, 730–759. [CrossRef]

17. Warnke, P.H.; Douglas, T.; Wollny, P.; Sherry, E.; Steiner, M.; Galonska, S.; Becker, S.T.; Springer, I.N.; Wiltfang, J.; Sivananthan, S. Rapid Prototyping: Porous Titanium Alloy Scaffolds Produced by Selective Laser Melting for Bone Tissue Engineering. *Tissue Eng. Part C Methods* **2009**, *15*, 115–124. [CrossRef] [PubMed]

18. Yan, C.; Hao, L.; Hussein, A.; Raymont, D. Evaluations of cellular lattice structures manufactured using selective laser melting. *Int. J. Mach. Tools Manuf.* **2012**, *62*, 32–38. [CrossRef]

19. Van Bael, S.; Kerckhofs, G.; Moesen, M.; Pyka, G.; Schrooten, J.; Kruth, J.P. Micro-CT-based improvement of geometrical and mechanical controllability of selective laser melted Ti6Al4V porous structures. *Mater. Sci. Eng. A* **2011**, *528*, 7423–7431. [CrossRef]

20. Shirazi, S.F.S.; Gharehkhani, S.; Mehrali, M.; Yarmand, H.; Metselaar, H.S.C.; Adib Kadri, N.; Osman, N.A.A. A review on powder-based additive manufacturing for tissue engineering: Selective laser sintering and inkjet 3D printing. *Sci. Technol. Adv. Mater.* **2015**, *16*, 1–20. [CrossRef] [PubMed]

21. Weißmann, V.; Bader, R.; Hansmann, H.; Laufer, N. Influence of the structural orientation on the mechanical properties of selective laser melted Ti6Al4V open-porous scaffolds. *Mater. Des.* **2016**, *95*, 188–197. [CrossRef]

22. Wauthle, R.; Vrancken, B.; Beynaerts, B.; Jorissen, K.; Schrooten, J.; Kruth, J.; Humbeeck, J. Van Effects of build orientation and heat treatment on the microstructure and mechanical properties of selective laser melted Ti6Al4V lattice structures. *Addit. Manuf.* **2015**, *5*, 77–84. [CrossRef]

23. Liu, Y.J.; Li, X.P.; Zhang, L.C.; Sercombe, T.B. Processing and properties of topologically optimised biomedical Ti-24Nb-4Zr-8Sn scaffolds manufactured by selective laser melting. *Mater. Sci. Eng. A* **2015**, *642*, 268–278. [CrossRef]

24. Ahmadi, S.M.; Hedayati, R.; Ashok Kumar Jain, R.K.; Li, Y.; Leeflang, S.; Zadpoor, A.A. Effects of laser processing parameters on the mechanical properties, topology, and microstructure of additively manufactured porous metallic biomaterials: A vector-based approach. *Mater. Des.* **2017**, *134*, 234–243. [CrossRef]

25. Jamshidinia, M.; Kovacevic, R. The influence of heat accumulation on the surface roughness in powder-bed additive manufacturing. *Surf. Topogr. Metrol. Prop.* **2015**, *3*, 1–10. [CrossRef]

26. Simchi, A.; Pohl, H. Effects of laser sintering processing parameters on the microstructure and densification of iron powder. *Mater. Sci. Eng. A* **2003**, *359*, 119–128. [CrossRef]
27. Oliver, W.C. An improved technique for determining hardness and elastic modulus using load displacement sensing indentation experiments. *J. Mater. Res.* **1992**, *7*, 1564–1583. [CrossRef]
28. Thijs, L.; Verhaeghe, F.; Craeghs, T.; Van Humbeeck, J.; Kruth, J.P. A study of the microstructural evolution during selective laser melting of Ti-6Al-4V. *Acta Mater.* **2010**, *58*, 3303–3312. [CrossRef]
29. Gu, D.; Hagedorn, Y.C.; Meiners, W.; Meng, G.; Batista, R.J.S.; Wissenbach, K.; Poprawe, R. Densification behavior, microstructure evolution, and wear performance of selective laser melting processed commercially pure titanium. *Acta Mater.* **2012**, *60*, 3849–3860. [CrossRef]
30. Chiumenti, M.; Neiva, E.; Salsi, E.; Cervera, M.; Badia, S.; Moya, J.; Chen, Z.; Lee, C.; Davies, C. Numerical modelling and experimental validation in Selective Laser Melting. *Addit. Manuf.* **2017**, *18*, 171–185. [CrossRef]
31. Abdul Aziz, M.S.; Furumoto, T.; Tanaka, R.; Hosokawa, A.; Alkahari, M.R.; Ueda, T. Thermal Conductivity of Metal Powder and Consolidated Material Fabricated via Selective Laser Melting. *Key Eng. Mater.* **2012**, *523–524*, 244–249.
32. Song, L.; Xiao, H.; Ye, J.; Li, S. Direct laser cladding of layer-band-free ultrafine Ti6Al4V alloy. *Surf. Coat. Technol.* **2016**, *307*, 761–771. [CrossRef]
33. Sun, G.; Zhou, R.; Lu, J.; Mazumder, J. Evaluation of defect density, microstructure, residual stress, elastic modulus, hardness and strength of laser-deposited AISI 4340 steel. *Acta Mater.* **2015**, *84*, 172–189. [CrossRef]

 *metals*

Article

Experiment of Process Strategy of Selective Laser Melting Forming Metal Nonhorizontal Overhanging Structure

Wentian Shi *, Peng Wang, Yude Liu and Guoliang Han

School of Materials Science and Mechanical Engineering, Beijing Technology and Business University, Beijing 100048, China; wangpeng_12321@126.com (P.W.); liuyd@th.btbu.edu.cn (Y.L.); hanguoliang163@163.com (G.H.)

* Correspondence: shiwt@th.btbu.edu.cn; Tel.: +86-138-1022-3727

Received: 9 March 2019; Accepted: 25 March 2019; Published: 27 March 2019

Abstract: To improve the precision of the nonhorizontal suspension structure and the forming quality of the overhanging surface by selective laser melting, the influence of laser power on the upper surface and the overhanging surface forming quality of 316L stainless steel at different forming angles was studied in the experiment. The influence of different scanning strategies, upper surface remelting optimization, and overhang boundary scanning optimization on the formation of overhanging structures was compared and analyzed. The forming effect of chromium–nickel alloy is better than 316L stainless steel below the limit forming angle in the overhanging structure. The better forming effect of chromium–nickel alloy can be obtained by narrowing the hatch space with the boundary optimization process. The experiment results show that the forming of the overhanging structure below the limit forming angle should adopt the chessboard scanning strategy. The smaller laser power remelting is beneficial to both the forming of the overhanging surface and the quality of upper surface forming. The minimum value of surface roughness using the 110 W power laser twice during surface remelting and boundary scanning 75° overhanging surface can reach 11.9 μm and 31.1 μm, respectively. The forming accuracy error range above the limit forming angle is controlled within 0.4 mm, and the forming quality is better. A boundary count scanning strategy was applied to this study to obtain lower overhanging surface roughness values. This research proposes a process optimization and improvement method for the nonhorizontal suspension structure formed by selective laser melting, which provides the process support for practical application.

Keywords: selective laser melting; nonhorizontal suspension structure; boundary remelting; surface roughness; forming accuracy

1. Introduction

As a rapid prototyping (RP) technique, selective laser melting (SLM) is a new technology and widely used in aerospace, automobile manufacturing, medical applications, industrial product design, architectural design, entertainment products, biotechnology, and other industrial fields [1,2]. SLM is an additive manufacturing technology and the principle is the discrete stacking which uses the high-energy laser beam to melt metal powder. The parts can be quickly formed without tools, fixtures, and molds with the advantages of short production cycle and high material utilization rate by using SLM [3–5].

Nonhorizontal suspension structure is the most basic and common structure encountered in the forming process, and is also the biggest geometric problem of forming in SLM experiments. Due to the inherent defects such as warpage, suspension, and sticky powder during processing, SLM cannot form parts with high quality and high precision. Therefore, if SLM can form overhanging structures with good quality, the technology will be improved and even promote the large-scale popularization and application of SLM. The support of forming is necessary and ensures the stability of the structure during the forming of suspension structures. At the same time, the excess heat will be transmitted by the support to prevent the structure from warping and deforming. However, the additional support will increase the time of processing, so it is necessary to investigate the strategy of forming suspension structures without support.

Kruth et al. [6] found that the forming level or near-level hanging surface can only be accurately formed by adding support. He [6,7] also proposed that adding monitoring and feedback devices in the optical system can flexibly change the laser power and improve the forming quality of the hanging surface. Yasa et al. [8] increased the feedback control of the forming process, laser surface remelting, laser etching, and other postprocessing methods to improve the forming effect of the suspension structure. From the microlevel view, the forming layer of the underside overhanging surface is a powder rather than metal entity, resulting in heat conduction, micromelting, adhesion, and other behaviors which are different from the solid structure. The behavior change of the molten pool can directly reflect the basic principle of SLM forming on the hanging surface. Therefore, many experts devote themselves to the forming mechanism of the microlevel suspension structure, and have obtained many achievements. Lott et al. [9] preliminarily expounded the dynamic behavior of the molten pool. Khan et al. [10] numerically simulated the molten pool instability in SLM forming, and concluded that the area of molten pool depended on the boundary conditions. Alkahari et al. [11] studied the molten pool behavior in the first-layer forming process under different laser powers, scanning speeds, and layer thicknesses by single-layer scanning experience.

The metal powder undergoes rapid melting, rapid cooling, and solidification during the SLM forming process due to the laser directly acting on the surface of the metal powder. The defects such as spheroidization, pores, cracks, dross, over-burning, warping, and so on are also the result of the varying of fast heat and rapid cooling. Some postprocessing methods (such as heat treatment, sanding, etc.) can improve the surface quality of the formed parts, but they are time-consuming and laborious and have limited effect, and for some complex structures (such as overhanging structures, hollow porous structures, etc.), it cannot be fundamentally solved by postprocessing methods. To eliminate the defect, it is necessary to decrease or even avoid the defect by adjusting the process parameters and other prior processes. Therefore, when forming parts such as a suspended structure or a complicated curved surface, a good forming effect should be obtained from the aspects of process parameters, scanning strategy, support added or not, and so on. In this study, the energy of the laser input was strictly controlled, and the process parameters, scanning strategy, support addition method, and auxiliary optimization process were changed. At the same time, the substrate was preheated by 100 °C to reduce the influence of temperature difference on the forming, and the optimal suspension structure was obtained with a good surface quality and high precision.

2. Materials and Methods

2.1. Materials and Experimental Setup

The experiment was carried out on the Renishaw AM400 (Renishaw plc, London, UK). The AM400 uses a continuous laser mode with a maximum power of 400 W and a 1075 nm Nd: YAG laser with a laser beam diameter of 70 μm. The working area (shown in Figure 1) can be machined to a maximum volume of 250 mm × 250 mm × 300 mm and provides a closed environment filled with argon as a shielding gas to maintain oxygen concentrations below 200 ppm. The experimental scanning strategy is based on the meander scan strategy, a schematic of which is shown in Figure 2; it can be seen that the angle between the Nth layer and the N+1th layer is 67°, where d is the point distance, δ is the hatch space, and Φ is the spot diameter.

There were two materials used in the experiment: one was the 316L stainless steel and the other was the chrome–nickel alloy steel. The 316L stainless steel had a particle diameter ranging from 5 μm to 41 μm and the average particle diameter was 17 μm. The chemical composition of the 316L stainless steel powder was Fe (Balance), Cr (16% to 18%), Ni (10% to 14%), Mo (2% to 3%), Mn (2% max), Si (1% max), N (0.1% max), O (0.1% max), P (0.045% max), C (0.03% max), S (0.03% max). The chrome–nickel alloy steel powder had an extremely high sphericity and a particle diameter ranging from 17 μm to 58 μm, and the average particle diameter was 31 μm. The chromium–nickel alloy steel included Fe (Balance), Cr (1% max), Mn (1% max), Ni (1% max), Mo (0.15% max), C (0.228% max), N (0.228% max). Electron microscopy (SEM) photographs and compositional contents are shown in Figure 3.

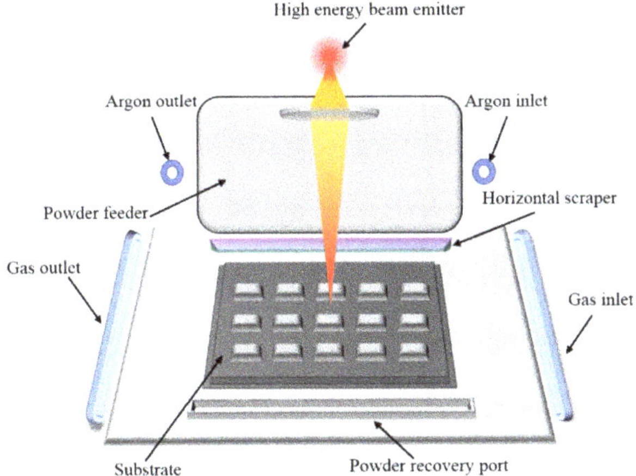

Figure 1. The equipment workspace of Renishaw AM400.

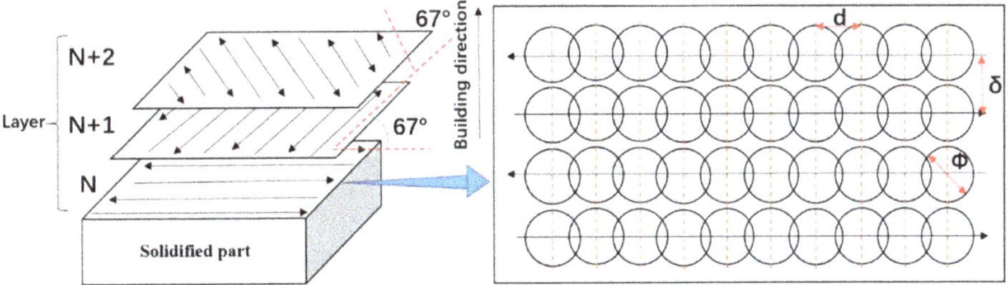

Figure 2. The schematic of forming experiment piece.

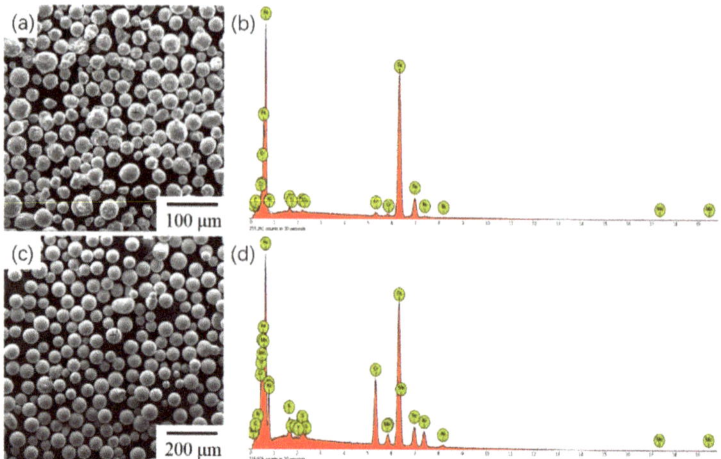

Figure 3. Powder characteristics and composition content: (**a**) morphology of 316L stainless steel powder; (**b**) composition content of 316L power; (**c**) morphology of chrome–nickel alloy steel powder; (**d**) composition content of chrome–nickel alloy steel.

2.2. Forming Experiment

Firstly, 316L stainless steel powder was used for 50 μm layer thickness experiment, and the forming quality under processing parameters such as different angles, different scanning strategies, support addition, different boundary scanning schemes, boundary scanning times, different upper surface remelting schemes, and upper surface remelting times were compared. The processing parameters and the combination of exposure time, point distance, hatch space, and scanning strategy are shown in Table 1. The laser scanning surface can be divided into upper surface, inner filling surface, and outer boundary scanning surface (as shown in Figure 4a). In the forming experiment, the main body was the noumenon, and three scanning strategies—meander, stripe, and chessboard—were adopted, and the boundary scan was solidified once. The boundary solidification scan occurred after the noumenon scan. The specific parameters were as follows: Table 1 (No. 1–3). There were nine kinds of optimization processes, which were divided into three categories: support addition (Table 1, No. 4), optimal matching boundary remelting processes with different power boundaries (Table 1, No. 5–7), and optimal matching remelting times of surface layers with different power levels (Table 1, No. 8–12). A total of 12 groups of experiments were carried out on noumenon formed by different scanning strategies, supported noumenon 1, noumenon 1 with boundary optimization, and noumenon 1 with upper surface optimization. The scheme of the block-forming experiment is shown in Table 2. Each block was inclined by 15°, 30°, 45°, 60°, 75° (as shown in Figure 4). The bottom of all samples was 5 mm * 5 mm and the length of the inclined plane was 10 mm. The experiment was carried out in terms of the presence or absence of support, the number of boundary remelting times, and the number of remelting of the upper surface layer. According to the 316L forming experiment, the chrome–nickel alloy powder was studied in the vicinity of the limit forming angle with the layer thickness of 35 μm, and the overhanging structures of 20°, 25°, 30°, and 40° were formed. The forming parameter tables are shown in Table 3. The overhanging portion of the formed overhanging structure is determined by the formula (1) [12]:

$$a = h \cdot \cot\theta \tag{1}$$

where a is the length of the overhanging portion, h is the layer thickness, and θ is the angle between the contour of the layer and the overhanging surface. In this experiment, the 316L stainless steel had a layer thickness of 50 μm and a spot diameter of 70 μm. The stable boundary forming can be obtained when the diameter of the spot is larger than the length of the overhanging portion. The forming limit inclination angle was 35.5° from the formula (1). Considering the formation of stainless steel, the thickness of the formed chrome–nickel alloy steel layer was 35 µm, and the limit forming angle was 26.6°.

Table 1. Block-forming test data sheet of 316L stainless steel.

Number	Test Plan	Laser Power (W)	Exposure Time (µs)	Point Distance (µm)	Hatch Space (µm)	Scanning Strategy	Remark
1	Noumenon1	200	80	40	110	Meander	The system scans the boundary once by default.
2	Noumenon2	200	80	40	110	Chessboard	
3	Noumenon3	200	80	40	110	Stripe	
4	Support						
5	Boundary1	110	100	20	100		Scan the boundary again according to the setting parameters.
6	Boundary2	160	100	20	100		
7	Boundary3	200	100	20	100		
8	Upper surface1	110	80	40	110	Meander	Scan once by setting parameters.
9	Upper surface2	160	80	40	110	Meander	
10	Upper surface3	200	80	40	110	Meander	
11	Upper surface4	110	80	40	110	Meander	Scan again by setting parameters.
12	Upper surface5	160	80	40	110	Meander	

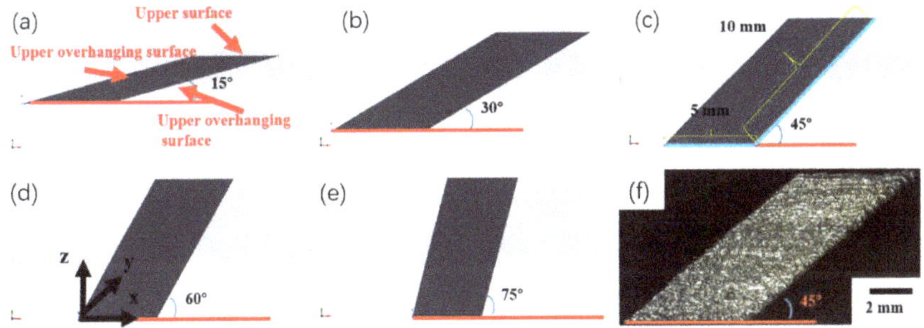

Figure 4. Schematic diagram and physical drawing of the 316L stainless steel: (**a**) 15°; (**b**) 30°; (**c**, **f**) 45°; (**d**) 60°; (**e**) 75°.

Table 2. Block-forming scheme of 316L stainless steel.

Number	Test Plan	Number	Test Plan
1	Noumenon1	2	Noumenon2
3	Noumenon3	4	Noumenon1 + Support
5	Noumenon1 + Boundary1	6	Noumenon1 + Boundary2
7	Noumenon1 + Boundary3	8	Noumenon1 + Upper surface1
9	Noumenon1 + Upper surface2	10	Noumenon1 + Upper surface3
11	Noumenon1 + Upper surface4	12	Noumenon1 + Upper surface5

Three scanning strategies, namely meander, chessboard, and stripe scanning, are shown in Figure 5. The angle between the layers was 67° counterclockwise. In the selective laser melting technology, in order to ensure the boundary quality during the melt-solidification forming, a boundary scan curing can be performed after each layer of scanning is completed. The boundary remelting is a boundary solidification operation after scanning the boundary. It is not a remelting operation according to the original boundary, but a solidification of the boundary inside the original boundary. Boundary counting includes boundary scan and boundary remelting. With the increase of counting times, boundary scanning and boundary remelting will be carried out separately (that is, boundary remelting will be carried out in the middle of two boundary scans). Different counting methods were used for forming the overhanging surface, as shown in Figure 6, and counting was performed multiple times to ensure the forming effect of the boundary.

Table 3. The parameters of chromium–nickel alloy steel block-forming experiment.

Number	Test Plan	Laser Power (W)	Exposure Time (μs)	Point Distance (μm)	Hatch Space (μm)	Scanning Strategy	Remark
1	Noumenon1	100	80	40	100	Meander	
2	Boundary1	80	80	40	40		Count 1 time
3	Boundary2	80	80	40	40		Count 2 times
4	Boundary3	80	80	40	40		Count 4 times
5	Boundary4	80	80	40	40		Count 10 times

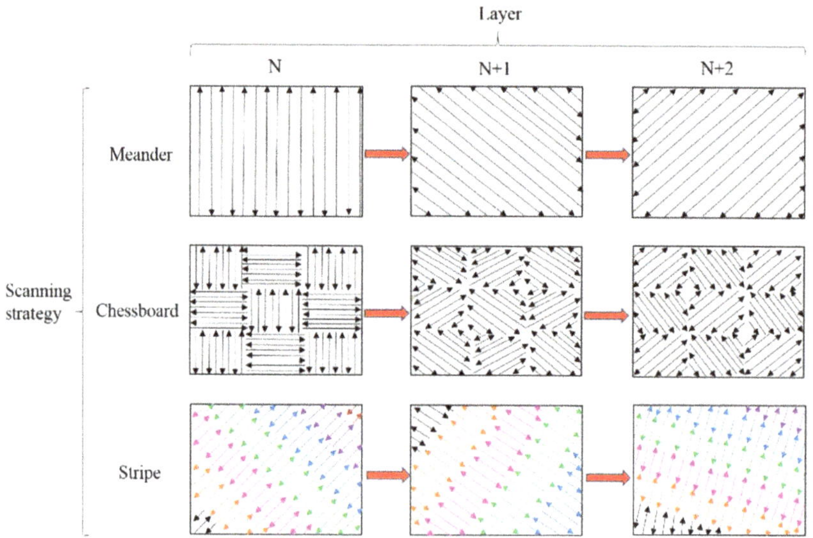

Figure 5. Schematic diagram of scanning strategy.

The measurement of the surface roughness was carried out using laser scanning confocal microscopy (VK-X100, Osaka, Japan) to obtain surface roughness values, and the roughness of the final surface was measured three times and averaged as the surface roughness of the final surface. The optical microscope (OM) (DM4000M; Leica, Wetzlar, Germany) was used to measure the forming precision of the sample with a measuring accuracy of 1 μm. The measuring method was to fit the microprofile of the part to an ideal straight line to measure the accurate dimension value, and the calculated average value was the accurate dimension value of the part.

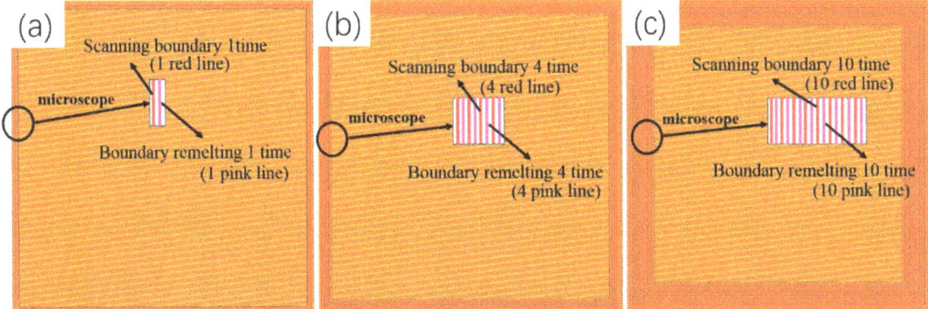

Figure 6. Boundary count (Corresponding Table 3, No. 1, 4, 5) (red line: scanning boundary; pink line: boundary remelting): (**a**) 1 time; (**b**) 4 times; (**c**) 10 times.

3. Results and Discussion

3.1. 316L Forming Surface Quality Analysis

The upper surface topography of different scanning strategies (a–o) and overhanging structures with additional support (p–t) is shown in Figure 7. Due to the lower forming angle of 15°–30°, other scanning strategies cannot be formed except for the additional support strategy. The chessboard scanning strategy can ensure the upper surface forming quality of 30°, and the missing of materials in the stripe strategy is the most serious.

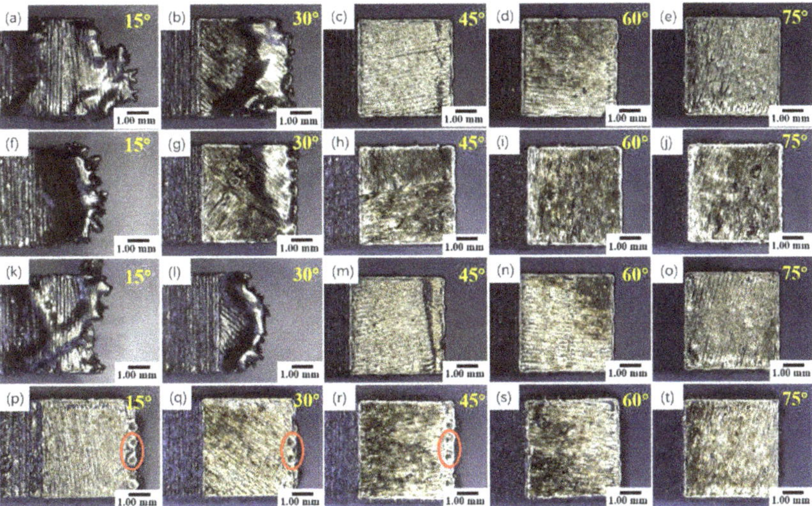

Figure 7. Different scanning strategies and support for adding surface features of the formed overhang structure (corresponding Table 2, No. 1–4): (**a–e**) meander; (**f–j**) chessboard; (**k–o**) stripe; (**p–t**) support.

As shown in Figure 8c, in the stripe scanning process, some scanning lines are not supported and rely entirely on the transverse bonding force of melted materials. In the case of insufficient bonding force, materials are prone to be missing, which results in the process defects of collapse. Due to the other scanning method, after lamination (as shown in Figure 8b), even if some areas are overhanging, the remaining areas can be supported by not only the transverse bonding force but also the materials of the bottom in the structure to ensure the lap joint effect, so the forming effect is excellent. The forming quality can be guaranteed by the additional support component from 15° to 30°. There are overhanging defect marks on the forming upper surface (red circle of Figure 7) because of the unreasonable additional support, and the defect can be eliminated by adjusting the position of the additional support.

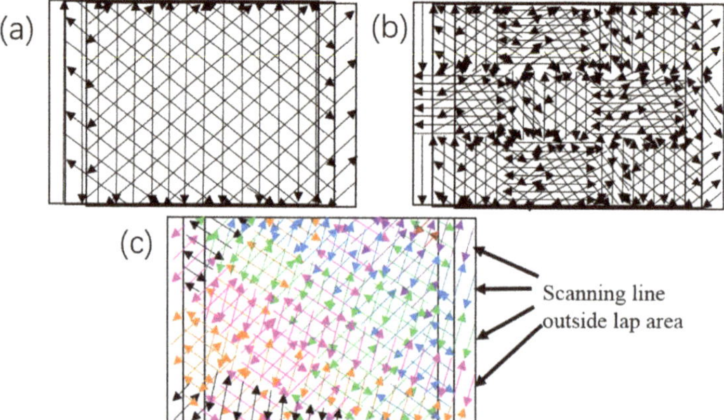

Figure 8. Effect diagram of different scanning strategies after lamination: (**a**) meander; (**b**) chessboard; (**c**) stripe.

The forming quality with the angle 45°–75° above the limit forming angle is excellent, and the meander scan strategy is better than the chessboard scan and stripe scan in the upper surface forming quality. Although the additional support can ensure good forming effect, the surface spheroidization is still serious. As shown in Figure 9, the spheroidizing defect happened by spatter. The irregular spheres spattered by excessive laser power during scanning forming solidify and are prone to forming the upper surface spheroidizing phenomenon.

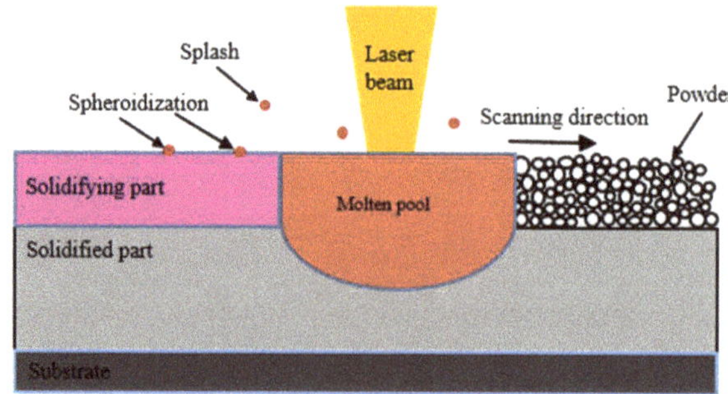

Figure 9. Surface spheroidization defect formation mechanism.

Warpage and slag hanging easily happen when SLM is used to form suspension structures. The reason for these defects lies in the vast quantities of heat generated by the interaction between the radiated laser and the powder material in the forming process. The heat conductivity of metal powders is not as good as that of solidified metals, so heat easily accumulates to form larger melting pools, resulting in slag hanging and overburning defects. So it is necessary to control the energy density of the laser when forming the surface of the hanging structure.

The low energy density is used from the suspension scanning laser, and the forming effect will be improved. In addition, rapid heat dissipation can prompt liquid metal to solidify in time, thus avoiding suspension defects and improving the forming effect of the suspension structure. Therefore, the control of laser scanning process parameters of the overhanging surface is very important to the forming quality of the suspension structure. Since the underside overhanging surface is directly in contact with the powder, a smaller energy density should be used to prevent burn-through, thereby a better quality of the underside overhanging surface can be obtained.

The topography of the overhanging structure is shown in Figure 10 for different scanning strategies. The quality of the side surface corresponds to the upper surface forming effect; with the inclination angle increased, the forming quality becomes better than before. The forming quality is poor under the minimum forming angle, in which the chessboard scan is best and the stripe scan is the worst. The forming effects of various scanning strategies above the limit forming angle are favorable, and the best forming quality can be obtained, wherein the meander scan strategy is optimal. Different scanning strategies for forming a 30° overhanging structure are shown in Figure 10b,g,l, and the chessboard scan forming effect is relatively favorable, followed by the meander scan, and the stripe scan forming effect is the worst.

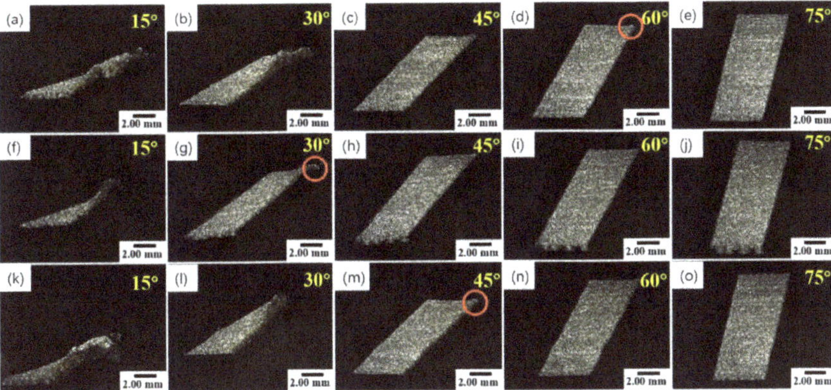

Figure 10. Side profile of shaped suspension structures with different scanning strategies (corresponding Table 2, No. 1–3): (**a–e**) meander; (**f–j**) chessboard; (**k–o**) stripe.

3.2. 316L Surface Roughness

The surface roughness of the upper overhanging surface formed by different optimization processes is shown in Figure 11. During the forming process with the fabrication angle from 15° to 75°, as the angle increases, the surface roughness decreases, and the surface forming quality is favorable. The forming quality and surface roughness is better when the forming angle of the overhanging surface is larger and the forming quality of the upper surface layer is also better. The values of upper surface roughness after optimization with the fabrication angle of 45°–75° are all below 30 μm in the experiment. With the increasing of the remelting power of the upper surface, the upper surface morphology deteriorates (Figure 12c–e), and the value of upper surface roughness increases. Due to the thickness of the forming layer being 50 μm, the higher laser power increases the splash phenomenon, resulting in an increase of surface spheroidization (Figure 12d), and the surface quality deteriorates. The number of laser scans also has an effect on the upper surface formation. The higher power scanning twice results in a poor surface topography, as shown in Figure 12f. The smaller power scanning twice has some improvement on the surface. As shown in Figure 13, the spheroidization phenomenon almost disappears with respect to the primary remelting surface. However, due to the instability of the process and the difference in the scanning directions, the surface is uneven, as shown in Figure 12e. The result shows that the best remelting technology is using 110 W power laser, and the upper surface roughness value can be as low as 11.9 μm.

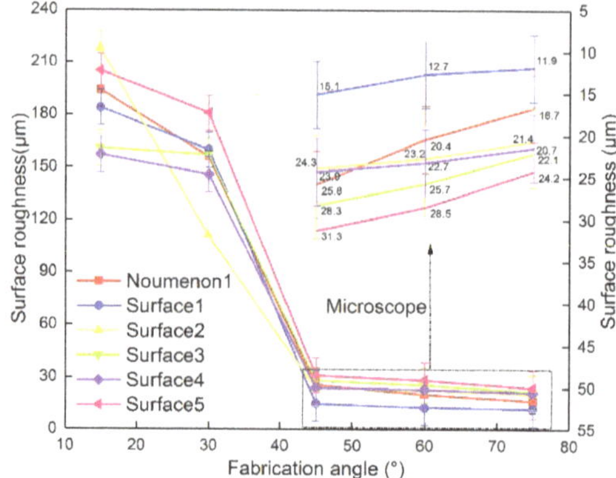

Figure 11. Different optimization schemes to form the surface roughness of the overhanging structure (corresponding Table 2, No. 1, 8–12).

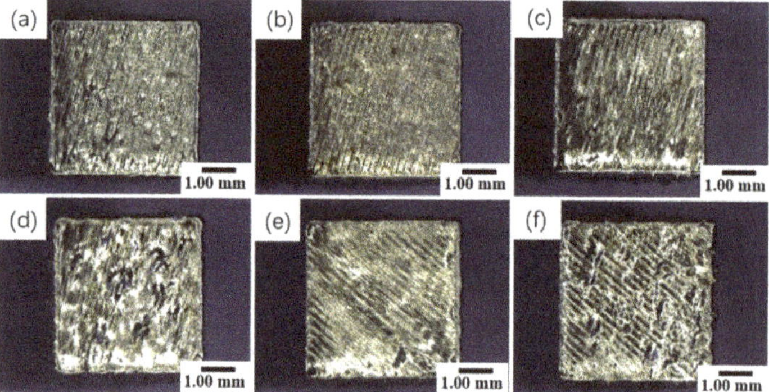

Figure 12. 75° upper surface topography (corresponding Table 2, No. 1, 8–12): (**a**) noumenon1; (**b**) surface1; (**c**) surface2; (**d**) surface3; (**e**) surface4; (**f**) surface5.

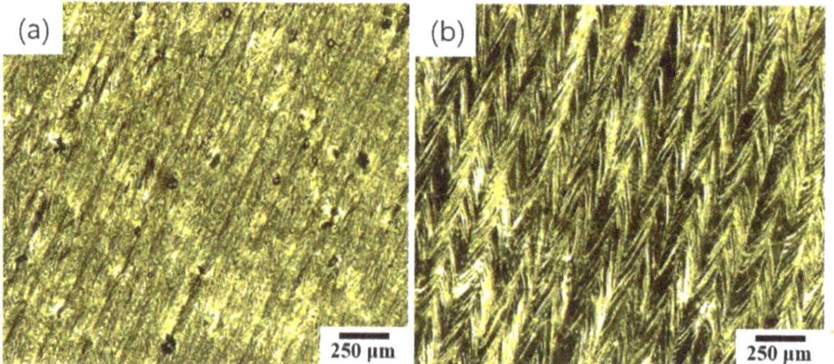

Figure 13. Remelted surface topography (corresponding Table 2, No. 8, 11): (**a**) 110 W power laser surface remelting once; (**b**) 110 W power laser surface remelting twice.

The surface roughness of the forming underside overhanging surface with different scanning strategies is shown in Figure 14. The overhanging structure is formed below the limit forming angle, and the chessboard scanning strategy is best for forming. The meander scanning above the extreme forming angle is more favorable for forming. Due to the unique scanning method of meander scanning, partition scanning, the board has a poor lap joint effect between the overhanging area and the other area when forming the overhanging structure. When the overhanging structure is formed below the forming angle using meander scanning or stripe scanning, the single line directly acts on the powder, and the chessboard scan has corresponding support and can obtain the better forming effect. The number of boundary scanning has a great influence on the underside overhanging surface of the shaped structure. The underside surface roughness value of the suspension structure with twice scanning 75° is reduced from 94.1 μm to 31.1 μm. The number of boundary scanning can improve the forming quality of the overhanging structure. The multiple scanning of the boundary is similar to the remelting of the narrow boundary by the laser and is beneficial to improve the forming quality. Thus multiple boundary scanning is beneficial to consolidate the forming quality of the underside overhanging surface. The lower laser power also has a good effect on the improvement of the underside overhanging surface. The surface roughness value of the underside overhanging surface increases with the increase of the laser power, and the boundary scanning with the 110 W power laser has the best underside surface roughness quality.

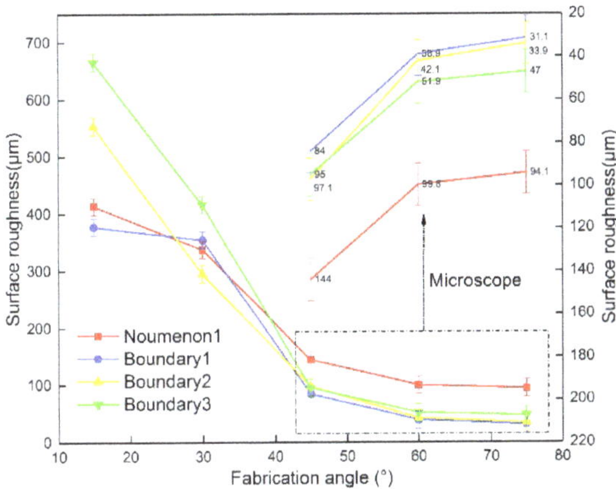

Figure 14. Different optimization schemes for forming overhanging surface roughness of overhanging structures (corresponding Table 2, No.1, 5–7).

The profile of the upper and underside overhanging surfaces of the overhanging structure is observed under the SEM. As shown in Figure 15, the defect of the sticky powdery phenomenon is more serious both in the upper and underside overhanging surfaces. Current research on the phenomenon of sticky powder shows that there is no better solution than using postprocessing to improve the quality of the overhanging surface. After sandblasting in the later process, it was found that the amount of powder adhering to the overhanging surface can be effectively reduced, and the forming quality of the overhanging surface greatly improved. The forming precision of the 45° suspension block formed in different directions of the x, y, and z (Figure 4) is shown in Table 4. The forming effect is good in this experiment, and the forming accuracy error (x1, y1, z1) is within 0.4 mm. The scanning with additional support has the best forming precision, and the remelting of boundary is beneficial to improve the forming precision. The effect of the meander scanning forming the overhang structure is better than the chessboard and the stripe scanning, and the forming precision is worse with the increasing of the remelting power.

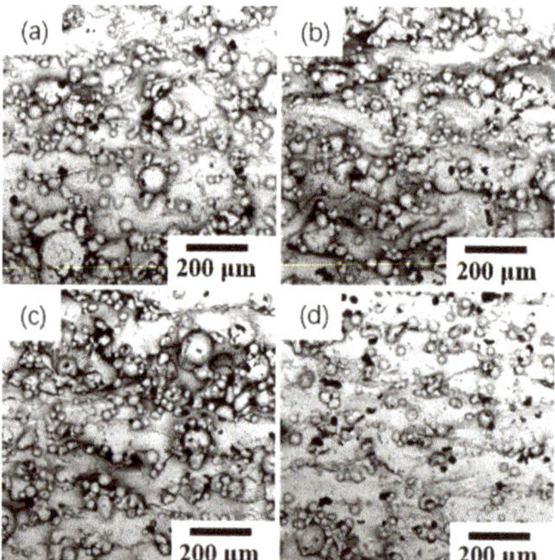

Figure 15. Different optimization schemes for forming overhanging surface roughness of overhanging structures: (a–b) upper overhanging surface; (c–d) underside overhanging surface.

Table 4. Block forming accuracy table.

Group	X (mm)	x′ (mm)	x_1 (mm)	y (mm)	y′ (mm)	y_2 (mm)	z (mm)	z′ (mm)	z_3 (mm)
1	5	5.09	0.09	5	5.08	0.08	7.07	7.15	0.08
2	5	5.15	0.15	5	5.12	0.12	7.07	7.29	0.12
3	5	5.11	0.11	5	5.09	0.09	7.07	7.18	0.11
4	5	5.04	0.04	5	5.05	0.05	7.07	7.08	0.01
5	5	5.06	0.06	5	5.07	0.07	7.07	7.13	0.06
6	5	5.14	0.14	5	5.16	0.16	7.07	7.35	0.28
7	5	5.21	0.21	5	5.22	0.22	7.07	7.43	0.36

3.3. Chrome–Nickel Alloy Steel Forming Surface Quality Analysis

Considering the result of the 316L stainless steel powder forming experiment, the chrome–nickel alloy steel powder is used to form 20°, 25°, 30°, 40° overhanging structure with a layer thickness of 35 μm. The forming effect at different angles as shown in Figure 16. Compared with 316L stainless steel, the minimum forming angle (20°, 25°) can ensure good forming, and no slag, warp, and so forth in the forming experiment of chrome–nickel alloy steel powder. The overhanging structure below the limit forming angle can also achieve a good forming effect. The warping deformation is caused by joint action of the thermal stress, the tissue stress, and the residual stress existing in the rapid laser forming process. Due to the force exceeding the strength of the material, the plastic deformation happened. The overhanging structure has a large warpage deformation due to the lack of support [7], so that it is difficult to avoid the defect of overhanging in the overhang structure experiment without the support. The single layer with the overhanging portion shrinks when the scanning is completed, and the volume of the molten powder shrinks during the process of solidification, causing the overhang portion to warp upward [13]. Due to the difference of temperature between the top and bottom of the scanning layer and the uneven distribution of thermal conductivity, the upper portion of the forming layer shrinks faster than the bottom, and is prone to forming the defect of overhanging [14]. As shown in Figure 8, in the red circle is a warp phenomenon, and the warped portion is scattered around each of the suspension members. In contrast, the overhanging structure of the chrome–nickel alloy steel powder has almost no warpage, and it can be seen that the chrome–nickel alloy steel powder has better forming effect than the 316L stainless steel powder. It can be seen that the adhesion of chrome–nickel alloy is better than that of 316L stainless steel.

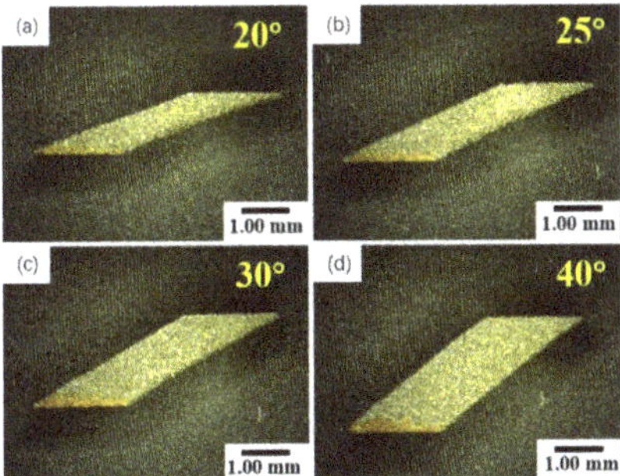

Figure 16. Side profile of the overhanging structure at different angles (corresponding Table 3, No. 1): (**a**) 20°; (**b**) 25°; (**c**) 30°; (**d**) 40°.

3.4. Chrome–Nickel Alloy Steel Surface Roughness

The trend curve of the influence of boundary counting strategy on overhanging surface roughness is shown in Figure 17. The different trends of the upper and underside overhanging surfaces are contrasted in the diagram. At the same time, it can be clearly found that with the increase of counting times, the surface roughness of both the upper and underside overhanging surfaces shows a downward trend. When the counting times are 10, the surface roughness value is the lowest and the forming effect is the best. The upper overhanging surface roughness value of the suspension surface on the 20° suspension structure is the lowest, which can reach 15.716 µm, and the underside surface roughness value of the suspension surface is the lowest, which is 30.716 µm.

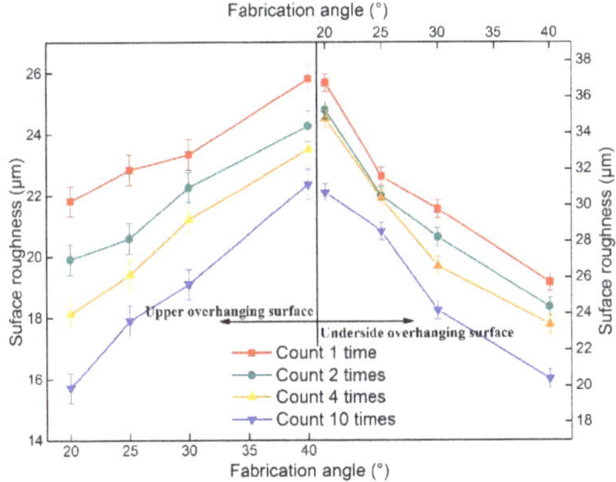

Figure 17. Boundary scanning strategy affects surface roughness trend (corresponding Table 3, No. 2–5).

4. Conclusions

In summary, the experimental optimization of forming overhanging structures of 316L stainless steel and chromium–nickel alloy steel was studied by selective laser melting. By using different scanning strategies, upper-surface-remelting processes, boundary-scanning powers, and boundary-counting strategies, the forming effects of the upper surface and the upper and underside overhang surface were analyzed, and the effects of the processing strategies on the surface roughness were also discussed. The conclusions are as follows:

1. The forming effect of chromium–nickel alloy is better than the 316L stainless steel below the limit forming angle in the overhanging structure.

2. Upper surface remelting can reduce the value of surface roughness, decrease the defect of surface spheroidization, and improve surface forming quality. Surface quality is better with the increase of the number of remelting, but increasing surface remelting power is not conducive to improving surface quality.

3. The optimization of boundary remelting is beneficial to the forming of suspension structure. Multiple boundary counting is beneficial to improve the forming quality of the suspension structure below the limit forming angle, and the lower scanning power is more beneficial to the forming of boundary.

4. By comparing different scanning strategies, it is found that the application of 316L stainless steel in the chessboard scanning strategy is beneficial to forming in a certain range below the ultimate forming angle, and the forming quality beyond a certain range is still not optimistic. The meander scanning strategy is suitable for forming suspension structures above the limit forming angle.

Author Contributions: Data curation, W.S. and P.W.; Formal analysis, W.S. and P.W.; Investigation, Y.L.; Methodology, P.W.; Resources, Y.L.; Supervision, Y.L.; Validation, G.H.; Visualization, G.H.; Writing—original draft, P.W.; Writing—review & editing, W.S., P.W., and Y.L.

Funding: This work was supported by grant 51505006 from the National Natural Science Foundation of China.

Conflicts of Interest: The authors declare no conflict of interest.

References

1. Raj, B.; Mudali, U.K. Materials development and corrosion problems in nuclear fuel reprocessing plants. *Prog. Nucl. Energy* **2006**, *48*, 283–313. [CrossRef]

2. Boyer, R.R. An overview on the use of titanium in the aerospace industry. *Mater. Sci. Eng.* **1996**, *213*, 103–114. [CrossRef]

3. Gu, D.D.; Wang, Z.Y.; Shen, Y.F.; Li, Q.; Li, Y.F. In-situ TiC particle reinforced Ti–Al matrix composites: Powder preparation by mechanical alloying and Selective Laser Melting behavior. *Appl. Surf. Sci.* **2009**, *255*, 9230–9240. [CrossRef]

4. Kothari, K.; Radhakrishnan, R.; Wereley, N.M. Advances in gamma titanium aluminides and their manufacturing techniques. *Prog. Aerosp. Sci.* **2012**, *55*, 1–16. [CrossRef]

5. Chianeh, V.A.; Hosseini, H.R.M.; Nofar, M. Micro structural features and mechanical properties of Al–Al 3 Ti composite fabricated by in-situ powder metallurgy route. *J. Alloys Compd.* **2009**, *473*, 127–132. [CrossRef]

6. Kruth, J.P.; Mercelis, P.; Vaerenbergh, J.V.; Froyen, L.; Rombouts, M. Binding mechanisms in selective laser sintering and selective laser melting. *Rapid Prototyping J.* **2005**, *11*, 26–36. [CrossRef]

7. Van Vaerenbergh, J. Benchmarking of SLS/SLM processes. In Proceedings of the Ims Rapid Manufacturing Network Rapid Product Development Session Enabling Processes, Stuttgart, Germany, 20 October 2005.

8. Yasa, E.; Craeghs, T.; Badrossamay, M.; Kruth, J.P. Rapid manufacturing research at the catholic university of leuven. In Proceedings of the US—TURKEY Workshop On Rapid Technologies, Istanbul, Turkey, 24–25 September 2009.

9. Lott, P.; Schleifenbaum, H.; Meiners, W.; Wissenbach, K.; Hinke, C.; Bültmann, J. Design of an optical system for the in situ process monitoring of selective laser melting (slm). *Physics Procedia.* **2011**, *12*, 683–690. [CrossRef]

10. Khan, M.; Sheikh, N.A.; Jaffery, S.H.I.; Ali, L.; Alam, K. Numerical simulation of meltpool instability in the selective laser melting (slm) process. *Lasers Eng.* **2014**, *28*, 319–336.

11. Alkahari, M.R.; Furumoto, T.; Ueda, T.; Hosokawa, A. Melt pool and single track formation in selective laser sintering/selective laser melting. *Adv. Mater. Res.* **2014**, *933*, 196–201. [CrossRef]

12. Yong-qiang, Y.; Jian-bin, L.; Di, W.; Zi-yi, L. A study of 316L stainless steel non-horizontal overhanging surface in selective laser melting. *Mater. Sci. Eng.* **2011**, *19*, 94–99.

13. Jun, H.; Yi-ping, T.; Bing-heng, L.; Dian-liang, W. Research on supporting design rules for non-horizontal undersurface of parts in rapid prototyping by light curing. *Chin. J. Mech. Eng.* **2004**, *40*, 155–159.

14. Tao, M. Research on several key problems in rapid prototyping technology of laser selective sintering. Ph.D. Thesis, Jiangsu University, Jiangsu, China, 2003.

Communication

Comparison of Nano-Mechanical Behavior between Selective Laser Melted SKD61 and H13 Tool Steels

Jaecheol Yun [1], Van Luong Nguyen [1], Jungho Choe [1], Dong-Yeol Yang [1], Hak-Sung Lee [2], Sangsun Yang [1] and Ji-Hun Yu [1,*]

[1] Powder & Ceramic Division, Korea Institute of Materials Science (KIMS), Changwon 51508, Korea; jaecheol@kims.re.kr (J.Y.); nguyenvanluong.hut@gmail.com (V.L.N.); drgb0443@kims.re.kr (J.C.); dyyang@kims.re.kr (D.-Y.Y.); nanoyang@kims.re.kr (S.Y.)

[2] Materials Processing Innovation Research Division, Korea Institute of Materials Science (KIMS), Changwon 51508, Korea; hslee@kims.re.kr

* Correspondence: jhyu01@kims.re.kr; Tel.: +82-55-280-3576; Fax: +82-55-280-3289

Received: 8 November 2018; Accepted: 4 December 2018; Published: 6 December 2018

Abstract: Using nanoindentation under various strain rates, the mechanical properties of a laser powder bed fusion (PBF) SKD61 at the 800 mm/s scan speed were investigated and compared to PBF H13. No obvious pile-up due to the ratio of the residual depth (h_f) and the maximum depth (h_{max}) being lower than 0.7 and no cracking were observed on any of the indenter surfaces. The nanoindentation strain-rate sensitivity (m) of PBF SKD61 was found to be 0.034, with hardness increasing from 8.65 GPa to 9.93 GPa as the strain rate increased between 0.002 s^{-1} and 0.1 s^{-1}. At the same scan speed, the m value of PBF H13 (m = 0.028) was lower than that of PBF SKD61, indicating that the mechanical behavior of PBF SKD61 was more critically affected by the strain rate compared to PBF H13. PBF processing for SKD61 therefore shows higher potential for advanced tool design than for H13.

Keywords: Laser powder bed fusion; selective laser melting; SKD61 tool steel; nanoindentation; strain-rate sensitivity

---

1. Introduction

SKD61 and H13 are types of hot work tool steels applied to casting molds, extrusion tools, forging dies, etc., and the hardness of SKD61 (229 BHN) is lower than that of H13 (235 BHN) [1]. Both steels are fabricated by conventional methods requiring expensive dedicated tools, and thus are inappropriate for small-scale or complex-shape productions [2–4]. Selective laser melting (SLM), is a laser powder-bed additive manufacturing process that is suitable for the processing of tool steels, including SKD61 and H13, because it offers the ability to not only reduce the amount of machining and hence wastage of this expensive material, but also to produce intricate molds with a nearly full density and a refined microstructure [3,4].

The mechanical behavior of hot work tool steels prepared by the PBF method is one of the most important characteristics [5,6]. It was reported that the hardness of PBF SKD61 fabricated from commercial powders was higher than that made from gas atomized spherical powders [4,7]. Very recently, a few studies have addressed the mechanical properties of tool steel processed by PBF using nanoindentation [3,8]. For example, the H13 PBF-processed at 200 mm/s scan speed exhibited the lowest creep resistance and highest hardness values. However, very few studies have dealt with the mechanical behavior of PBF SKD61 using nanoindentation. It is thus necessary to perform nanoindentation tests to probe the creep behavior of the PBF SKD61 and compare it with that of PBF H13 [9,10].

In a previous study, PBF processing at an 800 mm/s laser scan speed was applied successfully for SKD61 powders [4,11]. This study expands on those reported findings by using nanoindentation tests to investigate the mechanical properties of PBF SKD61. Results for PBF-processed H13 at the same laser scan speed sourced from the literature are used in the evaluation conducted here [2].

2. Experimental Procedure

The material used in this work was SKD61 powder prepared by a gas atomizing process (Hot Gas Atomization System, HEMMIGA 100/25, PSI Ltd., Leicester, UK), as shown in Figure 1. Specifically, SKD61 ingot obtained from SeAH Special Steel Corp. (Changwon, Korea) was melted into a graphite crucible at 1690 °C under a high purity argon atmosphere. The alloy liquid was then ejected through a spray nozzle under hot N_2 gas pressure of 50 bar. Upon gas atomization, the gas-atomized powders were collected and loaded onto a series of ASTM E11 standard sieves to obtain a specific particle size range of 10~45 μm. The elemental chemical compositions of the produced powders were confirmed by inductively coupled plasma (ICP, Spectro Arcos ICP-AES, Kleve, Germany). The powders were then processed by a PBF device (a Concept Laser Mlab-Cusing System, Lichtenfels, Germany) equipped with a 90 W Nd:YAG fiber laser to manufacture cuboid specimens (10 mm × 10 mm × 10 mm). Detailed preparation of the PBF samples was described in [3,4].

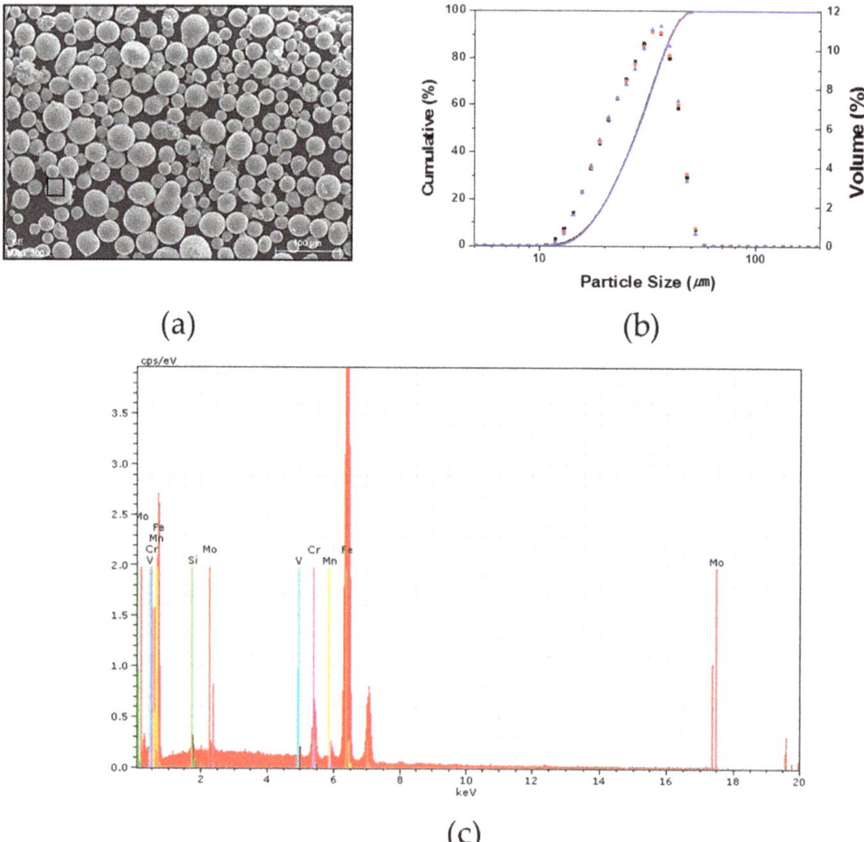

(a)   (b)

(c)

Figure 1. (**a**) SEM picture of the powder input; (**b**) particle size distribution; and (**c**) EDS result at the location indicated by the rectangle in (**a**).

The cuboid specimens prepared by PBF were mounted in epoxy resin, cross-sectioned, and then mechanically polished on an auto-disc polishing machine (Struers LaboForce-100, Ballerup, Denmark) to reach mirror-like surfaces. The specimens with a surface roughness of lower than 50 nm determined using a SV-3200S4 (Mitutoyo, Sakado, Japan) were used for the next steps. An optical microscope (OM, Nikon ECLIPSE MA200, Nikon, Tokyo, Japan) and a scanning electron microscope (SEM, JSM-5800, JEOL, Tokyo, Japan) were also used to observe the microstructure on the sample surfaces. Nanoindentation tests were then performed for those samples at room temperature at the same maximum load (500 mN) on a NanoTest nanoindenter (Micro Materials Ltd., Wrexham, UK) using a three-sided Berkovich diamond indenter. Specifically, all nanoindentation tests were carried out at loading rates of 50 mN/s, 25 mN/s, 16.67 mN/s, 12.5 mN/s, 10 mN/s, 5 mN/s, and 1 mN/s. The indenter was then held at the maximum load for 5 s, which was followed by unloading at a rate of 50 mN/s for all tests. At least 10 indentation points at each loading rate were carried out and the results were averaged.

3. Results and Discussion

Figure 2a shows the OM and SEM microstructure result of the SKD61 sample prepared by PBF at the 800 mm/s scanning speed, in which the inset is the SEM result. In the OM image, the melt pool confirmed the scan direction and some pores. Figure 2b shows a SEM result at the location of an indenter point. Figure 2a reveals some black pores with irregular shapes and small volumes observed over the sample. The volume fraction of the black phase was calculated, using the image analysis software, to be 1.4%, which is lower than that (5.2%) for PBF H13 at the same scan speed [3]. Black pores, indicating imperfections in the microstructure, may stem from insufficient laser energy to melt powder materials within the melt pool, causing a short cooling (faster solidification) time and insufficient melting of the substrate and powders [3].

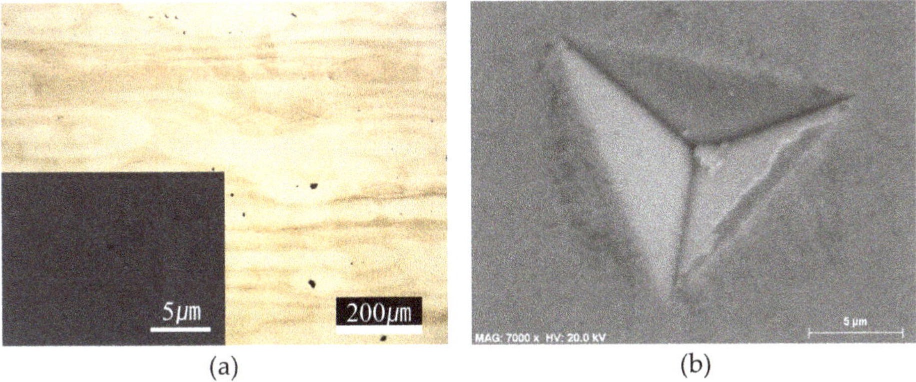

(a)   (b)

Figure 2. (a) Microstructure of the PBF- processed SKD61 at 800 mm/s scanning speed; (b) SEM graph of an indenter point after nanoindentation tests.

The widely used Oliver and Pharr method is accurate only for nanoindents that do not show significant pile-up deformation, which is related to the ratio of the residual depth (h_f) and the maximum depth (h_{max}) [12–14]. In the SEM results of indenter points on all sample surfaces, no obvious pile-up was observed, as shown in Figure 2b. This result is similar to that of the PBF H13 material, where it was confirmed that $\frac{h_f}{h_{max}}$ in all nanoindentation tests was lower than 0.7 [3]. Therefore, the Oliver and Pharr method can be adopted to extract the hardness values of the material here. Moreover, no cracking in Figure 2b is observed, indicating that the material has low crack sensitivity.

In this study, a constant rate of loading (CRL) method, where a steady loading rate is used until the tip depth rate becomes nearly constant, was used to determine the strain rate ($\dot{\varepsilon}$) because the CRL allows for simple calculation of strain rates, and also is proven to be more suitable to correlate with the conventional constant strain rate tests [12,15,16]. In fact, both indentation hardness and strain rate are not constant under CRL loading, causing difficulty in determining a consistent creep exponent from a single CRL nanoindentation experiment [9,10,17,18]. However, if only the indentation hardness and indentation strain rate at the maximum load point are used, the strain-rate sensitivity can be determined from a group of CRL nanoindentation tests under different loading rates [15]. Accordingly, the above loading rates corresponded to strain rates of 0.1 s^{-1}, 0.05 s^{-1}, 0.033 s^{-1}, 0.025 s^{-1}, 0.02 s^{-1}, 0.01 s^{-1}, and 0.002 s^{-1}, respectively.

Figure 3 is plotted to compare the hardness of SKD61 and H13 processed by PBF at an 800 mm/s scan speed. Error bars on the data reflect the standard deviation calculated for the hardness from multiple indentations for each sample. Experimental results from nanoindentation tests of H13 prepared by PBF at 800 mm/s, obtained from another report in the literature, are included for comparison [3]. As shown in Table 1, the hardness increases in ranges of (8.65–9.99) GPa and (7.91–8.84) GPa for the SKD61 and H13 materials, respectively, prepared by PBF at the same scan speed of 800 mm/s at nanoindentation strain rates in the range of 0.002 s^{-1} and 0.1 s^{-1}. This shows the PBF SKD61 has about 10% higher hardness values on average than the PBF H13 material.

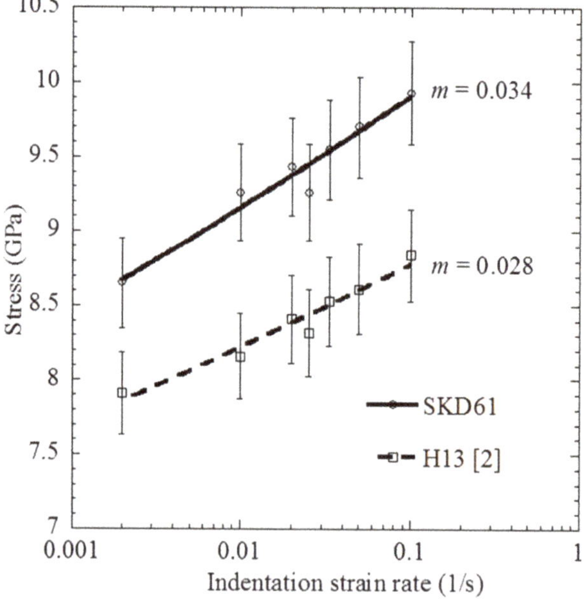

Figure 3. Indentation stress as a function of the strain rate for PBF processed by SKD61 and H13.

Table 1. Hardness values of SKD61 and H13 prepared by the PBF process at the same scan speed of 800 mm/s.

Strain Rate (s^{-1})	Hardness (GPa)	
	SKD61	H13 [2]
0.002	8.65	7.91
0.01	9.26	8.16
0.02	9.43	8.41
0.025	9.26	8.32
0.033	9.55	8.53
0.05	9.70	8.61
0.1	9.93	8.84

By assuming the indentation strain rate and hardness are proportional to the flow strain rate and stress, respectively, it follows that the strain rate sensitivity (SRS) can be defined as:

$$m = \left(\frac{\partial \ln H}{\partial \ln \dot{\varepsilon}} \right) \tag{1}$$

The slope of the straight line in Figure 3 represents the strain-rate sensitivity exponent (*m*), with a value of 0.034 and 0.028 for the PBF SKD61 and H13, respectively. The strain-rate sensitivity of PBF SKD61 was higher than that of PBF H13, indicating that the mechanical behavior of H13 material is less susceptible to the strain rate than SKD61.

In recent studies, the optimal laser scan speed for the PBF process was around 200 mm/s for H13 [3], and the hardness increased from 8.61 GPa to 9.29 GPa; these values are lower than those of SKD61 in the present study at the same strain rate range. The laser energy density (E) can be defined as:

$$E = \frac{P}{vht} \tag{2}$$

where *P* is the laser power (W), *v* is the scan speed (mm/s), *h* is the hatch distance (μm), and *t* is the layer thickness (μm). The laser energy density was calculated by Equation (2), and it was 156.2 kW·h/m^3. Because the laser energy density is inversely proportional to the scan speed [19–22], the optimal condition of the PBF process for SKD61 consumes less energy than for H13. In the case of the same process conditions, PBF SKD61 showed less pores then PBF H13 in the SEM images. It is not shown in this manuscript because the results of PBF SKD61 and H13 are included in various process conditions. We considered optimal process conditions and then the indentation stress of two specimens, in which PBF SKD61 and H13 were compared. Moreover, PBF SKD61 showed a better microstructure (less pores) and higher hardness than PBF H13. Therefore, PBF SKD61 shows higher potential for advanced tool design than PBF H13.

4. Conclusions

This study demonstrates successful printing of SKD61 by the PBF method at an 800 mm/s laser speed. From the results, no obvious pile-up (due to $\frac{h_f}{h_{max}} < 0.7$) and no cracking on all indenter surfaces were observed. Hardness increased from 8.65 GPa to 9.93 GPa as the strain rate increased in a range of 0.002 s^{-1} and 0.1 s^{-1}. At the same scan laser speed, the *m* value of PBF SKD61 (*m* = 0.034) was higher than that of PBF H13 (*m* = 0.028), indicating that the mechanical behavior of the PBF SKD61 was more critical to the strain rate compared to PBF H13. As a result, PBF processing for SKD61 shows higher potential in practical application than for H13 due to the superior microstructure and mechanical behavior of the former, and less laser energy consumption at the optimal condition.

Author Contributions: writing-original draft preparation, J.Y.; conceived and designed the experiments, V.L.N. and J.Y.; performed the experiments, J.C.; analyzed the date, D.-Y.Y., H.-S.L. and S.Y.; edited the paper, J.-H.Y.

Conflicts of Interest: The authors declare no conflict of interest.

References

1. Astmsteel, Hot Work Steel Comparison: H13 vs 1.2344 vs SKD61. Available online: http://www.astmsteel. com/steel-knowledge/hot-work-steel-h13-1-2344-skd61/ (accessed on 27 November 2018).
2. Sander, J.; Hufenbach, J.; Giebeler, L.; Wendrock, H.; Kuhn, U.; Eckert, J. Microstructure and properties of FeCrMoVC tool steel produced by selective laser melting. *Mater. Des.* **2016**, *89*, 335–341. [CrossRef]
3. Nguyen, V.L.; Kim, E.A.; Lee, S.R.; Yun, J.C.; Choe, J.H.; Yang, D.Y.; Lee, H.S.; Lee, C.W.; Yu, J.H. Evaluation of strain-rate sensitivity of selective laser melted H13 tool steel using nanoindentation tests. *Metals* **2018**, *8*, 589. [CrossRef]

4. Yu, J.H.; Yang, D.Y.; Lee, H.S.; Kim, S.W.; Yang, S.S.; Lim, T.S.; Lee, C.W. Characteristics of SLM printed SKD61 alloy by using of gas atomized spherical powders. *KSME Spring Conf.* **2016**, 1143–1144.

5. Dirk, H.; Vanessa, S.; Eric, W.; Claus, E. Additive manufacturing of metals. *Acta Mater.* **2016**, *117*, 1–22.

6. Gu, D.D.; Meiners, W.; Wissenbach, W.; Proprawe, R. Laser additive manufacturing of metallic components: Materials, processes and mechanisms. *Int. Mater. Rev.* **2012**, *57*, 133–164. [CrossRef]

7. Geldart, D.; Abdullah, E.C.; Hassnapour, A.; Nwoke, L.C.; Wouters, I. Characterization of powder flowability using measurement of angle of repose. *China Particuol.* **2006**, *4*, 104–107. [CrossRef]

8. Duan, Z.; Pei, W.; Gong, X.; Chen, H. Superplasticity of annealed H13 steel. *Materials* **2017**, *10*, 870. [CrossRef] [PubMed]

9. Attar, H.; Ehtemam-Haghighi, S.; Kent, D.; Dargusch, M.S. Recent developments and opportunities in additive manufacturing of titanium-based matrix composited: A review. *Int. J. Mach. Tool Manu.* **2018**, 85–102. [CrossRef]

10. Attar, H.; Ehtemam-Haghighi, S.; Kent, D.; Okulov, I.V.; Wendrock, H.; Bönisch, M.; Volegov, A.S.; Calin, M.; Eckert, J.; Dargusch, M.S. Nanoindentation and wear properties of Ti and Ti-TiB composite materials produced by selective laser melting. *Mater. Sci. Eng. A* **2017**, 20–26. [CrossRef]

11. Yun, J.; Choe, J.; Lee, H.; Kim, K.-B.; Yang, S.; Yang, D.-Y.; Kim, Y.-J.; Lee, C.-W.; Yu, J.-H. Mechanical Property Improvement of the H13 Tool Steel Sculptures built by Metal 3D Printing Process via Optimum Conditions. *J. Korean Powder Metall. Inst.* **2017**, *24*, 195–201. [CrossRef]

12. Oliver, W.C.; Pharr, G.M. An improved technique for determining hardness and elastic modulus using load and displacement sensing indentation experiments. *J. Mater. Res.* **1992**, *7*, 1564–1583. [CrossRef]

13. Gu, S.T.; Chai, G.Z.; Wu, H.P.; Bao, Y.M. Characterization of local mechanical properties of laser-cladding H13-TiC composite coatings using nanoindentation and finite element analysis. *Mater. Des.* **2012**, *39*, 72–80. [CrossRef]

14. Yadroitsev, I.; Bertrand, P.; Smurov, I. Parametric analysis of the selective laser melting process. *Appl. Surf. Sci.* **2007**, *253*, 8064–8069. [CrossRef]

15. Du, Y.; Liu, X.H.; Fu, B.; Shaw, T.M.; Lu, M.; Wassick, T.A.; Bonilla, G.; Lu, H. Creep characterization of solder bumps using nanoindentation. *Mech. Time-Depend Mater.* **2017**, *21*, 287–305. [CrossRef]

16. Peykov, D.; Martin, E.; Chromik, R.R.; Gauvin, R.; Trudeau, M. Evaluation of strain rate sensitivity by constant load nanoindentation. *J. Mater. Sci.* **2012**, *47*, 7189–7200. [CrossRef]

17. Hu, J.J.; Zhang, Y.S.; Sun, W.M.; Zhang, T.H. Nanoindentation-induced pile-up in the residual impression of crystalline Cu with different grain size. *Crystals* **2018**, *8*, 9. [CrossRef]

18. Liu, C.Z.; Chen, J. Nanoindentation of lead-free solders in microelectronic packaging. *Mater. Sci. Eng. A* **2007**, *448*, 340–344. [CrossRef]

19. Simchi, A. Direct laser sintering of metal powders: Mechanism, kinetics and microstructural features. *Mater. Sci. Eng. A* **2006**, *428*, 148–158. [CrossRef]

20. Carter, L.N.; Wang, X.; Read, N.; Khan, R.; Aristizabal, M.; Essa, K.; Attallah, M.M. Process Optimisation of Selective Laser Melting using Energy Density Model for Nickel-based Superalloys. *Matter. Sci. Technol.* **2016**, *32*, 657–662. [CrossRef]

21. Körner, C. Additive manufacturing of metallic components by selective electron beam melting—A review. *Int. Mater. Rev.* **2016**, *61*, 361–377. [CrossRef]

22. Kasperovich, G.; Haubrich, J.; Gussone, J.; Requena, G. Correlation between porosity and processing parameters in TiAl6V4 produced by selective laser melting. *Mater. Des.* **2016**, *105*, 160–170. [CrossRef]

Article

Additive Manufactured A357.0 Samples Using the Laser Powder Bed Fusion Technique: Shear and Tensile Performance

Lucia Denti

Department of Engineering "Enzo Ferrari", University of Modena and Reggio Emilia, Via Vivarelli 10, 41125 Modena, Italy; lucia.denti@unimore.it; Tel.: +39-059-205-6100

Received: 26 July 2018; Accepted: 24 August 2018; Published: 27 August 2018

Abstract: New aluminium alloys, with lower silicon content than in the first-developed formulations, have recently been introduced in the field of Additive Manufacturing and are dedicated to automotive applications. As they are relatively new, mechanical characterization under standard protocols of the automotive field are of utmost scientific as well as industrial relevance. The paper addresses the mechanical properties and microstructure of A357.0. Static tensile and shear tests of samples built by Laser Powder Bed Fusion, with different orientations in the machine work volume, have been performed. The aim was to identify possible anisotropy in the tensile and shear behaviour of this innovative alloy. Particularly for shear, the effect of adhesion between the layers onto shear strength was studied. Results analysis, by means of statistical tools, allows for the affirmation that no tensile modulus or yield strength anisotropy is observed. Instead, a small (yet statistically significant) increase in both shear- and tensile strength and a decrease in ductility are obtained as the direction of the specimens approaches the growth direction. Scanning Electron Microscope (SEM) observation of the failure mechanisms assisted in the interpretation of the results, by relating different failure modes to the relative orientation of loads versus the directions of inherent anisotropy in Laser Powder Bed Fusion processes.

Keywords: additive manufacturing; laser powder bed fusion; A357.0; mechanical performance

1. Introduction

Over the next few years, the Additive Manufacturing (AM) industry is expected to continue to grow strongly. Oerlikon group, one of the major suppliers of metal powders for AM, forecasted that by 2021 the sale of AM products and services will be around 26.5 billion USD (United States dollars). AM has great potential, and is gaining every day in broader usage. In order to accelerate the adoption of AM, improvements in areas such as new material formulations and more solid knowledge of the expected mechanical behaviour will need to take place [1].

The formulations of aluminium (Al) alloys that are available for AM, and more specifically for Laser Powder Bed Fusion (LPBF), are still very few. Al alloys have been introduced for AM applications recently because, if compared for example to Ti alloys, they are easier to machine by means of conventional subtractive processes. However, since Al is comparatively cheaper to purchase, the material saving that is allowed by AM becomes less impressive. As a consequence, the production of Al parts by LPBF has less immediate commercial appeal [2].

Other obstacles to the use of Al alloys are:

- The presence of some heavily applied hardenable alloy elements (such as Zn), which leads to turbulent melt pools, splatter, and porosity [2]. Many formulations are not suitable for or easy to be used in LPBF;

- High reflectivity;
- Poor flowability, which can impede the deposition of a smooth thin layer [3];
- The low viscosity of the molten material limits the melt pool sizes.

Currently, the most common aluminium alloy for LPBF is AlSi10Mg. It is used to produce parts with high strength and high dynamic loadability. The use of such parts in the aerospace and automotive industries is quickly spreading [4].

As previously mentioned, much is expected from the market in terms of a wider range of aluminium alloys to be processed by LPBF. For example, a specific alloy has been introduced recently: Scalmalloy®RP is the Airbus Group's unique aluminium-magnesium-scandium alloy (AlMgSc). It was developed for additive manufacturing and displays high strength [5]. On the side of high-performance alloys, the new A20X™ was patented in 2010 and gained full aerospace approval in 2012 [6]. The formulation consists of an aluminium copper base, with the addition of intermetallic titanium diboride. The resulting ultra-fine microstructure accomplishes strength values as high as 477 MPa, together with the possibility to operate at high temperatures. Initially developed for casting, the alloy in now available for use in powder bed AM.

Also, a new formulation of an Al-Si-Mg alloy has been marketed of late: A357.0, in which the Si content is reduced to 7 wt % [7–9]. The alloy proposed by CITIM GmbH ensures, as an example, a nominal strength of 415 MPa and a total extension at fracture of 5% in the as-built state [9]. A357.0 is of special interest for automotive applications since the same alloy is commonly used for many car body parts produced by traditional processes. Typical examples are brackets, carters, and stands. The quality standards of high-performance cars require the adoption of expensive solutions, mainly gravity and low-pressure permanent mould casting, with the use of cores. While outstanding as to mechanical and dimensional quality of parts, as well to absence of defects, such processes suffer from high investment costs, long time-to-market, and very low flexibility. Such process characteristics mismatch production types characterized by very small batches. In the case of small batch, variable cost, still high, can be much easier attributed than fixed ones. LPBF processes are an increasingly frequent choice for many aluminium components traditionally obtained by casting. Yet, LPBF is still at the early stages in this field as the reliability and predictability of a technology that only recently included aluminium alloys among the options is still unproven. The choice of an innovative process with a known material means the same target material can be kept. As an advantageous consequence, the complex structural validations of a part design (including static, dynamic, durability, and non-linear crash load cases) do not need to be rerun.

Data on the mechanical performance of Al-based materials used in LPBF are often not comparable directly due to the non-homogeneity of the characterization standards that are used and the effect of different process parameters that cause a great dispersion of data. Table 1 lists the mechanical properties available in the datasheets for the as-built condition.

Moreover, the difficulty in data collection increases for the most recently marketed materials. AlSi10Mg has been studied for a while, and at least its static properties are quite consolidated. Some authors also addressed the optimization of process parameters, providing information such as the effect of power on the build rate and the combination of power and volumetric energy density for best process performance [13]. The higher-silicon alloy Al-12Si has been studied by Prashanth et al., who provided an extremely interesting focus on mechanical properties and their relation to the microstructure in different build orientations [14]. Prashanth and colleagues observed a remarkably fine grain structure, along with outstanding strength, at least twice that of the cast material. Low silicon aluminium alloys are still quite new in the field of LPBF and very little tested [6,7]. For most of the major system suppliers, A357.0 is still under development and is expected to reach full commercial availability soon.

Table 1. Mechanical performance specified by the suppliers of Aluminium alloys used for Laser Powder Bed Fusion (LPBF) in the as-built condition.

Mechanical Performance	Tensile Strength [MPa]		Yield Strength [MPa]		Modulus of Elasticity [GPa]	Total Extension at Fracture [%]	
Direction	XY direction	Z direction	XY direction	Z direction		XY direction	Z direction
AlSi10Mg [10]	460 ± 20	470 ± 20	270 ± 20	230 ± 20	70 ± 10	9 ± 2	6 ± 2
AlSi10Mg [9]	410 ± 40		240 ± 40		65 ± 5	5 ± 2	
AlSi10Mg [11]	397 ± 11		227 ± 11		64 ± 10	6 ± 1	
AlSi9Cu3 [9]	380 ± 40		200 ± 40		62 ± 10	2.5 ± 1.0	
AlSi9Cu3 [11]	415 ± 15		236 ± 8		57 ± 5	5 ± 1	
AlSi12 [12] [1] AlSi10Mg [12] [1]	329 ± 4	344 ± 2	211 ± 4	205 ± 3	75	9 ± 1	6 ± 1
AlSi12 [11]	409 ± 20		211 ± 20			5 ± 3	
AlSi7Mg [9]	410 ± 50		270 ± 50		70 ± 15	5 ± 3	
AlSi7Mg [11]	375 ± 17		211 ± 15		59 ± 21	8 ± 2	
Scalmalloy® [5]	520		470			13	
A20X™ [6]	477		415			13	

[1] After stress-relieving.

This paper is focused on the characterization of ISO A357.0 parts (ASM, ASTM, and SAE 356.0 specifications) produced by LPBF. This alloy is traditionally used for applications in which good weldability, pressure tightness, and good resistance to corrosion are required. Some of the many examples of its use may include pump parts of aircrafts, automotive transmission cases, aircraft fittings and control parts, as well as water-cooled cylinder blocks. Many of the listed applications involve the production of big parts, so the present contribution was focused specifically on verifying the properties expected for large components to be produced relatively quickly. As a consequence, the following choices were made in the research plan, namely: (i) to use one of the machines with the biggest build chamber on the market; (ii) to select high-productivity process parameters in order for the process to be economically viable; (iii) to use a sufficiently high plate temperature so that distortion even of big parts would not be an issue and stress-relieving would not be required; and iv) to disregard any possible subsequent densification steps, as for example hot isostatic pressing, because it would hamper the industrial benefits of the solution under investigation.

The aim of the research is to evaluate the tensile and shear behaviour of this new alloy and to verify if different build orientations cause a quantitative variation of the mechanical properties and a qualitative change of the failure modes.

2. Materials and Methods

A357.0 powder was used to produce tensile and shear specimens. Of the powder, the nominal chemical composition versus the experimental one is listed in Table 2, and the particle size measured by the supplier by laser granulometry is shown in Figure 1 [12].

Table 2. Nominal chemical analysis of the A357.0 powder [12].

Weight %	Si	Mg	Fe	Zn	Ti	Al
	7.2	0.51	0.19	0.02	0.09	bal.

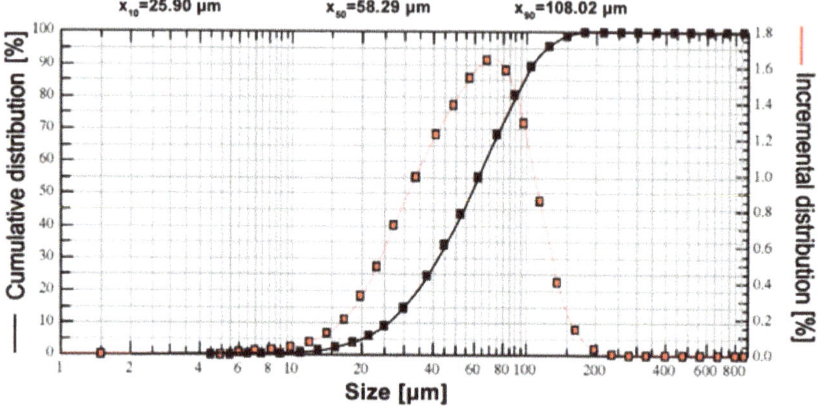

Figure 1. Particle size measured by laser granulometry [12].

The A357.0 powder used in this paper was observed by scanning electron microscope, SEM (ESEM, Quanta-200, FEI, Thermo Fisher Scientific, Eindhoven, The Netherland), equipped with an X-ray energy dispersion spectroscopy system (X-EDS, INCA Oxford Instruments, Abingdon, UK) in order to check the particle shape and the chemical composition.

The specimens were manufactured by means of an X Line 2000 R (Concept Laser GmbH, Lichtenfels, Germany) by using the following process parameters:

- Laser power: 950 W;
- Laser velocity: 2000 mm/s;
- Laser spot: 400 µm;
- Hatch distance: 0.2 mm;
- Inert gas: nitrogen;
- Platform temperature: preheated at 200 °C;
- Scan strategy: skin-core;

The machine has one of the largest build chambers on the market ($800 \times 400 \times 500$ mm^3) and is specifically dedicated to the manufacture of large functional components with extremely high productivity. The system is equipped with two fibre lasers that each output 1000 W, an automated powder transport, and a rotating mechanism so that two build modules can be used reciprocally, without any downtime.

Heating of the platform acts as an 'in-process' ageing treatment [7] and allows for the limitation of the cooling rates and the residual stresses as compared to unheated builds [6], hence the parts do not require any stress relieving treatment. All the specimens in this study were tested and observed in the as-built condition.

Shear tests were performed according to ASTM B831-05, by adopting a test velocity of 1 mm/min (the standard cross head speed rate limit is 19.1 mm/min). Such speed allowed for a stress rate lower than 689 MPa/min, which is the standard loading rate limit on the cross-section. The shear specimens were built in the LPBF machine as rectangular parallelepipeds and then machined to the final geometries, as in Figure 2A.

Tensile tests were performed according to UNI EN ISO 6892-1 by adopting a strain rate of 0.00025 ± 0.00005 s^{-1}, which was achieved using a test velocity of 0.5 mm/min. The mechanical tests were preceded by a calibration test with an initial velocity of 1.0 mm/min. Strain was measured by an extensometer with a gauge length of 25 mm. The tangent tensile modulus was automatically calculated by the machine software. The tensile specimens were built in the LPBF machine as 14 mm diameter and 80 mm long cylinders and were then machined to the final geometry shown in Figure 2B.

Tensile and shear specimens were built in different orientations relative to the LPBF process. Figure 3 shows the build orientations of samples parallel to the layers (0°), parallel to the growth direction (90°), and inclined at 45° with respect to both previous directions (45°). In the case of 45° shear specimens, specific attention has been paid to test interlayer shear strength; the geometry and orientation of the specimens allowed for the application of the shear loads on a surface which is parallel to the powder bed (Figure 3). Figure 3 also shows the direction along which the recoater blade spreads the powder in the bed, and the direction of the inert gas flux on the powder bed. The latter is very important in order to blow the melting slags away from the build area.

The number of specimens used in each test is shown in Table 3.

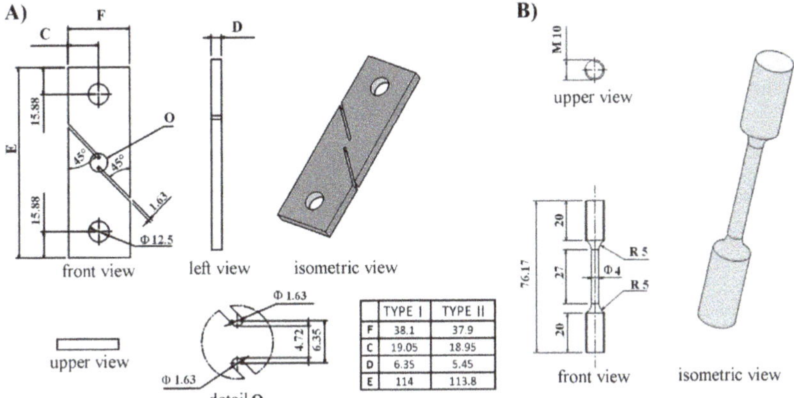

Figure 2. Geometry of the used specimens: (**A**) shear, ASTM B831-05 type I and type II; (**B**) tensile, UNI EN ISO 6892-1.

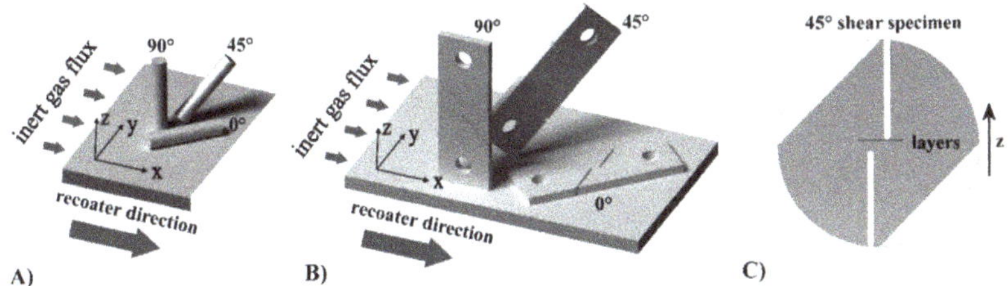

Figure 3. Orientation of the built specimens: (**A**) tensile specimens; (**B**) shear specimens; (**C**) detail of 45° shear specimen.

Table 3. Number of specimens used in the tests.

Specimens Tested	No. of Specimens in 0° Direction	No. of Specimens in 45° Direction	No. of Specimens in 90° Direction
ASTM B831-05 Figure 2 type I	9	9	-
ASTM B831-05 Figure 2 type II	8	8	17
UNI EN ISO 6892-1	17	17	17

Brinell hardness HBW 2.5/62.5/15 hardness (ball diameter 2.5 mm, applied force 62.5 kgf, duration time 15 s) was measured according to standard EN ISO 6506-1. The tests were carried out on a surface parallel to the powder bed as well as on one parallel to the growth direction. Metallographic sections of the specimens, parallel to the XY, XZ, and YZ planes, were prepared by means of microcutting, embedding in epoxy resin, and polishing with a 3 μm diamond suspension. Residual porosity was measured by image analysis on polished sections using an optical microscope, OM (Eclipse LV150N, Nikon, Tokyo, Japan). Metallurgical structures were observed after etching with the Dix-Keller reactant (HF 2% vol, HCl 1.5% vol, HNO_3 2.5% vol; water bal.). The etched surfaces were observed by an OM as well as by an Nova NanoSEM 450 FEG-SEM (FEI, Thermo Fisher Scientific, Eindhoven, The Netherland) in immersion lens mode.

Rupture surfaces of all specimens were observed by SEM with the aid of X-ray energy dispersion spectroscopy (X-EDS) analysis (X-EDS, INCA Oxford Instruments, Abingdon, UK).

3. Results and Discussion

Figure 4 shows SEM images of the A357.0 powder used in this paper, sampled from the build chamber after the job completion. The sphericity of powder particles is not as regular or repetitive as that of other raw materials for LPBF. Irregular shape and poor flowability are among the well-known intrinsic difficulties in working with aluminium [3,7]. The same is not true, for example, for titanium or Ni-based alloys. The shape of the A357.0 particles may have an effect on the flowability of the powder and lead to a lack of continuity in the distributed powder layer. Table 4 shows the chemical composition measured on the powder by means of an X-ray energy dispersion spectroscopy system. No traces of contaminants were detected, even when several powder specimens were observed and a careful investigation of possible impurities was carried out. The chemical composition was consistent with the nominal one listed in Table 2. As far as the relatively high oxygen content is concerned, it should be considered that the powder was not kept in an inert atmosphere after extraction from the machine, during SEM specimen preparation. In order to cross-check the oxygen percentage in the consolidated specimens, a sample was ground and polished in such a way as to remove the outer oxidised layers and was immediately introduced in the SEM chamber, ensuring the shortest possible contact with air. The X-EDS spectrum and the chemical composition detected on this sample are shown in Figure 5, where an oxygen percentage of 1.3% is reported.

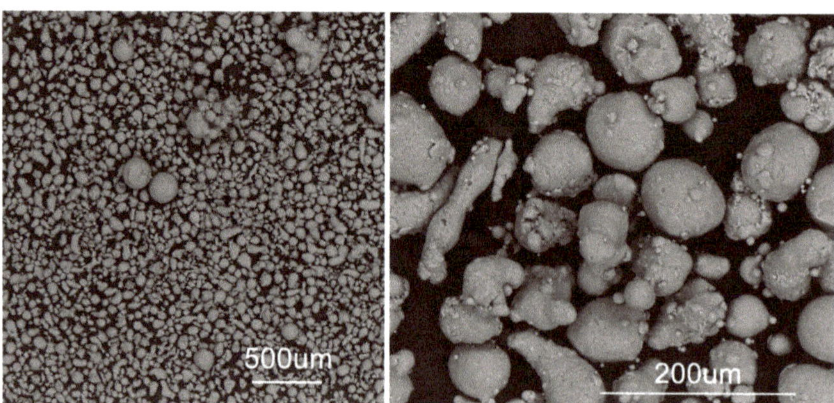

Figure 4. A357.0 powder.

Table 4. Experimental chemical analysis of the A357.0 powder after job completion.

Weight %	Si	Mg	Fe	Zn	Ti	O	Al
	8.7	0.5				6.3	bal.

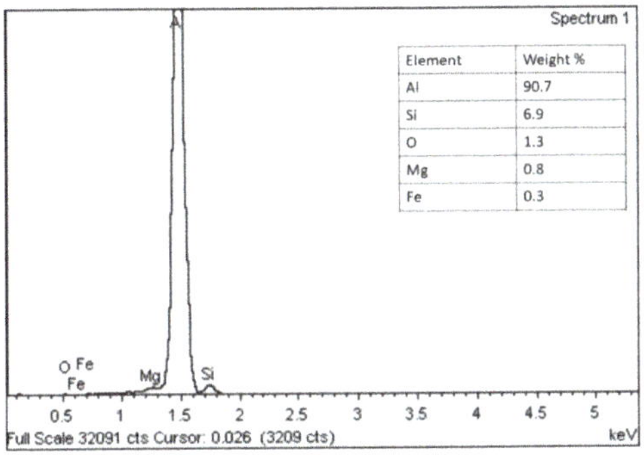

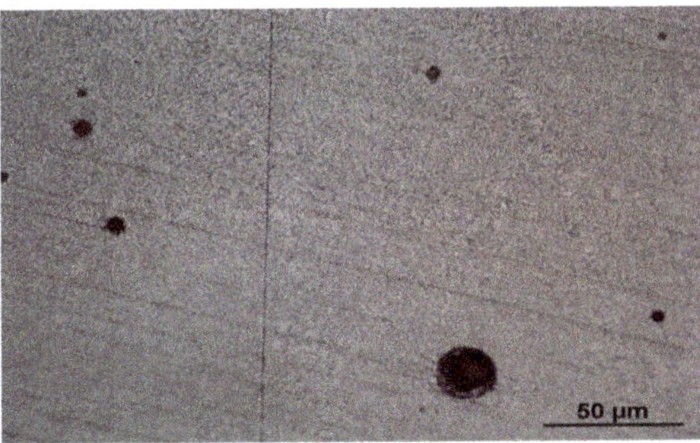

Figure 5. X-ray energy dispersion spectroscopy (X-EDS) spectrum and quantitative element percentage of a freshly polished surface of a tensile specimen.

The residual porosity measured on five images, such as the example in Figure 6, is on average 0.69%, with a standard deviation of 0.28%. A final density of 99.3% has thus been achieved, which is consistent with common industrial specifications. The microstructure obtained in the three orientations can be appreciated in Figure 7, where OM images are illustrated next to FEG-SEM ones. A very fine cellular structure is observed, equiaxed in the XY plane (Figure 7A) and elongated in the Z direction (Figure 7B,C), comparable to the 'fibre texture' observed by Prashanth et al. for Al-12Si [14] and by Hadadzadeh et al. for AlSi10Mg [15]. Hadadzadeh and colleagues produced extremely interesting proofs that when the building direction changes from vertical to horizontal, a columnar to equiaxed transition is obtained in the AlSi10Mg alloy produced by LPBF. The microstructure observed in this study could be consistent with this claim.

HBW hardness was 95.2 (Standard Deviation, SD 1.3) on the surface parallel to the powder bed and 102.6 (SD 0.8) on the one parallel to the Z direction.

Figure 6. Optical microscope (OM) image of a polished section for the measurements of residual porosity.

The results of all the mechanical tests are listed in Table 5 and shown in Figures 8 and 9. They include: elastic modulus (E), yield strength ($R_{p0.2}$), tensile strength (R_m), percentage total extension at fracture (A_t), and shear strength (S).

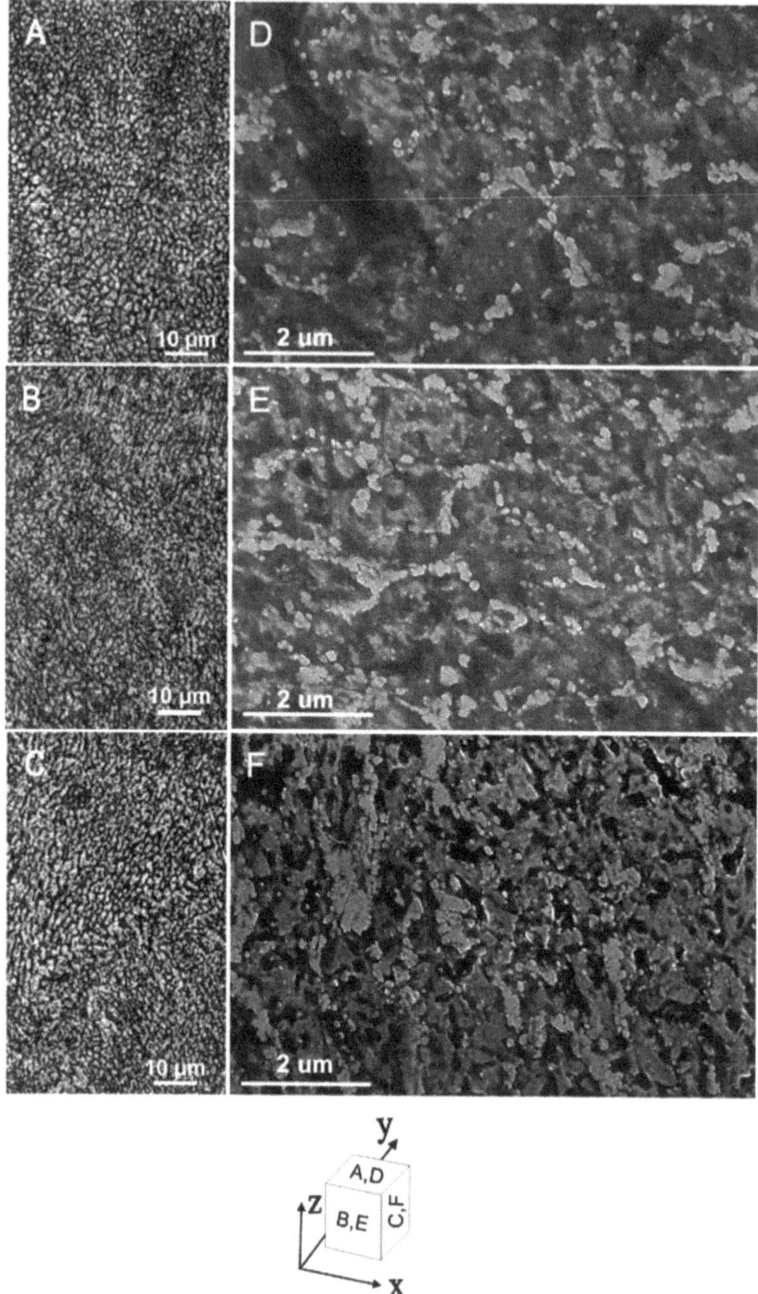

Figure 7. OM (**A–C**) and FEG-SEM (**D–F**) images of etched sections parallel to: the *XY* plane (**A,D**); the *XZ* plane (**B,E**) and the *YZ* plane (**C,F**).

No distinction in the mechanical behaviour was observed between shear specimens produced with the dimensions labelled as Type I and II in Figure 2. Hence, the results of the shear tests and the observation of the rupture surfaces are presented and discussed jointly in this section. The yield-, tensile-, and shear strength (the obtained values are the average of 17 measurements) rise when the direction of the specimens approaches the growth direction, but the deviation between different orientations is quite small. The opposite trend is observed for extension, but in this case the variation is much bigger. The elastic modulus does not show any relevant trend for varying orientation. The results for strength are in contrast with those obtained by other authors for Al-12Si; in that case, the effect of texture proved negligible on the yield strength, which was found to be unaffected by the build orientation [14].

Table 5. Results of the tensile and shear tests. Standard deviation is between brackets. E = elastic modulus, $R_{p0.2}$ = yield strength, R_m = tensile strength, A_t = percentage total extension at fracture, and S = shear strength.

Tensile and Shear Test Results	0° Mean (SD)	45° Mean (SD)	90° Mean (SD)
E [GPa]	75.0 (4.1)	77.8 (6.8)	74.8 (5.3)
$R_{p0.2}$ [MPa]	184 (17)	195 (17)	192 (17)
R_m [MPa]	284 (19)	298 (17)	305 (15)
A_t [%]	6.5 (0.8)	5.1 (0.8)	3.5 (0.7)
S [MPa]	165 (10)	173 (9)	192 (9)

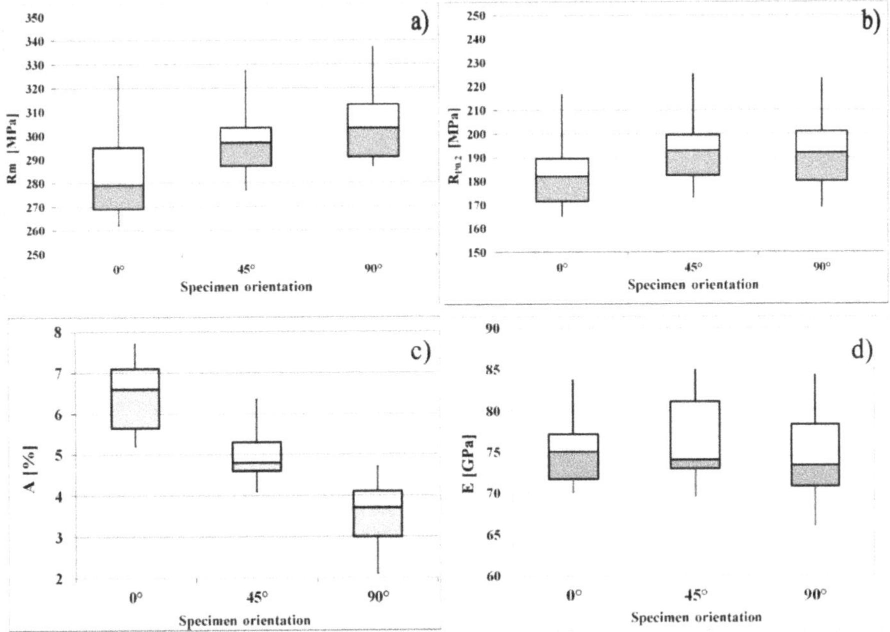

Figure 8. Tensile test results: (**a**) tensile strength values; (**b**) yield strength values; (**c**) percentage total extension at fracture values; (**d**) elastic modulus values.

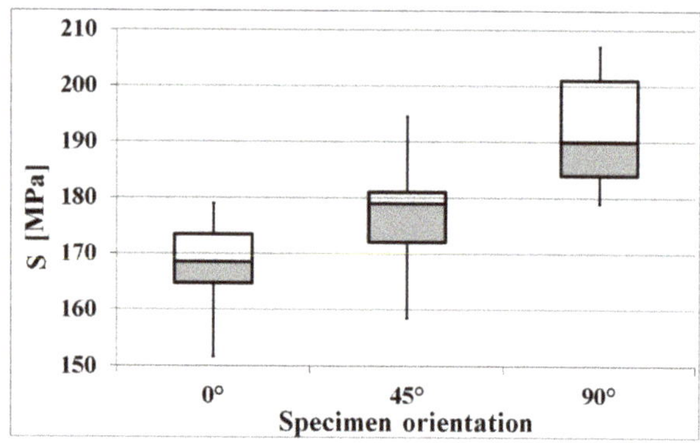

Figure 9. Shear strength results.

For the 90° specimens, the yield and tensile strength values are in good agreement with those obtained by other authors who produced density optimised samples [16]. Rao et al. found a yield stress of 187 MPa (vs 192 ± 17 MPa in this research) and an ultimate stress of 290 MPa (vs 305 ± 15 MPa). A comparison with the nominal datasheets should take into account that all the suppliers declare the mechanical properties either in the as-built condition (as summed in Table 1), or after the T6 heat treatment. The use of parts in the as-built state is unfeasible for many industrial applications, especially in the automotive field, due to excessive anisotropy, scarce ductility, and undue frozen stresses or post-build deformations. In the present paper, the aim was to study process conditions capable of minimizing shrinkage-related deformation even in the case of big components. Consequently, the temperature of the build plate was chosen so as to allow a continuous stress relief of the part during its construction. Therefore, comparatively lower values of yield and tensile strengths, against the data in Table 1, are not surprising. Conversely, the results of the tensile tests in this research are encouragingly comparable to the typical values of the mechanical properties of A357.0 obtained by casting and T6 heat treatment, shown in Table 6.

Anisotropy of mechanical performance, i.e., possible significant differences in the values that have been measured for the three orientations, was evaluated by statistical tools (Statistica 8, Statsoft, Hamburg, Germany). A correlation analysis was done of the independent variable 'orientation' on all the measured mechanical characteristics listed in Table 4. Table 7 registers the results, expressed in terms of correlation coefficients and probability values (p-values). When lower than 0.05, the p-values can be taken as a decision to reject the null hypothesis of insignificance of the correlation. A significant, even if slight, effect of orientation onto tensile and shear strength was observed. The same applies to total extension at break, but in this case a much greater variation was measured: ductility was nearly halved from 0° to 90°. For elastic modulus and yield strength, instead, the variations are within the scattering of experimental data, since the associated p-values are much higher than 0.05. As anticipated above, the same is true for shear strength of specimens produced with the two different thickness values (Types I and II): no differences are observed.

Failure mechanisms on rupture surfaces are expected to reflect the described anisotropy of strength and ductility.

Table 6. Typical mechanical properties of A357.0 aluminium alloy obtained by casting [17].

Alloy	Heat Treatment	Rm [MPa]	$Rp_{0.2}$ [MPa]	A [%]	S [MPa]
A357.0	T6	315	250	3	275

Table 7. Correlation coefficients and *p*-values resulting from the correlation analysis for the variables E, $Rp_{0.2}$, Rm, A, and S versus build orientation. Records below the level of significance of 0.05 are bold.

Correlation Analysis	E	$Rp_{0,2}$	Rm	A [%]	S
Correlation coefficient	−0.02	0.19	0.51	−0.85	0.78
p-value	0.882	0.189	2.26×10^{-4}	5.53×10^{-14}	6.41×10^{-8}

The rupture surfaces have been observed by means of a SEM. Rupture surfaces obtained in the shear tests show that for specimens produced in all directions, shear rupture generates two distinct failure morphologies. Very fine dimple areas are adjacent to zones of shear flow where the material is dragged onto the surface (Figure 10). For 0° and 90° specimens, the dimples are parallel to the rupture surface, that is to say parallel to the images (Figure 10d,f), whereas specimens built at 45°, show only dimples pointing out of the rupture surface (Figure 10e). For these specimens, shear stresses are applied onto a plane that is parallel to the layers. A further distinction can be observed between 0° and 90° specimens. For 90°, dimples are parallel not only to the rupture surface, but also to the direction of the shear load. In addition, for 0° specimens numerous areas of lack of fusion are detected, where unmolten powder particles are still present (Figure 11). An extremely fine fibre microstructure is clearly visible, composed of thin grains elongated by the Z direction. The described morphology accounts for the differences measured for the shear strength. Namely, specimens built in the 0° direction are the weakest, due to a combination of lack of fusion defects and ineffective orientation of the fibre structure versus the shear stresses. For 45° specimens, the shear stresses act perpendicularly to the texture obtained by LPBF. Specimens produced in the 90° orientation are the strongest since they benefit from shear loads acting along the direction of the fine elongated grain structure. An additional effect of higher percentage of columnar grains in 90° specimens as compared to 0° ones, similar to the findings by Hadadzadeh et al. [15], cannot be excluded.

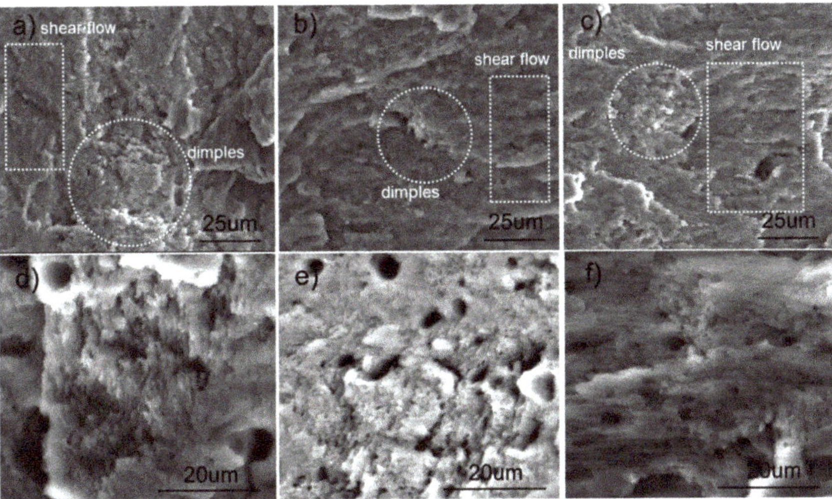

Figure 10. SEM observations of the rupture surface of A357.0 shear test specimens. The failure morphology changes with the growth direction of the specimen: (**a**,**d**) 0°, (**b**,**e**) 45°, and (**c**,**f**) 90°.

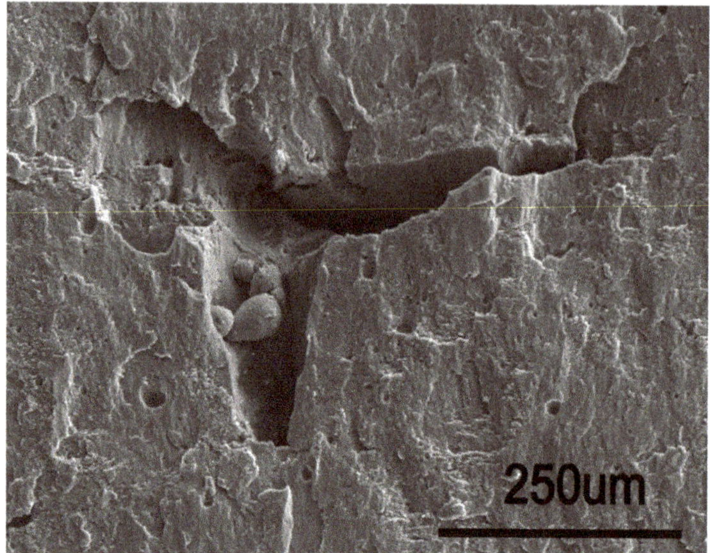

Figure 11. Lack of fusion onto the shear rupture surface of a 0° specimen.

Other authors previously observed a mixed quasi-cleavage and dimpled morphology on the fracture surfaces of A356 tensile samples, in the as-cast condition or after T6, as well as of A357 produced by LPBF [18,19]. Jiang Wen-Ming et al., found that the dimpled morphology is dispersed uniformly under the T6 condition, compared to the as-cast condition.

The rupture surfaces obtained in the tensile tests exhibit the presence of numerous large voids (Figure 12), which could apparently be ascribed to a much higher porosity than that measured on metallographic sections like the one in Figure 5. Actually, crack propagation proceeds through pores located at different layers, so the number of open pores that can be observed on the rupture surfaces is not a direct measurement of the percentage porosity. Differently from previous results for Al-12Si [14], no step-like morphology was observed in this study, which corroborates the achievement of a fully densified metal structure in which hatch overlaps are successfully evened out by a correct choice of laser parameters. Dimpled rupture is the prevailing failure mode for the specimens in the present study, but it acts synergistically with other mechanisms. The dimples are equiaxed in those regions of the specimens where a tensile stress component is predominant. However, traces of plastic micro-deformation and parallel slip bands can be also detected on the surfaces. Parallel slip bands are clearly visible around the large void in Figure 12a, and they are shown at higher magnification in Figure 12d. Shear band areas have been observed before on the rupture surfaces of A356 cast tensile test specimens [18], but the additive manufactured specimens in this study show much larger slip band areas than do the cast specimens.

The rupture surfaces in Figure 12 may be ascribed to cellular fracture [14]. Cellular fracture is typical of multi-phase materials, where the various phases have different mechanical properties, as in this case α-Al and Si precipitates that segregate along the cellular boundaries. On the cellular fracture surface, the features of both brittle and ductile fracture are present simultaneously. Each particular phase constituent cracks according to its proper decohesion mechanism. Very often in the boundaries between brittle and ductile phases, the continuity is preserved, as in the surfaces observed in Figure 12.

Some authors [20,21] have affirmed that the fracture morphology of an aluminium alloy is influenced to a great extent by the type of test and by the specimen geometry; these variables change the local stress state and consequently the failure mode. The observations in this paper confirm these findings, since the statistical proof of anisotropy in the mechanical response is combined with the distinct failure modes that are identified depending on the combination between the build orientation, which determines the orientation of the columnar structure in the sample, and the type of load (tensile or shear) that is applied macroscopically.

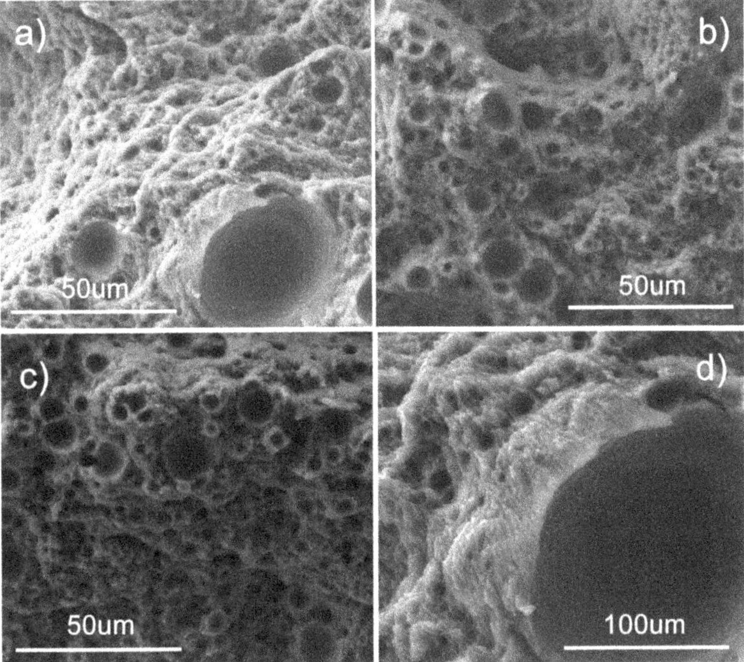

Figure 12. Rupture surfaces of A357.0 tensile test specimens: (**a,d**) 0°,(**b**) 45°, and (**c**) 90°. Dimpled rupturing is an important failure mode for these specimens, but it acts synergistically with the formation of parallel slip bands.

4. Conclusions

The paper reports the characterization of A357.0 specimens produced by means of LPBF and contributes to the knowledge base of this very recent alloy. The results allow for the assertion of the following:

- The measured yield- and tensile strength values are: (i) in good agreement with those reported in literature; (ii) encouragingly comparable to those typically measured for cast parts after T6 treatment; and (iii) consistent with those reported in the datasheets by AM system producers, if the heat-treatment conditions are considered.

- Both the tensile- and the shear strength show a small (but statistically significant) raise as the direction of the specimens approaches the growth direction. The opposite trend is observed for percent total extension at fracture that is almost halved in the Z- versus in the XY direction. Vertical samples are the strongest and least ductile.

- Tensile modulus and yield strength do not show statistically significant anisotropy. Variations for different building orientations are comparable to the scatter of experimental data.

- The rupture morphology of A357.0 is influenced to a great extent by the load condition, by the specimen geometry, and by the relative orientation of loads versus the directions of inherent anisotropy in LPBF processes. Specimens show an extremely fine fibre microstructure made of thin grains that extend along the Z direction. Both the shear- and the tensile rupture morphologies account for the anisotropy measured for the mechanical properties, since the fibre microstructure varies its reinforcing effectiveness depending on its relative orientation as compared to that of the applied stresses.

- Both for shear loads and for tensile ones, dimpled rupture is an important failure mode, but it acts synergistically with the formation of parallel slip bands. Regardless of the type of load applied macroscopically to the part, when the local stress state on the elongated grains is tensile, the rupture mode is a very fine dimple coalescence. When the fibre structure is locally subjected to shear stresses, which act tangentially to the elongated grains, rupture occurs by shear plastic flow, in which the material is spread over the surface. The extension of the plastic flow area increases as shear stresses become parallel to the build layers.

Funding: This research received no external funding.

Acknowledgments: Grateful acknowledgements are made by the author to Eng. Andrea Comin (Maserati S.pA) for the technical support and to Andrea Pasquali, general manager of ZARE Srl (Boretto, RE, Italy), for the technical support and for the construction of the specimens.

Conflicts of Interest: The author declares no conflict of interest.

References

1. Oerlikon, Additive Manufacturing the Next Industrial Revolution. Factsheet, 12.2017. Available online: https://www.oerlikon.com/ecoma/files/Oerlikon-AM-Factsheet_II.pdf (accessed on 8 August 2018).
2. Herzog, D.; Seyda, V.; Wycisk, E.; Emmelmann, C. Additive manufacturing of metals. *Acta Mater.* **2016**, *117*, 371–392. [CrossRef]
3. Sercombe, T.B.; Li, X. Selective laser melting of aluminium and aluminium metal matrix composites: Review. *Mater. Technol.* **2016**, *31*, 77–85. [CrossRef]
4. Wu, J.; Wang, X.Q.; Wang, W.; Attallah, M.M.; Loretto, M.H. Microstructure and strength of selectively laser melted AlSi10Mg. *Acta Mater.* **2016**, *117*, 311–320. [CrossRef]
5. APWORKS. Scalmalloy®—Aluminum Powder at Its Best. Available online: https://www.apworks.de/en/scalmalloy/ (accessed on 8 August 2018).
6. Aeromet International Ltd., A20X™ POWDER. Available online: http://a20x.com/powder/ (accessed on 8 August 2018).
7. Trevisan, F.; Calignano, F.; Lorusso, M.; Pakkanen, J.; Ambrosio, E.P.; Mariangela, L.; Pavese, M.; Manfredi, D.; Fino, P. Effects of Heat Treatments on A357 Alloy Produced by Selective Laser Melting. In Proceedings of the European Congress and Exhibition on Powder Metallurgy, Hamburg, Germany, 9–13 October 2016; The European Powder Metallurgy Association: Brussels, Belgium, 2016; pp. 1–6.
8. Aversa, A.; Lorusso, M.; Trevisan, F.; Ambrosio, E.P.; Calignano, F.; Manfredi, D.; Biamino, S.; Fino, P.; Lombardi, M.; Pavese, M. Effect of process and post-process conditions on the mechanical properties of an A357 alloy produced via laser powder bed fusion. *Metals* **2017**, *7*, 68. [CrossRef]
9. CITIM GmbH. Metal Additive Manufacturing: Aluminum AlSi9Cu3, Aluminum AlSi7Mg, Aluminum AlSi10Mg. Available online: http://www.citim.de/en/metal-additive-manufacturing (accessed on 15 November 2017).
10. *EOS Titanium Ti64*; Material data sheet; EOS GmbH—Electro Optical Systems: Krailling, Germany, 2017.
11. SLM Solutions, 3D Metals: AlSi10Mg, AlSi12, AlSi7Mg0.6, AlSi9Cu3. Available online: https://slm-solutions.com/download-center (accessed on 27 August 2018).
12. Concept Laser: CL 30AL, CL 31AL. Available online: https://www.concept-laser.de/en/products/materials.html (accessed on 15 November 2017).
13. Kretzschmar, N. Economic Validation of Metal Powder Bed Based AM Processes. Master's Thesis, Aalto University, Helsinki, Finland, May 2015.
14. Prashanth, K.G.; Scudino, S.; Klauss, H.J.; Surreddi, K.B.; Löber, L.; Wang, Z.; Chaubey, A.K.; Kühn, U.; Eckert, J. Microstructure and mechanical properties of Al-12Si produced by selective laser melting: Effect of heat treatment. *Mater. Sci. Eng. A* **2014**, *590*, 153–160. [CrossRef]

15. Hadadzadeh, A.; Amirkhiz, B.S.; Li, J.; Mohammadi, M. Columnar to equiaxed transition during direct metal laser sintering of AlSi10Mg alloy: effect of building direction. *Addit. Manuf.* **2018**, *23*, 121–131. [CrossRef]
16. Rao, H.; Giet, S.; Yang, K.; Wu, X.; Davies, C.H.J. The influence of processing parameters on aluminium alloy A357. *Mater. Des.* **2016**, *109*, 334–346. [CrossRef]
17. Kaufman, J.G.; Rooy, E.L. *Aluminum Alloy Castings Properties, Processes, and Applications*; ASM International: Almere, The Netherlands, 2004.
18. Jiang, W.; Fan, Z.; Liu, D. Microstructure, tensile properties and fractography of A356 alloy under as-cast and T6 obtained with expendable pattern shell casting process. *Trans. Nonferrous Met. Soc. China* **2012**, *22*, s7–s13. [CrossRef]
19. Rao, J.H.; Zhang, Y.; Fang, X.; Chen, Y.; Wu, X.; Davies, C.H.J. The origins for tensile properties of selective laser melted aluminium alloy A357. *Addit. Manuf.* **2017**, *17*, 113–122. [CrossRef]
20. Warmuzek, M. *Aluminum-Silicon Casting Alloys: Atlas of Microfractographs*; ASM International: Almere, The Netherlands, 2004.
21. Bhandarkar, M.D.; Lisagor, W.B. *Metallurgical Characterization of the Fracture of Several High Strength Aluminum Alloys*; National Aeronautics and Space Administration: Washington, DC, USA, 1977.

MDPI

St. Alban-Anlage 66

4052 Basel

Switzerland

Tel. +41 61 683 77 34

Fax +41 61 302 89 18

www.mdpi.com

Metals Editorial Office

E-mail: metals@mdpi.com

www.mdpi.com/journal/metals